Multiple-Valued Computing in Quantum Molecular Biology

This book mainly focuses on the design methodologies of various quantum circuits, DNA circuits, DNA-quantum circuits, and quantum-DNA circuits. In this text, the author has compiled various design aspects of multiple-valued logic DNA-quantum and quantum-DNA sequential circuits, memory devices, programmable logic devices, and nanoprocessors. *Multiple-Valued Computing in Quantum Molecular Biology: Sequential Circuits, Memory Devices, Programmable Logic Devices, and Nanoprocessors* is Volume 2 of a two-volume set, and consists of four parts.

This book presents various design aspects of multiple-valued logic DNA-quantum and quantum-DNA sequential circuits, memory devices, programmable logic devices, and nanoprocessors. Part I discusses multiple-valued quantum and DNA sequential circuits such as D flip-flop, SR latch, SR flip-flop, JK flip-flop, T flip-flop, shift register, ripple counter, and synchronous counter, which are described, respectively, with the applications and working procedures. After that, multiple-valued quantum-DNA and DNA-quantum sequential circuits such as D flip-flop, SR flip-flop, JK flip-flop, T flip-flop, shift register, ripple counter, and synchronous counter circuits are explained with working procedures and architecture. Part II discusses the architecture and design procedure of memory devices such as random access memory (RAM), read-only memory (ROM), programmable read-only memory (PROM), and cache memory, which are sequentially described in multiple-valued quantum, DNA, quantum-DNA, and DNA-quantum computing. In Part III, the author examines the architectures and working principles of programmable logic devices such as programmable logic array (PLA), programmable array logic (PAL), field programmable gate array (FPGA), and complex programmable logic device (CPLD) in multiple-valued quantum, DNA, quantum-DNA, and DNA-quantum computing. Multiple-valued quantum, DNA, quantum-DNA, and DNA-quantum nanoprocessors are designed with algorithms in Part IV. Furthermore, the basic components of ternary nanoprocessors such as T-RAM, ternary instruction register, ternary incrementor circuit, ternary decoder, ternary multiplexer, ternary accumulator in quantum, DNA, quantum-DNA, and DNA-quantum computing are also explained in detail.

This book will be of great help to researchers and students in quantum computing, DNA computing, quantum-DNA computing, and DNA-quantum computing.

Multiple-Valued Computing in Quantum Molecular Biology

Sequential Circuits, Memory Devices, Programmable Logic Devices, and Nanoprocessors, Volume 2

Hafiz Md. Hasan Babu

CRC Press
Taylor & Francis Group
Boca Raton London New York

CRC Press is an imprint of the
Taylor & Francis Group, an **informa** business

First edition published 2024
by CRC Press
6000 Broken Sound Parkway NW, Suite 300, Boca Raton, FL 33487-2742

and by CRC Press
4 Park Square, Milton Park, Abingdon, Oxon, OX14 4RN

CRC Press is an imprint of Taylor & Francis Group, LLC

© 2024 Hafiz Md. Hasan Babu

Library of Congress Control Number: 2023942122

ISBN: 978-1-032-46487-9 (hbk)
ISBN: 978-1-032-46490-9 (pbk)
ISBN: 978-1-003-38192-1 (ebk)

DOI: 10.1201/9781003381921

Typeset in Nimbus Roman
by KnowledgeWorks Global Ltd.

To my respected great parents and also to my lovely wife, daughter, and son, who made it possible to write this book

Contents

Author		xxi
Preface		xxiii
Acknowledgments		xxv
Acronyms		xxvii
Introduction		xxix
Fundamentals of Multiple-Valued Quantum Molecular Biology		**1**
0.1	Introduction	1
0.2	Multiple-Valued Logic	1
0.3	Multiple-Valued Computing	2
0.4	Molecular Biology	2
0.5	Quantum Molecular Biology	3
0.6	Fundamental Operations of Multiple-Valued Quantum Computing	3
0.7	Quantum Ternary Logic	4
	0.7.1 Why Ternary Logic in Quantum Computing?	4
0.8	Quantum Fundamental Gates in Multi-Valued Logic	5
	0.8.1 Quantum Ternary Shift Gates	5
	0.8.2 Quantum Ternary C^2 NOT Gate	8
0.9	Quantum Multi-Valued Basic Logic Operations	8
	0.9.1 Ternary Quantum-AND Operation	9
	0.9.2 Ternary Quantum-NAND Operation	10
	0.9.3 Ternary Quantum-OR Operation	11
	0.9.4 Ternary Quantum-NOR Operation	12
	0.9.5 Ternary Quantum XOR Operation	13
	0.9.6 Ternary Quantum XNOR Operation	14
0.10	Fundamental Operations in Multiple-Valued DNA Computing	15
	0.10.1 Ternary DNA-NOT Operation	16
	0.10.2 Ternary DNA-AND Operation	17
	0.10.3 Ternary DNA NAND Operation	17
	0.10.4 Ternary DNA-OR Operation	18
	0.10.5 Ternary DNA-NOR Operation	20
	0.10.6 Ternary DNA-XOR Operation	21
	0.10.7 Ternary DNA XNOR Operation	21

0.11 Summary   22
Bibliography   23

I Multi-Valued Sequential Circuits in Quantum Molecular Biology   25

Overview   **27**

1 Multiple-Valued Sequential Circuits in Quantum Computing   **29**
  1.1 Introduction   29
  1.2 Multi-Valued Quantum D Flip Flop   30
     1.2.1 General Organization of Multi-Valued D Flip Flop   31
     1.2.2 Circuit Architecture of Multi-Valued Quantum D Flip-Flop   32
     1.2.3 Working Principle of Multi-Valued Quantum D Flip-Flop   32
     1.2.4 Example   34
  1.3 Multi-Valued Quantum SR Latch   34
     1.3.1 General Organization of Multi-Valued Quantum SR Latch   35
     1.3.2 Circuit Architecture of Multi-Valued Quantum SR Latch .   35
     1.3.3 Working Principle of Multi-Valued Quantum SR Latch . .   36
     1.3.4 Example   37
  1.4 Multi-Valued Quantum SR Flip Flop   38
     1.4.1 General Organization of Multi-Valued Quantum SR Latch   38
     1.4.2 Circuit Architecture of Multi-Valued Quantum SR Flip
            Flop   39
     1.4.3 Working Principle of Multi-Valued Quantum SR Flip Flop   40
     1.4.4 Example   41
  1.5 Multi-Valued Quantum JK Flip Flop   42
     1.5.1 General Organization of Multi-Valued Quantum JK Flip
            Flop   42
     1.5.2 Circuit Architecture of Multi-Valued Quantum JK Flip
            Flop   44
     1.5.3 Working Principle of Multi-Valued Quantum JK Flip Flop   44
     1.5.4 Example   46
  1.6 Multi-Valued Quantum T Flip Flop   46
     1.6.1 General Organization of Multi-Valued Quantum T Flip
            Flop   47
     1.6.2 Circuit Architecture of Multi-Valued Quantum T Flip Flop   48
     1.6.3 Working Principle of Multi-Valued Quantum T Flip Flop .   49
     1.6.4 Example   49
  1.7 Multi-Valued Quantum Shift Register   50
     1.7.1 General Organization of Multi-Valued Quantum Shift
            Register   51
     1.7.2 Circuit Architecture of Multi-Valued Quantum Shift
            Register   52
     1.7.3 Working Principle of Multi-Valued Quantum Shift Register   53

	1.7.4	Example	54
1.8	Multi-Valued Quantum Ripple Counter		55
	1.8.1	General Organization of Multi-Valued Quantum Ripple Counter	55
	1.8.2	Circuit Architecture of Multi-Valued Quantum Ripple Counter	57
	1.8.3	Working Principle of Multi-Valued Quantum Ripple Counter	58
	1.8.4	Example	59
1.9	Multi-Valued Quantum Synchronous Counter		60
	1.9.1	General Organization of Multi-Valued Quantum Synchronous Counter	60
	1.9.2	Circuit Architecture of Multi-Valued Quantum Synchronous Counter	62
	1.9.3	Working Principle of Multi-Valued Quantum Synchronous Counter	63
	1.9.4	Example	64
1.10	Summary		65
Bibliography			65

2 Multi-Valued Sequential Circuits in DNA Computing 67
2.1	Introduction		67
2.2	Multi-Valued DNA D Flip Flop		68
	2.2.1	General Organization of Multi-Valued DNA D Flip Flop	69
	2.2.2	Circuit Architecture of Multi-Valued DNA D Flip Flop	70
	2.2.3	Working Principle of Multi-Valued DNA D Flip Flop	71
	2.2.4	Example	71
2.3	Multi-Valued DNA SR Latch		72
	2.3.1	General Organization Multi-Valued DNA SR Latch	72
	2.3.2	Circuit Architecture of Multi-Valued DNA SR Latch	73
	2.3.3	Working Principle of Multi-Valued DNA SR Latch	74
	2.3.4	Example	75
2.4	Multi-Valued DNA SR Flip Flop		76
	2.4.1	General Organization of Multi-Valued DNA SR Flip Flop	77
	2.4.2	Circuit Architecture of Multi-Valued DNA SR Flip-Flop	78
	2.4.3	Working Principle of Multi-Valued DNA SR Flip-Flop	79
	2.4.4	Example	80
2.5	Multi-Valued DNA JK Flip-Flop		81
	2.5.1	General Organization of Multi-Valued DNA JK Flip Flop	81
	2.5.2	Circuit Architecture of Multi-Valued DNA JK Flip-Flop	83
	2.5.3	Working Principle of Multi-Valued DNA JK Flip Flop	84
	2.5.4	Example	85
2.6	Multi-Valued DNA T Flip-Flop		85
	2.6.1	General Organization of Multi-Valued DNA T Flip-Flop	86
	2.6.2	Circuit Architecture of Multi-Valued DNA T Flip-Flop	87

		2.6.3	Working Principle of Multi-Valued DNA T Flip-Flop . . .	88
		2.6.4	Example	89
	2.7	Multi-Valued DNA Shift Register	89
		2.7.1	General Organization of Multi-Valued DNA Shift Register	90
		2.7.2	Circuit Architecture of Multi-Valued DNA Shift Register .	92
		2.7.3	Working Principle of Multi-Valued DNA Shift Register . .	93
		2.7.4	Example	94
	2.8	Multi-Valued DNA Ripple Counter	95
		2.8.1	General Organization of Multi-Valued DNA Ripple Counter	95
		2.8.2	Circuit Architecture of Multi-Valued DNA Ripple Counter	97
		2.8.3	Working Principle of Multi-Valued DNA Ripple Counter .	98
		2.8.4	Example	99
	2.9	Multi-Valued DNA Synchronous Counter	100
		2.9.1	General Organization of Multi-Valued DNA Synchronous Counter	101
		2.9.2	Circuit Architecture of Multi-Valued DNA Synchronous Counter	101
		2.9.3	Working Principle of Multi-Valued DNA Synchronous Counter	103
		2.9.4	Example	104
	2.10	Summary	105
	Bibliography	105
3	**Multi-Valued Sequential Circuits in Quantum-DNA Computing**			**107**
	3.1	Introduction	107
	3.2	Multi-Valued Quantum-DNA D Flip Flop	108
		3.2.1	General Organization of Multi-Valued Quantum-DNA D Flip-Flop	109
		3.2.2	Circuit Architecture of Multi-Valued Quantum-DNA D Flip-Flop	110
		3.2.3	Working Principle of Multi-Valued Quantum-DNA D Flip-Flop	112
		3.2.4	Example	113
	3.3	Multi-Valued Quantum-DNA SR Latch	114
		3.3.1	General Organization Multi-Valued Quantum-DNA SR Latch	115
		3.3.2	Circuit Architecture of Multi-Valued Quantum-DNA SR Latch	116
		3.3.3	Working Principle of Multi-Valued Quantum-DNA SR Latch	117
		3.3.4	Example	119
	3.4	Multi-Valued Quantum-DNA SR Flip Flop	120

3.4.1   General Organization of Multi-Valued Quantum-DNA SR
Flip-Flop 120

3.4.2   Circuit Architecture of Multi-Valued Quantum-DNA SR
Flip-Flop 122

3.4.3   Working Principle of Multi-Valued Quantum-DNA SR
Flip-Flop 123

3.4.4   Example 125

3.5   Multi-Valued Quantum-DNA JK Flip-Flop 125

3.5.1   General Organization of Multi-Valued Quantum-DNA JK
Flip-Flop 126

3.5.2   Circuit Architecture of Multi-Valued Quantum-DNA JK
Flip-Flop 127

3.5.3   Working Principle of Multi-Valued Quantum-DNA JK
Flip-Flop 129

3.5.4   Example 130

3.6   Multi-Valued Quantum-DNA T Flip-Flop 131

3.6.1   General Organization of Multi-Valued Quantum-DNA T
Flip-Flop 132

3.6.2   Circuit Architecture of Multi-Valued Quantum-DNA T
Flip-Flop 134

3.6.3   Working Principle of Multi-Valued Quantum-DNA T Flip-
Flop 135

3.6.4   Example 137

3.7   Multi-Valued Quantum-DNA Shift Register 138

3.7.1   General Organization of Multi-Valued Quantum-DNA
Shift Register 139

3.7.2   Circuit Architecture of Multi-Valued Quantum-DNA Shift
Register 140

3.7.3   Working Principle of Multi-Valued Quantum-DNA Shift
Register 142

3.7.4   Example 143

3.8   Multi-Valued Quantum-DNA Ripple Counter 144

3.8.1   General Organization of Multi-Valued Quantum-DNA Ripple Counter 145

3.8.2   Circuit Architecture of Multi-Valued Quantum-DNA Ripple Counter 146

3.8.3   Working Principle of Multi-Valued Quantum-DNA Ripple
Counter 148

3.8.4   Example 150

3.9   Multi-Valued Quantum-DNA Synchronous Counter 151

3.9.1   General Organization of Multi-Valued Quantum-DNA
Synchronous Counter 151

3.9.2   Circuit Architecture of Multi-Valued Quantum-DNA Synchronous Counter 153

3.9.3   Working Principle of Multi-Valued Quantum-DNA Synchronous Counter   155
3.9.4   Example   158
3.10   Summary   158
Bibliography   158

4   Multi-Valued Sequential Circuits in DNA-Quantum Computing   161
4.1   Introduction   161
4.2   Multi-Valued DNA-Quantum D Flip Flop   162
4.2.1   General Organization of Multi-Valued DNA-Quantum D
Flip Flop   163
4.2.2   Circuit Architecture of Multi-Valued DNA-Quantum Using
D Flip-Flop   164
4.2.3   Working Principle of Multi-Valued DNA-Quantum D Flip-
Flop   165
4.2.4   Example   167
4.3   Multi-Valued DNA-Quantum SR Latch   167
4.3.1   General Organization Multi-Valued DNA-Quantum SR
Latch   168
4.3.2   Circuit Architecture of Multi-Valued DNA-Quantum SR
Latch   169
4.3.3   Working Principle of Multi-Valued DNA-Quantum SR
Latch   171
4.3.4   Example   172
4.4   Multi-Valued DNA-Quantum SR Flip Flop   173
4.4.1   General Organization of Multi-Valued DNA-Quantum SR
Flip Flop   174
4.4.2   Circuit Architecture of Multi-Valued DNA-Quantum SR
Flip-Flop   175
4.4.3   Working Principle of Multi-Valued DNA-Quantum SR
Flip-Flop   176
4.4.4   Example   178
4.5   Multi-Valued DNA-Quantum JK Flip-Flop   178
4.5.1   General Organization of Multi-Valued DNA-Quantum JK
Flip-Flop   179
4.5.2   Circuit Architecture of Multi-Valued DNA-Quantum JK
Flip-Flop   181
4.5.3   Working Principle of Multi-Valued DNA-Quantum JK
Flip-Flop   182
4.5.4   Example   184
4.6   Multi-Valued DNA-Quantum T Flip-Flop   185
4.6.1   General Organization of Multi-Valued DNA-Quantum T
Flip-Flop   186
4.6.2   Circuit Architecture of Multi-Valued DNA-Quantum T
Flip-Flop   187

4.6.3   Working Principle of Multi-Valued DNA-Quantum T Flip-
          Flop 189
4.6.4   Example 191
4.7   Multi-Valued DNA-Quantum Shift Register 191
4.7.1   General Organization of Multi-Valued DNA-Quantum
          Shift Register 192
4.7.2   Circuit Architecture of Multi-Valued DNA-Quantum Shift
          Register 193
4.7.3   Working Principle of Multi-Valued DNA-Quantum Shift
          Register 195
4.7.4   Example 197
4.8   Multi-Valued DNA-Quantum Ripple Counter 198
4.8.1   General Organization of Multi-Valued DNA-Quantum Rip-
          ple Counter 198
4.8.2   Circuit Architecture of Multi-Valued DNA-Quantum Rip-
          ple Counter 199
4.8.3   Working Principle of Multi-Valued DNA-Quantum Ripple
          Counter 201
4.8.4   Example 203
4.9   Multi-Valued DNA-Quantum Synchronous Counter 203
4.9.1   General Organization of Multi-Valued DNA-Quantum
          Synchronous Counter 204
4.9.2   Circuit Architecture of Multi-Valued DNA-Quantum Syn-
          chronous Counter 205
4.9.3   Working Principle of Multi-Valued DNA-Quantum Syn-
          chronous Counter 207
4.9.4   Example 209
4.10   Summary 210
Bibliography 210

II   Multiple-Valued Memory Devices in Quantum Molecular Biology   211

Overview   213

5   Multiple-Valued Quantum Memory Devices   215
5.1   Introduction 215
5.2   Multiple-Valued Quantum Random Access Memory 216
5.2.1   Basic Definitions 218
5.2.2   Block Diagram 218
5.2.3   Architecture of Basic Components 218
5.2.4   Circuit Architecture of Multiple-Valued Quantum RAM
          Memory 226
5.3   Multiple-Valued Quantum Read-Only Memory 228
5.3.1   Basic Definitions 229

     5.3.2   Block Diagram of Multiple-Valued Quantum Read-Only
            Memory 230
     5.3.3   Circuit Architecture of Multiple-Valued Quantum ROM . 230
  5.4 Multiple-Valued Quantum Programmable Read-Only Memory . . 237
     5.4.1   Basic Definitions 238
     5.4.2   Block Diagram 238
     5.4.3   Circuit Architecture of Multiple-Valued Quantum PROM 238
  5.5 Multiple-Valued Quantum Cache Memory 244
     5.5.1   Basic Definitions 245
     5.5.2   Block Diagram 245
     5.5.3   Architecture of Basic Components 246
     5.5.4   Circuit Architecture of Multiple-Valued Quantum Cache
            Memory 248
  5.6 Summary 252
  Bibliography 252

6 Multiple-Valued DNA Memory Devices   **255**
  6.1 Introduction 255
  6.2 Multiple-Valued DNA Random Access Memory 256
     6.2.1   Block Diagram 257
     6.2.2   Architecture of Basic Components 259
     6.2.3   Circuit Architecture of Multiple-Valued DNA RAM . . . 266
     6.2.4   Working Principle of Multi-Valued DNA RAM 267
  6.3 Multiple-Valued DNA Read-Only Memory 268
     6.3.1   Basic Definition 270
     6.3.2   Block Diagram 271
     6.3.3   Circuit Architecture of Multiple-Valued DNA ROM . . . 272
     6.3.4   Design Procedure 274
     6.3.5   Working Principle 275
  6.4 Multiple-Valued DNA Programmable Read-Only Memory . . . 279
     6.4.1   Basic Definitions 280
     6.4.2   Block Diagram 280
     6.4.3   Circuit Architecture of Multiple-Valued DNA PROM . . . 280
     6.4.4   Design Procedure 281
     6.4.5   Working Principle 283
  6.5 Multiple-Valued DNA Cache Memory 286
     6.5.1   Block Diagram 287
     6.5.2   Architecture of Basic Components 287
     6.5.3   Circuit Architecture of Multiple-Valued DNA Cache
            Memory 289
     6.5.4   Working Principle 289
  6.6 Summary 292
  Bibliography 292

7 Multiple-Valued Quantum-DNA Memory Devices **295**

7.1   Introduction 295

7.2   Multiple-Valued Quantum-DNA Random Access Memory 296

     7.2.1   General Organization of Multiple-Valued Quantum-DNA RAM 297

     7.2.2   Architecture of Basic Components of Multiple-Valued Quantum-DNA RAM 297

     7.2.3   Circuit Architecture of Multi-Valued Quantum-DNA RAM 298

     7.2.4   Working Principle 298

7.3   Multiple-Valued Quantum-DNA Read-Only Memory 298

     7.3.1   Basic Definitions 300

     7.3.2   General Organization of Multiple-Valued Quantum-DNA ROM 300

     7.3.3   Architecture of Basic Components of Multiple-Valued Quantum-DNA ROM 302

     7.3.4   Circuit Architecture of Multiple-Valued Quantum-DNA ROM 302

     7.3.5   Working Principle 304

7.4   Multiple-Valued Quantum-DNA Programmable Read Only Memory 307

     7.4.1   Basic Definitions 307

     7.4.2   General Organization of Multiple-Valued Quantum-DNA PROM 308

     7.4.3   Basic Components of Multiple-Valued Quantum-DNA PROM Memory 309

     7.4.4   Circuit Architecture of Multiple-Valued Quantum-DNA PROM 309

     7.4.5   Working Principle 310

7.5   Multiple-Valued Quantum-DNA Cache Memory 314

     7.5.1   General Organization of Multiple-Valued Quantum-DNA Cache Memory 314

     7.5.2   Circuit Architecture of Multiple-Valued Quantum-DNA Cache Memory 314

     7.5.3   Working Principle 314

7.6   Summary 316

Bibliography 317

8 Multiple-Valued DNA-Quantum Memory Devices **319**

8.1   Introduction 319

8.2   Multiple-Valued DNA-Quantum Random Access Memory 319

     8.2.1   Circuit Architecture of Multiple-Valued DNA-Quantum RAM 319

     8.2.2   Working Principle 320

8.3   Multiple-Valued DNA-Quantum Read Only Memory 322

8.3.1   Circuit Architecture of Multiple-Valued DNA-Quantum
ROM 322
8.3.2   Design Procedure 323
8.3.3   Working Principle 325
8.4   Multiple-Valued DNA-Quantum Programmable Read Only
Memory 329
8.4.1   Circuit Architecture of Multiple-Valued DNA-Quantum
PROM 330
8.4.2   Design Procedure 330
8.4.3   Working Principle 332
8.5   Multiple-Valued DNA-Quantum Cache Memory 335
8.5.1   Working Principle of Multiple-Valued DNA-Quantum
Cache Memory 336
8.6   Summary 338
Bibliography 338

III   Multiple-Valued Programmable Logic Devices in Quantum Molecular Biology   339

Overview   341

9   Multiple-Valued Programmable Logic Devices in Quantum Computing 343

9.1   Introduction 343
9.2   Multiple-Valued Quantum Programmable Logic Array 343
9.2.1   General Organization of Multi-Valued Quantum PLA . . . 343
9.2.2   Circuit Architecture of Multi-Valued Quantum PLA 344
9.2.3   Working Principle 345
9.3   Multiple-Valued Quantum Programmable Array Logic 347
9.3.1   General Organization of Multi-Valued Quantum PAL . . . 347
9.3.2   Circuit Architecture of Multi-Valued Quantum PAL 349
9.3.3   Working Principle 351
9.4   Multi-Valued Quantum Field Programmable Gate Array 351
9.4.1   General Organization of Multi-Valued Quantum FPGA . . 352
9.4.2   Circuit Architecture of Basic Components 353
9.4.3   Circuit Architecture of Multi-Valued Quantum FPGA . . . 355
9.4.4   Working Principle 357
9.5   Multi-Valued Quantum Complex Programmable Logic Devices . . 359
9.5.1   General Organization of Multi-Valued Quantum CPLD . . 361
9.5.2   Circuit Architecture of Multi-Valued Quantum CPLD . . . 362
9.5.3   Working Principle 362
9.6   Summary 364
Bibliography 364

10 Multiple-Valued Programmable Logic Devices in DNA Computing **367**
    10.1 Introduction 367
    10.2 Multiple-Valued DNA Programmable Logic Array 367
          10.2.1 General Organization of Multiple-Valued DNA PLA . . . 368
          10.2.2 Circuit Architecture of Multi-Valued DNA PLA 369
          10.2.3 Working Principle 369
    10.3 Multiple-Valued DNA Programmable Array Logic 371
          10.3.1 General Organization Multiple-Valued DNA PAL 372
          10.3.2 Architecture of Ternary DNA PAL 372
          10.3.3 Working Principle 373
    10.4 Multi-Valued DNA Field Programmable Gate Array 375
          10.4.1 Architecture of Basic Components 375
          10.4.2 General Organization of Multi-Valued DNA FPGA 378
          10.4.3 Circuit Architecture of Multi-Valued DNA FPGA 379
          10.4.4 Working Procedure 381
    10.5 Multi-Valued DNA Complex Programmable Logic Devices . . . 384
          10.5.1 General Organization of Multi-Valued DNA CPLD 384
          10.5.2 Circuit Architecture of Multi-Valued DNA CPLD 385
          10.5.3 Working Procedure 387
    10.6 Summary 389
    Bibliography 390

11 Multiple-Valued Programmable Logic Devices in Quantum-DNA
   Computing **391**
    11.1 Introduction 391
    11.2 Multiple-Valued Quantum-DNA Programmable Logic Array . . . 391
          11.2.1 General Organization of Multi-Valued Quantum-DNA
                 PLA 392
          11.2.2 Circuit Architecture of Multiple-valued Quantum-DNA
                 PLA 393
          11.2.3 Working Principle 395
    11.3 Multiple-Valued Quantum-DNA Programmable Array Logic . . . 395
          11.3.1 General Organization of Multi-Valued Quantum-DNA PAL 396
          11.3.2 Circuit Architecture of Multi-Valued Quantum-DNA PAL 397
          11.3.3 Working Principle 398
    11.4 Multi-Valued Quantum-DNA Field Programmable Gate Array . . 400
          11.4.1 Circuit Architecture of Multi-Valued Quantum-DNA FPGA
                 400
          11.4.2 Working Principle 401
    11.5 Multi-Valued Quantum-DNA Complex Programmable Logic De-
         vices 403
          11.5.1 Circuit Architecture of Multi-Valued Quantum-DNA CPLD
                 403
          11.5.2 Working Principle 406
    11.6 Summary 407
    Bibliography 407

**12 Multiple-Valued Programmable Logic Devices in DNA-Quantum
Computing   409**
12.1 Introduction 409
12.2 Multiple-Valued DNA-Quantum Programmable Logic Array . . . 409
    12.2.1 General Organization of Multiple-Valued DNA-Quantum
        PLA 410
    12.2.2 Circuit Architecture of Multi-Valued DNA-Quantum PLA 411
    12.2.3 Working Principle 412
12.3 Multiple-Valued DNA-Quantum Programmable Array Logic . . . 413
    12.3.1 General Organization of Multi-Valued DNA-Quantum PAL 413
    12.3.2 Circuit Architecture of Multi-Valued DNA-Quantum PAL 415
    12.3.3 Working Principle 416
12.4 Multi-Valued DNA-Quantum Field Programmable Gate Array . . 417
    12.4.1 Circuit Architecture of Multiple-Valued DNA-Quantum
        FPGA 418
    12.4.2 Working Procedure 419
12.5 Multi-Valued DNA-Quantum Complex Programmable Logic Devices 422
    12.5.1 Circuit Architecture of Multiple-Valued DNA-Quantum
        CPLD 422
    12.5.2 Working Procedure 426
12.6 Summary 428
Bibliography 428

**IV Multiple-Valued Nano-Processors in Quantum Molecular
Biology   431**

Overview   433

13 Multiple-Valued Quantum Nano Processor   435
13.1 Introduction 435
13.2 Basic Definitions 436
13.3 Block Diagram of Complete 2-Qubit Ternary Nano Processor . . . 436
13.4 Basic Components of Quantum Ternary Nano Processor 438
    13.4.1 Quantum Ternary RAM 438
    13.4.2 Quantum Ternary Instruction Register (IR) 442
    13.4.3 Quantum Ternary Program Counter 446
    13.4.4 Ternary 2-Qubit Incrementor 447
    13.4.5 Ternary Quantum Decoder 448
    13.4.6 Ternary Quantum ALU 448
    13.4.7 Ternary Quantum Multiplexer 448
    13.4.8 Ternary Quantum Accumulator 451
13.5 Applications 452
13.6 Summary 452
Bibliography 452

14 Multiple-Valued DNA Nano Processor **455**
  14.1   Introduction 455
  14.2   Basic Definitions 455
  14.3   Block Diagram of Ternary 2-Bit DNA Nano Processor 456
  14.4   Basic Components of Ternary DNA Nano Processor 457
        14.4.1   Ternary DNA RAM 458
        14.4.2   Ternary 2-Bit DNA Instruction Register (IR) 461
        14.4.3   Ternary DNA Program Counter 465
        14.4.4   Ternary 2-Bit DNA Incrementor Circuit 466
        14.4.5   Ternary DNA Decoder 466
        14.4.6   Ternary DNA ALU 467
        14.4.7   Ternary DNA Multiplexer 467
        14.4.8   Ternary DNA Accumulator 467
  14.5   Applications 470
  14.6   Summary 470
  Bibliography 471

15 Multiple-Valued Quantum-DNA Nano-Processor **473**
  15.1   Introduction 473
  15.2   Basic Definitions 474
  15.3   Block Diagram of Complete Ternary 2-Qubit Quantum-DNA Nano
         Processor 474
  15.4   Basic Components of Ternary Quantum-DNA Nano Processor . . 477
        15.4.1   Quantum Ternary RAM 477
        15.4.2   Ternary DNA CPU 478
  15.5   Applications 478
  15.6   Summary 478
  Bibliography 479

16 Multiple-Valued DNA-Quantum Nano-Processor **481**
  16.1   Introduction 481
  16.2   Basic Definitions 482
  16.3   Block Diagram of Complete Ternary 2-Bit DNA-Quantum Nano
         Processor 482
  16.4   Basic Components of Ternary DNA-Quantum Nano Processor . . 484
        16.4.1   Ternary DNA RAM 485
        16.4.2   Ternary Quantum CPU 485
  16.5   Applications 485
  16.6   Summary 485
  Bibliography 486

Final Remarks **487**

Index **489**

Author

Dr. Hafiz. Md. Hasan Babu is currently working as a Professor in the Department of Computer Science and Engineering, University of Dhaka as well as the Dean in the Faculty of Engineering and Technology of the University of Dhaka, Bangladesh. In addition, at present, he is a member (part-time) of Bangladesh Accreditation Council, Ministry of Education of the Government of the People's Republic of Bangladesh. He is also the Director of the Board of Directors of Bangladesh Submarine Cable Company Limited. Dr. Hasan Babu was the Chairman of the Department of Computer Science and Engineering at the University of Dhaka from 2003–2006 and Pro-Vice-Chancellor of the National University of Bangladesh from 2016–2020. He was also a professor and Founding Chairman of the Department of Robotics and Mechatronics Engineering, University of Dhaka. Dr. Hasan Babu obtained his PhD in electronics and computer science from Japan under the Japanese Government Scholarship and received his MSc in computer science and engineering from the Czech Republic under the Czech Government Scholarship. He also received the DAAD Research Fellowship from Germany.

Dr. Hasan Babu was awarded the Dr. M.O. Ghani Memorial Gold Medal by the Bangladesh Academy of Sciences in 2017 for his excellent research in the progress of physical sciences in Bangladesh. In addition, he was awarded the UGC Gold Medal Award-2017 in the mathematics, statistics, and computer science categories for his research on quantum multiplier-accumulator devices. He is currently an associate editor of the research journal *IET Computers and Digital Techniques* published by the Institution of Engineering and Technology of the United Kingdom. He was a member of Prime Minister's ICT Task Force in Bangladesh.

Dr. Hasan Babu was also President of the Bangladesh Computer Society from 2017–2020. Presently, he is President of the International Internet Society, Bangladesh chapter.

Professor Dr. Hasan Babu has published more than 100 research papers. Three of his research papers have received the best research awards in international conferences.

In addition, he has published the following four textbooks by three famous UK and US publishers for graduate and post-graduate students:

1. Hafiz Md. Hasan Babu, *Quantum Computing: A Pathway to Quantum Logic Design*, IOP (Institute of Physics) Publishing, 2020, Bristol, UK

2. Hafiz Md. Hasan Babu, *Reversible and DNA Computing*, Wiley Publishers, 2021, UK

3. Hafiz Md. Hasan Babu, *VLSI Circuits and Embedded Systems*, CRC Press/Taylor & Francis, 2022, USA

4. Md. Jahangir Alam, Guoqing Hu, Hafiz Md. Hasan Babu and Huazhong Xu, *Control Engineering Theory and Applications*, CRC Press/Taylor & Francis, 2022, USA

Preface

Multiple-valued logic improves the computation power. It provides more security, performs high-speed computations, and provides high-storage capacity, etc. over two-valued quantum computing. Quantum computing achieves tremendous processing power, low energy consumption, and exponential speed above traditional computers by regulating the behavior of minuscule physical things such as atoms, electrons, photons, and other microscopic particles. With the advent of nanotechnology, quantum computing vibrates an incredibly immense role in developing more compact and less power-consuming computers. The quantum computer is a completely new notion from regular computing, and it does not employ binary logic. Multiple-valued quantum computing is an exciting topic for multiple-valued computing systems since it addresses many problematic multiple-valued traditional computer concerns.

Multiple-valued DNA (deoxyribonucleic acid) computational systems are used to introduce many forms of biological computing systems. DNA encodes a biological organism's genetic information. It is constituted of polymer chains, which are commonly referred to as DNA strands. The information in DNA is stored as a code made up of four chemical bases: adenine (A), guanine (G), cytosine (C), and thymine (T). When compared to silicon-based conventional computers, DNA operations consume less power, and DNA-based logic gates can be reused once they have completed an operation. DNA computing is a new branch of computing that replaces traditional electronic computing with DNA, biochemistry, and molecular biology hardware. DNA computing carries the promise of cheap, huge, accessible data storage and an exponential increase in computing power and speed. Multiple-valued computing in quantum molecular biology can be introduced to get the advantages of both quantum and DNA computing.

This book, *Multiple-Valued Computing in Quantum Molecular Biology* starts from Part V, Chapter 20. Volume 1 of this book consists of four parts and 19 chapters. This book is Volume 2 where another four parts will explain the rest of the topics. Each part contains four chapters.

At the beginning of Part V, multiple-valued quantum and DNA sequential circuits such as D flip-flop, SR latch, SR flip-flop, JK flip-flop, T flip-flop, shift register, ripple counter, and synchronous counter are described respectively with applications and working procedures.

After that, multiple-valued quantum-DNA and DNA-quantum sequential circuits such as D flip-flop, SR flip-flop, JK flip-flop, T flip-flop, shift register, ripple counter, and synchronous counter circuits are explained with working procedure and architecture.

In Part VI, the architecture and design procedure of memory devices such as Random Access Memory (RAM), Read-Only Memory (ROM), Programmable Read-Only Memory (PROM), and cache memory are sequentially described in multiple-valued quantum, DNA, quantum-DNA, and DNA-quantum computing.

In Part VII, programmable logic devices such as Programmable Logic Array (PLA), Programmable Array Logic (PAL), Field Programmable Gate Array (FPGA), and Complex Programmable Logic Device (CPLD) in multiple-valued quantum, DNA, quantum-DNA, and DNA-quantum computing are described with their architecture and working principles.

Multiple-valued quantum, DNA, quantum-DNA, and DNA-quantum nano processors are designed with algorithms in Part VIII. The basic components of ternary nano processors, such as T-RAM, Ternary Instruction Register, Ternary Program Counter, ternary incrementor circuit, ternary decoder, ternary multiplexer, Ternary Arithmetic Logic Unit, ternary accumulator in quantum, DNA, quantum-DNA, and DNA-quantum computing, are also explained in detail in Part VIII.

Acknowledgments

I would like to express my sincerest gratitude and special appreciation to the various researchers in the field of multiple-valued computing in quantum molecular biology. The contents in this book have been compiled from a wide variety of research, where the researchers are pioneers in their respective fields. All research articles related to the contents are listed at the end of each chapter.

I am grateful to my great parents and dear family members for their endless support. Most of all, I want to thank my lovely wife, Mrs. Sitara Roshan, sweet daughter, Ms. Fariha Tasnim, and sweet son, Md. Tahsin Hasan, for their invaluable cooperation to help me complete this book.

Finally, I am also thankful to all of those, especially my beloved students Nitish Biswas, Md. Tareq Hasan, and Rownak Borhan Himel, who have provided immense support and valuable time to help me finish this book.

Acronyms

ALU	arithmetic logic unit
BCD	binary coded decimal
CLB	configurable logic block
CPLD	complex programmable logic device
CPU	central processing unit
CU	control unit
DNA	deoxyribonucleic acid
EMR	electron magnetic resonance
FPGA	field programmable gate array
IR	instruction register
LUT	look up table
MUX	multiplexer
MVL	multiple-valued logic
NMR	nuclear magnetic resonance
NTI	negative ternary inverter
PAL	programmable array logic
PC	program counter
PCR	polymerase chain reaction
PLA	programmable logic array
PROM	programmable read-only memory
PTI	positive ternary inverter
RAM	random access memory
ROM	read only memory
SIPO	serial-in parallel-out
SISO	serial-in serial-out
SPLD	simple programmable logic devices
STI	standard ternary inverter
TG	toffoli gate

Introduction

Multiple-valued quantum computing is a blend of quantum physics and theoretical computer science that allows this computing to be more unique and faster than current silicon-based multiple-valued computing systems. Multiple-valued quantum computing is an exciting topic for multiple-valued computing systems since it addresses many problematic multiple-valued traditional computer concerns. In classical computing, Shor's factorization method and Grover's search algorithm are challenging tasks that multiple-valued quantum computing handles in polynomial time.

Multiple-valued DNA (deoxyribonucleic acid) computational systems are used to introduce many forms of biological computing systems. DNA encodes a biological organism's genetic information. It is constituted of polymer chains, which are commonly referred to as DNA strands. The information in DNA is stored as a code made up of four chemical bases: adenine (A), guanine (G), cytosine (C), and thymine (T). Human DNA consists of about 3 billion bases, and more than 99 percent of those bases are the same in all people. The order, or sequence, of these bases, determines the information available for building and maintaining an organism, similar to the way in which letters of the alphabet appear in a certain order to form words and sentences. DNA contains features that make it possible to imitate traditional reasoning operations. Single-stranded DNA naturally migrate toward complementary sequences to create double-stranded complexes, whereas double-stranded DNA want to be in a double-stranded state. DNA computers have clear advantages over conventional computers when applied to problems that can be divided into separate, non-sequential tasks. The reason is that DNA strands can hold so much data in memory and conduct multiple operations at once, thus solving decomposable problems much faster. DNA has a massive data storage capacity and could store information for a long period. It is a stable molecule until it is subjected to extreme conditions. When compared to silicon-based conventional computers, DNA operations consume less power, and DNA-based logic gates can be reused once they have completed an operation.

In a magnetic field, the nucleus absorbs and emits electromagnetic radiation, which is known as nuclear magnetic resonance (NMR). Radio frequency (RF) is used in NMR to create a powerful magnetic field that excites the nuclei of molecules and causes them to exist in superposition. Quantum coherence and quantum entanglement are based on the superposition concept. Coherence in NMR refers to a physical state in which several spins line up and circle at the same speed around the direction of the magnetic field. The reversal of the NMR process is known as NMR relaxation. NMR relaxation qubits are set to the ground state to get the original molecule. In this case, EMR production must be stopped, and the molecules must lose energy to return to their ground state configuration. Reverse Nuclear Magnetic Resonance (RNMR) is a method that is performed at two temperatures: room temperature and zero kelvin. NMR and RNMR may aid in the conversion of a DNA molecular sequence to a qubit and a qubit to a DNA molecular sequence. Multiple-valued computing in quantum molecular biology can be introduced to get the advantages of both quantum and DNA computing.

Fundamentals of Multiple-Valued Quantum Molecular Biology

0.1 Introduction

The study of quantum computing is concerned with the development of computer-based technologies based on the ideas of quantum theory. Quantum theory explains how energy and matter behave on the quantum (atomic and subatomic) level. Quantum computing uses a mixture of qubits to carry out certain computational operations. All of these are completed much more quickly than with their conventional computing equipment. On the other hand, DNA computing uses biological molecules to do computations rather than standard silicon chips. Instead of using the binary alphabet (1 and 0) used by conventional computers, DNA computing uses the four-character genetic alphabets (A-adenine, G-guanine, C-cytosine, and T-thymine). The ability to produce tiny DNA molecules with any random sequence makes this possible. Any DNA operation's input can be represented by DNA molecules having a certain sequence. By performing laboratory procedures on the molecules, the instructions are carried out, and the outcome is characterized as a characteristic of the finished collection of molecules. Massively parallel computations and significant connections between computers and biological systems are promised by DNA computing. DNA computing is capable of doing millions of tasks at once.

Combining the benefits of DNA computing and quantum computing will maximize their benefits. In order to create two new computing processes that can be referred to as quantum molecular biology, these two computing systems can be combined. Quantum-DNA computing and DNA-quantum computing are also components of this quantum molecular biology. It is already known that using these two together is advantageous for a two-valued system. Therefore, these two processes can also be combined in a multiple-valued system. Multiple-valued quantum, multiple-valued DNA, multiple-valued quantum-DNA, and multiple-valued DNA-quantum computing will all be covered in this book.

0.2 Multiple-Valued Logic

Nearly all of the digital circuits used in modern technology are based on binary logic. Although widely utilized, binary circuits have several drawbacks; hence, solutions like Multiple-Valued Logic (MVL) have been sought after. Non-classical logics include many-valued logics. They adopt the truth-functionality principle, which states that the truth of a compound sentence is determined by the truth values of its component sentences, making them similar to classical logic (and so remains unaffected when one of its component sentences is replaced by another sentence with the same truth value). However, they are fundamentally different from classical logic in that they permit a wider range of truth degrees and do not limit the number of truth values to just two.

The multiple-valued system is more difficult to implement than the two-valued system. It is more costly, more space-consuming, requires more computational time, and generates more heat than the two-valued system. Quantum and DNA computing can be used to implement in the multiple-valued system. The reason quantum and DNA computers can solve complicated problems is that they generate multiple viable solutions at once. This is known as *parallel processing*. Quantum and DNA computing are already used to implement two-valued operations. Now, this chapter will explain how DNA computing constructs the operations for multi-valued systems.

0.3   Multiple-Valued Computing

Designing with multiple-valued logic has drawn a lot of attention in the previous three decades. Multiple-valued logic (MVL) emerged as a separate study in the early 1920s, thanks to a Polish philosopher named Lukasiewicz, whose goal was to add a third value to the binary system. The Lukasiewicz system is the result of this investigation. Emil Post, an American mathematician, invented multiple-valued algebra, sometimes known as post algebra, in response to this technique. Computations like addition, multiplication, etc. can be done in multi-valued logic, and this is actually multi-valued computing.

0.4   Molecular Biology

Molecular biology is a branch of biology that studies the chemical structures and processes of biological phenomena involving molecules, the basic units of life. The study of molecular biology is concerned with nucleic acids (such as DNA and RNA) and proteins—macromolecules that are critical to biological processes—and how they interact and behave within cells. In the 1930s, molecular biology emerged from the associated disciplines of biochemistry, genetics, and biophysics, and it is still strongly tied with those fields today. For molecular biology, various methodologies have been established, while researchers in the discipline may also use procedures and techniques that are native to genetics and other closely related fields. Molecular biology, in particular, uses techniques like *X-ray diffraction* and *electron microscopy* to better understand the three-dimensional structure of biological macromolecules. Molecular biologists study the molecular basis of genetic processes, mapping the location of genes on specific chromosomes, associating these genes with specific characteristics of an organism, and isolating, sequencing, and modifying specific genes using genetic engineering (recombinant DNA technology). Techniques including *Polymerase Chain Reaction* (PCR), *western blotting, and microarray analysis* can be used in these approaches. The field of molecular biology began in the 1940s intending to reveal the basic three-dimensional structure of proteins. In the early 1950s, as knowledge of protein structure grew, the structure of deoxyribonucleic acid (DNA), the genetic blueprint found in all living things, as described in 1953. With more investigation, scientists were able to learn more about not only DNA and ribonucleic acid (RNA), but also the chemical sequences within these substances that tell cells and viruses how to build proteins. Molecular biology remained a pure science with few practical applications until the 1970s when enzymes that could cut and reassemble portions of DNA in bacteria's chromosomes were found. Because it allows manipulation of the genetic sequences that determine the basic features of organisms, recombinant DNA technology has became one of the most active disciplines of molecular biology.

0.5 Quantum Molecular Biology

Quantum physics and molecular biology jointly form quantum molecular biology. Quantum computing and DNA computing can be combined to gain the advantages of both. This combination is called quantum molecular biology. Quantum molecular biology is defined as quantum-DNA computing and DNA-quantum computing. Quantum molecular biology is the field of study that investigates processes in living organisms that cannot be accurately described by the classical laws of physics. This means that quantum theory has to be applied to understand those processes. All matter, including living matter, is subject to the laws of physics. Here quantum molecular biology is defined as computations using qubits or DNA sequences. DNA molecular sequences or qubits are used here to represent information. In quantum computing, qubits are used to represent information or data, and in DNA computing, DNA molecules are used to represent information or data. So, quantum molecular biology is a kind of computing technology which can be quantum-DNA computing or DNA-quantum computing.

0.6 Fundamental Operations of Multiple-Valued Quantum Computing

Multi-Valued quantum computing is a challenging field for researchers. Two-valued quantum computing is not so old, but multi-valued quantum computing is a new topic to discuss. The benefits of working with other number systems over the binary number system are known. It is possible to work with more data with less effort in a multiple-valued system. Possibilities in which there are no middle options between true and false are familiar to computer scientists, computer engineers, applied mathematicians, and physicists. The soft logic of probability is familiar to statisticians, while the logic of uncertainty is familiar to physicists. When attempting to determine whether the status of a computer system is go, wait, or stop, the lack of better options is inconvenient and critical. These intermediate choices are the focus of multiple-valued logic. The difficulty in checking, correcting, or modifying over complicated flowcharts created by computer programmers is a severe disadvantage.

0.7   Quantum Ternary Logic

The qubits are known to all and have two states – |0> and |1>. The information unit in a three-valued quantum system (ternary quantum system) is termed as *qutrit*. The ternary quantum system represents one type of three-dimensional quantum system with the basis states |0>, |1>, and |2>. These basis states are called qutrit states and can be represented by 3×1 vectors: $|0> = \begin{bmatrix} 1 \\ 0 \\ 0 \end{bmatrix}$, $|1> = \begin{bmatrix} 0 \\ 1 \\ 0 \end{bmatrix}$, and $|2> = \begin{bmatrix} 0 \\ 0 \\ 1 \end{bmatrix}$

In a ternary quantum system, a qutrit can be defined as a linear superposition of the above-mentioned basis by the following equation:

$$\psi = \alpha |0> + \text{ß} |1> + \gamma |2>;$$

where α, β, and γ are the complex quantities to represent the probability amplitudes of the basis states and ψ is the wave function.'

0.7.1   Why Ternary Logic in Quantum Computing?

A ternary computer (sometimes known as a trinary computer) is a computer that does computations using ternary logic (three possible values) rather than binary logic (two possible values). In ternary quantum computing the superposition will be formed in the range of |0> to |2>. Ternary computing has many basic benefits over binary computing; some are given below:

1. Higher data throughput
2. Access to additional instructions
3. Back-compatibility with legacy binary codes
4. Preventing malware and viruses
5. Providing more security

However, the usual aim of ternary computers has not met with overwhelming success thus far, because they are not as efficient as binary computers in computing binary codes, which are widely used and appear to be nearly ubiquitous. Besides, constructing the ternary operational circuits is much more difficult than the binary operational circuit.

0.8 Quantum Fundamental Gates in Multi-Valued Logic

In the two-valued quantum computing, several quantum fundamental gates for reversible logic are learned which include the Pauli gate, Pauli-X gate, Hadamard gate, Toffoli gate, Fredkin gate, Deutsch gate, and Swap gate. All of them will not work in the ternary quantum system. The fundamental quantum gates which will work in ternary quantum computing are as follows:

1. Quantum Ternary Shift Gates

2. Quantum Ternary Toffoli Gates

3. Quantum Ternary C2 NOT Gate

These three forms of quantum ternary gates are discussed in the next section.

0.8.1 Quantum Ternary Shift Gates

Six ternary permutation matrices are widely used which are termed Quantum Ternary Shift Gates:

1. $Z(+0)$ is the primary state. Its columns correspond to 0, 1, and 2, respectively.

2. Transform $Z(+1)$ shifts the qutrit states by 1.

3. Transform $Z(+2)$ shifts the qutrit states by 2.

4. Transform $Z(12)$ swaps (permutes) the qutrit states $|1>$ and $|2>$.

5. Transform $Z(01)$ swaps the qutrit states $|0>$ and $|1>$.

6. Transform $Z(02)$ swaps the qutrit states $|0>$ and $|2>$.

Those transformations are depicted as follows.

$$(\text{1-Qutrit Ternary Permutation Transformations})$$

$$Z_3(+0) = \begin{bmatrix} 1 & 0 & 0 \\ 0 & 1 & 0 \\ 0 & 0 & 1 \end{bmatrix} Z_3(+1) = \begin{bmatrix} 0 & 0 & 1 \\ 1 & 0 & 0 \\ 0 & 1 & 0 \end{bmatrix} Z_3(+2) = \begin{bmatrix} 0 & 1 & 0 \\ 0 & 0 & 1 \\ 1 & 0 & 0 \end{bmatrix}$$

$$Z_3(12) = \begin{bmatrix} 1 & 0 & 0 \\ 0 & 0 & 1 \\ 0 & 1 & 0 \end{bmatrix} Z_3(01) = \begin{bmatrix} 0 & 1 & 0 \\ 1 & 0 & 0 \\ 0 & 0 & 1 \end{bmatrix} Z_3(02) = \begin{bmatrix} 0 & 0 & 1 \\ 0 & 1 & 0 \\ 1 & 0 & 0 \end{bmatrix}$$

TABLE 0.1

Operations of 1-Qutrit Permutation Gates/Ternary Shift Gates

A	Z(0)=A	Z(+1)=A+1	Z(+2)=A+2	Z (12)=2A	Z (1)=2A+1	Z (02)=2A+2
0	0	1	2	1	1	2
1	1	2	0	0	0	1
2	2	0	1	2	2	0

Table 0.1 shows the result of the operations.

From Table 0.1 and the matrix transformations, it is easy to understand the operations in the ternary shift gates. Figure 0.1 shows the general diagram of the ternary shift gates.

Figure 0.1 (a) shows the Z(+1) operations, where the given input will be shifted to the next one (i.e. |0> will become |1>, |1> will become |2>, and |2> will become to |0>). Figure 0.1 (b) shows the Z(+2) operations, where the given input will be shifted by +2 (i.e. |0> will become |2>, |1> will become |0>, and |2> will become to |1>). Figure 0.1 (c) shows the Z (02) operations, where if the given input is |0> it will be swapped by |2> and vice versa. Note that input |1> will not affect this gate. That means input |1> will pass as |1> itself (i.e. no swapping). Figure 0.1 (d) shows the Z (01) operations, in the same way, if the given input is |0> it will be swapped by |1> and vice versa. And, now input |2> will not affect this gate. That means input |2> will pass as |2> itself (i.e. no swapping). And Figure 0.1 (e) shows the Z (12) operations, where if the given input is |1> it will be swapped by |2> and vice versa. Eventually, input |0> will not affect this gate. That means input |0> will pass as |0> itself (i.e. no swapping).

0.8.1.1   Quantum Ternary Toffoli Gates

The Ternary Toffoli gate is another quantum ternary gate. Its inputs are A, B, and C, where A and B are the controlling inputs and C is the controlled input. Its outputs are

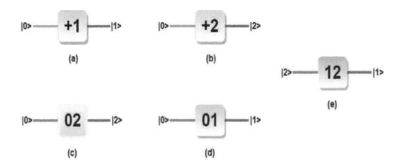

FIGURE 0.1

The General Architecture of the Ternary Permutation Gates

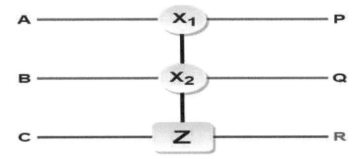

FIGURE 0.2
Symbol of Generalized 3-Qutrit Ternary Permutation Operations

$P = A$
$Q = B$
$R = Z$ transforms of C; if $A = X_1$ and $B = X_2$
$R = C$; otherwise

The symbol of the generalized 3-qutrit Ternary Permutation/Shift operations is shown in Figure 0.2.

As shown in Figure 0.2, the outputs of the ternary Toffoli gate are P, Q, and R. The outputs P and Q are equal to A and B, and the output R is equal to the Z transform of C if the inputs A and B are equal to X1 and X2, where $Z = \{+1, +2, 01, 02, 12\}$. Otherwise, the output R is equal to the input C.

Figure 0.3 shows the Toffoli gate for one controlled input. Here, the gate will open only if the controlled qutrit is $|2>$.

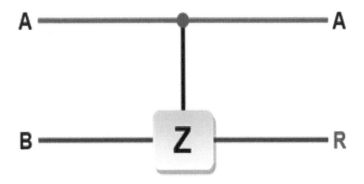

FIGURE 0.3
Ternary 1-bit Controlled Operation

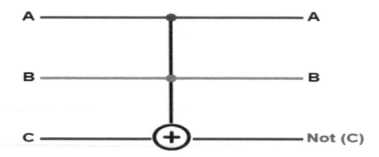

FIGURE 0.4
The General Structure Ternary 3-qutrit C^2 NOT Gate

0.8.2   Quantum Ternary C^2 NOT Gate

In ternary logic, several multi-qutrit control operations are possible. A 3-qutrit C^2NOT which is used for realizing the simplification rules for ternary minterms as

     C^2 NOT (A, B, C) = NOT(C); if A != B and A,B != 0
     C^2 NOT (A, B, C) = C; otherwise

where A, B are control inputs and C is the target input. Here, the final NOT (C) will provide the (+1) operation of the given input.

Figure 0.4 shows the symbol of the ternary C^2 NOT gate.

There might be confusion in this gate because it is called C^2 NOT gate but isn't doing the actual NOT operation. This gate will only activate if the input A = |1> and B = |2> or input A = |2> and B = |1> (see the logical expression). And if the control inputs fulfill the condition and input C is |0>, it will perform a Z (+1) operation and produce output |1>, and for input C as |1> and |2>, it will produce |2> and |0>, respectively.

0.9   Quantum Multi-Valued Basic Logic Operations

Quantum ternary logic functions are those functions that have significance if a third value is acquainted with the quantum binary logic. Quantum computing in the ternary logic system is quite interesting. Here, |0>, |1>, and |2> denote the ternary levels for basic logic gates to represent false, undefined, and true, respectively. The basic operations of quantum ternary logic can be defined as follows:

$$y_{OR} = \max(x, y)$$
$$y_{NOR} = \overline{\max(x, y)}$$
$$y_{AND} = \min(x, y)$$
$$y_{NAND} = \overline{\min(x, y)}$$
$$y_{XOR} = \operatorname{sum}(x, y)$$
$$y_{XNOR} = \overline{\operatorname{sum}(x, y)}$$

Where, $x, y = \{0, 1, 2\}$

TABLE 0.2

Truth Table for Ternary AND, NAND, OR, NOR, XOR, XNOR

Input 1	Input 2	AND	NAND	OR	NOR	XOR	XNOR
0	0	0	2	0	2	0	2
0	1	0	2	1	1	1	1
0	2	0	2	2	0	2	0
1	0	0	2	1	1	1	1
1	1	1	1	1	1	2	0
1	2	1	1	2	0	0	2
2	0	0	2	2	0	2	0
2	1	1	1	2	0	0	2
2	2	2	0	2	0	1	1

The truth table of those operations is shown in Table 0.2. Assume every operation is in the quantum system. For convenience, classical names are used.

The truth Table of Quantum Ternary AND, NAND, OR, NOR, XOR, and XNOR logic operations for ternary logic is presented in the above Table. For AND logic operation its output value depends on the minimum value of its inputs. Similarly, in the case of the OR logic operation, its output value depends on the maximum value of its inputs. For the XOR operation, its output value is the sum of the value of its inputs.

Therefore, the outputs of Quantum Ternary NAND, NOR, and XNOR logic operations become the inverted of quantum ternary AND, OR, and XNOR logic operations.

0.9.1 Ternary Quantum-AND Operation

Quantum ternary AND operation is defined as $Y_{QAND} = \min(X, Y)$, where X and Y are the input from $\{|0>, |1>, |2>\}$. The truth table of quantum ternary AND operation is given in Table 0.3.

TABLE 0.3

Truth Table of Quantum Ternary AND Operations

	|0>	|1>	|2>
|0>	|0>	|0>	|0>
|1>	|0>	|1>	|1>
|2>	|0>	|1>	|2>

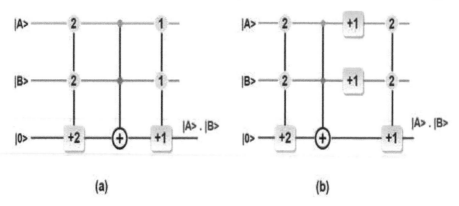

(a)   (b)

FIGURE 0.5
Circuit Diagram of Quantum Ternary AND Operation

0.9.1.1   Design Architecture of Ternary Quantum-AND Operation

The circuit diagram of the Quantum Ternary AND operation is shown in Figure 0.5; both (a) and (b) show the circuit diagram of the Quantum Ternary AND operation. Any of those can be used.

The Quantum Ternary AND operation require the shifting operations that include Z (+1) and Z (+2) shifting. Where one Z (+2) operation is controlled by both inputs and will open only if the value of both inputs is |2>. And following a C^2 NOT gate is used which will work only if the inputs A and B met conditions of A! =B and A, B! = |0>. And the following Z (+1) operation is controlled by both inputs and will open only if the value of both inputs is |1> (Figure 0.5 (a)). This will give the expected outputs that were shown in the truth table. Figure 0.5 (b) is the equivalent of Figure 0.5 (a), but requires more quantum ternary gates.

0.9.2   Ternary Quantum-NAND Operation

As it is known how to design the quantum ternary AND operation, it is understandable that only a NOT operation is required to get the result of quantum NAND operation. Quantum NAND operation is defined as $Y_{QNAND} = \overline{(\min(X, Y))}$ where X and Y are the input from {|0>, |1>, |2>}. Table 0.4 shows the truth table of Quantum Ternary NAND Operations.

From the truth table, it can be understood that this operation will not affect the quantum ternary AND output value |1>. If the output value |0> it will convert it to the |2> and vice versa.

TABLE 0.4

Truth Table of Quantum Ternary NAND Operations

	|0>	|1>	|2>
|0>	|2>	|2>	|2>
|1>	|2>	|1>	|1>
|2>	|2>	|1>	|0>

0.9.2.1 Design Architecture of Ternary Quantum-NAND Operation

The quantum ternary NAND operation is nothing but the $\overline{\text{AND}}$ operation. Therefore, the only need is to invert the output of the quantum ternary AND operation outputs. The circuit diagram of Quantum Ternary NAND operation is shown in Figure 0.6.

The output can be inverted in two ways. Figures 0.6 (a) and 0.6 (b) both can perform quantum ternary NAND operations. Because Z(02) swapping is the same as quantum NOT operation.

0.9.3 Ternary Quantum-OR Operation

Quantum OR operation is defined as $Y_{QOR} = \max (X, Y)$, where X and Y are the input from {\|0>, \|1>, \|2>}. The operations in the quantum ternary OR operation are shown in Table 0.5.

0.9.3.1 Design Architecture of Ternary Quantum-OR Operation

The Quantum Ternary OR operation requires the shifting operations that include Z(+1) and Z(+2) shifting. Where two Z(+1) operation is not controlled, and two Z(+1) operation is controlled by two input. And two Z(+2) operation is controlled

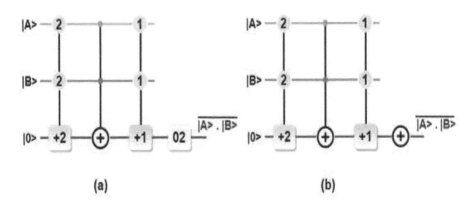

(a)   (b)

FIGURE 0.6

Circuit Diagram of Quantum Ternary NAND Operation

TABLE 0.5

Truth Table of Quantum Ternary OR
Operations

	|0>	|1>	|2>
|0>	|0>	|1>	|2>
|1>	|1>	|1>	|2>
|2>	|2>	|2>	|2>

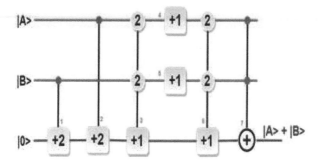

FIGURE 0.7

The Circuit Diagram of Quantum Ternary OR Operation

by respective one input (input |B> for the first Z(+2), and input |A> for the second
Z(+2)). And, a two-input controlled C^2 NOT gate is also required to get the expected
result that was shown above in the truth table. The circuit diagram of The Quantum
Ternary OR operation is shown in Figure 0.7.

0.9.4   Ternary Quantum-NOR Operation

Quantum Ternary NOR operation is defined as $Y_{QNOR} = \overline{(\max(X, \ Y))}$, where X and
Y are the input from {|0>, |1>, |2>}.

The operations in the quantum ternary NOR operation are shown in Table 0.6.

TABLE 0.6

Truth Table of Quantum Ternary NOR
Operations

	|0>	|1>	|2>
|0>	|2>	|1>	|0>
|1>	|1>	|1>	|0>
|2>	|0>	|0>	|0>

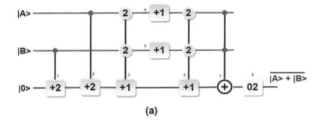

(a)

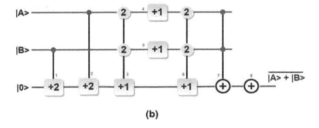

(b)

FIGURE 0.8
The Circuit Diagram of Quantum Ternary NOR Operation

0.9.4.1 Design Architecture of Ternary Quantum-NOR Operation

The Quantum Ternary NOR NOR operation is nothing but the $\overline{\text{OR}}$ operation. Therefore, it is needed to invert the output of the quantum ternary OR operation's outputs. Therefore the circuit diagram will be the same as the quantum ternary OR operation with an extra NOT operation at the end. The circuit diagram of Quantum Ternary NOR operation is shown in Figure 0.8.

0.9.5 Ternary Quantum XOR Operation

Quantum Ternary XOR operation is defined as Y_{QXOR} = sum (X, Y), where X and Y are the input from {|0>, |1>, |2>}. The operations in the quantum ternary XOR operation are shown in Table 0.7.

TABLE 0.7
Truth Table of Quantum Ternary XOR Operation

	|0>	|1>	|2>
|0>	|0>	|1>	|2>
|1>	|1>	|2>	|0>
|2>	|2>	|0>	|1>

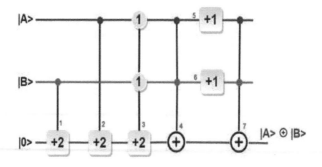

FIGURE 0.9
The Circuit Diagram of Quantum Ternary XOR Operation

TABLE 0.8
Truth Table of Quantum Ternary XNOR Operations

	$|0>$	$|1>$	$|2>$
$|0>$	$|2>$	$|1>$	$|0>$
$|1>$	$|1>$	$|0>$	$|2>$
$|2>$	$|0>$	$|2>$	$|1>$

0.9.5.1 Design Architecture of Ternary Quantum XOR Operation

The circuit diagram of the Quantum Ternary XOR operation is shown in Figure 0.9. The Quantum Ternary XOR operation requires the shifting operations that include Z (+1) and Z (+2) shifting.

Where two Z (+1) operation is not controlled, and two Z (+2) operation is controlled by respective one input. And one Z (+2) operation is controlled by both inputs. And, two two-input controlled C^2 NOT gate is also required to get the expected result.

0.9.6 Ternary Quantum XNOR Operation

Quantum Ternary XNOR operation is defined as $Y_{QXNOR} = \overline{\text{sum}(X, Y)}$, where X and Y are the input from $\{|0>, |1>, |2>\}$. The operations in the quantum ternary XOR operations are shown in Table 0.8.

0.9.6.1 Design Architecture of Ternary Quantum XNOR Operation

The Quantum Ternary XNOR operation is nothing but the ternary $\overline{\text{QXOR}}$ operation. Therefore, it is needed to invert the output of the quantum ternary XOR operation's outputs. In order to do that, an inverter is needed to add to the QXOR's circuit. The circuit diagram of the quantum ternary XNOR operation is shown in Figure 0.10.

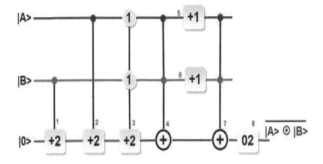

FIGURE 0.10
The Circuit Diagram of Quantum Ternary XNOR Operation

Z (O2) shift operation is added which will invert the output value of the quantum ternary XNOR operations. Quantum NOT operation can be added instead of Z (02) as shown earlier.

0.10 Fundamental Operations in Multiple-Valued DNA Computing

In this section, the fundamental operations will be implemented in ternary logic in DNA computing. NOT, OR, NOR, AND, NAND, XOR, XNOR, etc. will be implemented in DNA computing with ternary logic system.

DNA ternary logic functions are those functions that have significance if a third value is acquainted with the DNA binary logic. Here, ACCTAG, CAAGCT, and TG-GATC will represent 0, 1, and 2 in ternary logic which denotes the ternary levels for basic logic operations to represent false, undefined, and true, respectively. The basic operations of DNA ternary logic can be defined as follows:

$$y_{OR} = \max(x, y)$$
$$y_{NOR} = \overline{\max(x, y)}$$
$$y_{AND} = \min(x, y)$$
$$y_{NAND} = \overline{\min(x, y)}$$
$$y_{XOR} = \text{sum}(x, y)$$
$$y_{XNOR} = \overline{\text{sum}(x, y)}$$

where, $x, y = \{0, 1, 2\}$.

But in this case, 0, 1, and 2 are used for simplicity. The actual values are the DNA strands which are ACCTAG, CAAGCT, and TGGATC as mentioned before.

The truth table of those DNA operations is shown in Table 0.9. Assume every operation is in the quantum system. For convenience, classical names are used.

TABLE 0.9

Truth Table for Ternary AND, NAND, OR, NOR, XOR, XNOR Operations

Input 1	Input 2	AND	NAND	OR	NOR	XOR	XNOR
0	0	0	2	0	2	0	2
0	1	0	2	1	1	1	1
0	2	0	2	2	0	2	0
1	0	0	2	1	1	1	1
1	1	1	1	1	1	2	0
1	2	1	1	2	0	0	2
2	0	0	2	2	0	2	0
2	1	1	1	2	0	0	2
2	2	2	0	2	0	1	1

The truth table of DNA Ternary AND, NAND, OR, NOR, XOR, and XNOR logic operations for ternary logic is depicted in the Table 0.1. For AND logic operation its output value depends on the minimum value of its inputs. Similarly, in the case of the OR logic operation, its output value depends on the maximum value of its inputs. For the XOR operation, its output value is the sum of the value of its inputs.

Therefore, the outputs of DNA Ternary NAND, NOR, and XNOR logic operations become the inverted of quantum ternary AND, OR, and XNOR logic operations.

0.10.1 Ternary DNA-NOT Operation

It is known about the NOT operation principle that the NOT operation will simply invert the input value. Here in ternary DNA logic, the invert of ACCTAG is TGGATC and vice versa. And the invert value of CAAGCT is CAAGCT itself.

DNA-AND operation is defined as $Y_{DNOT}(X) = \overline{X}$, where X is the input from {ACCTAG, CAAGCT, TGGATC}.

0.10.1.1 Design Procedure of DNA Ternary NOT Operation

Figure 0.11 shows the architecture of the DNA ternary NOT operation. As it is said earlier, the architecture will not be complex in DNA ternary logic implementation. Here no base sequence is needed in the test tube. The fluorescent level will deter-

TABLE 0.10

Truth Table of DNA Ternary Standard NOT Operations

X	$\overline{X}$
ACCTAG	TGGATC
CAAGCT	CAAGCT
TGGATC	ACCTAG

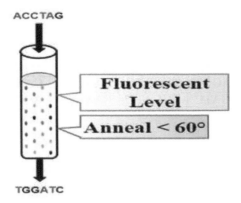

FIGURE 0.11
The Operational Diagram of DNA Ternary NOT Operation

mine the output. The Annealing temperature should be more than 60°C for the DNA ternary NOT operations.

0.10.2 Ternary DNA-AND Operation

DNA ternary AND operation is defined as $Y_{DTAND} = \min(X, Y)$, where X and Y are the input from {0, 1, 2}. Table 0.11 shows the truth table of DNA ternary AND operations.

0.10.2.1 Design Procedure of DNA Ternary AND Operation

Figure 0.12 shows the architecture of the DNA ternary AND operation. Where two input sequences will add to the test tube and the fluorescence level will produce the output based on the input sequence. The annealing temperature is more than 60°C for the DNA ternary AND operations.

0.10.3 Ternary DNA NAND Operation

DNA ternary NAND operation is defined as $Y_{DTNAND} = \overline{\min(X, Y)}$, where X and Y are the input from {0, 1, 2}. The truth table of DNA ternary NAND operations is shown in Table 0.12.

TABLE 0.11
Truth Table of DNA Ternary AND Operations

	ACCTAG	CAAGCT	TGGATC
ACCTAG	ACCTAG	ACCTAG	ACCTAG
CAAGCT	ACCTAG	CAAGCT	CAAGCT
TGGATC	ACCTAG	CAAGCT	TGGATC

TABLE 0.12

Truth Table of DNA Ternary NAND Operations

	ACCTAG	CAAGCT	TGGATC
ACCTAG	TGGATC	TGGATC	TGGATC
CAAGCT	TGGATC	CAAGCT	CAAGCT
TGGATC	TGGATC	CAAGCT	ACCTAG

0.10.3.1   Design Procedure of DNA Ternary NAND Operation

From Table 0.12, it is found that the DNA ternary NAND operation is the inverted output of the DNA ternary AND operation. The DNA ternary NOT operation is designed which inverts an input. Therefore, the DNA ternary NAND operation can be constructed by adding STI to the DNA ternary AND operation which is shown in Figure 0.13.

Figure 0.13 shows that the DNA ternary NAND is constructed using one DNA ternary AND operation followed by one DNA ternary NOT operation.

0.10.4   Ternary DNA-OR Operation

DNA ternary OR operation is defined as $Y_{DTOR} = \max(X, Y)$, where X and Y are the input from {0, 1, 2}. Table 0.13 shows the truth table of DNA ternary OR operations.

0.10.4.1   Design Procedure of DNA Ternary OR Operation

Figure 0.14 shows the architecture of the DNA ternary OR operation, where two input sequences will mix into the test tube and the fluorescence level will produce

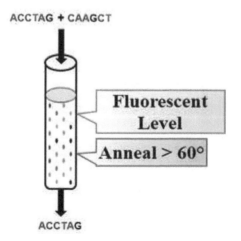

FIGURE 0.12

The Operational Diagram of DNA Ternary AND operation

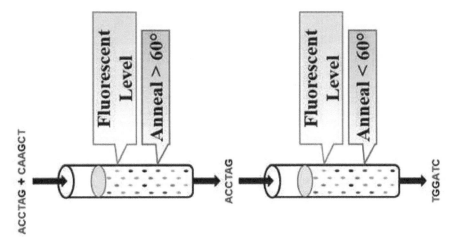

FIGURE 0.13
The Operational Diagram of DNA Ternary NAND Operation

TABLE 0.13
Truth Table of DNA Ternary OR Operations

	ACCTAG	CAAGCT	TGGATC
ACCTAG	ACCTAG	CAAGCT	TGGATC
CAAGCT	CAAGCT	CAAGCT	TGGATC
TGGATC	TGGATC	TGGATC	TGGATC

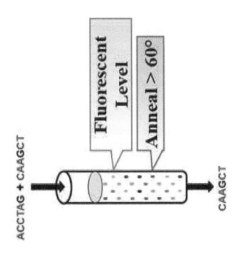

FIGURE 0.14
The Operational Diagram of DNA Ternary OR operation

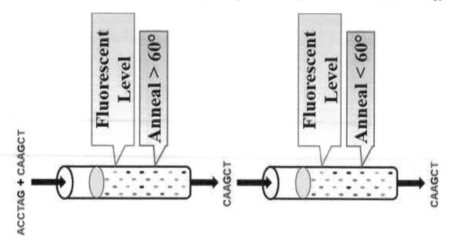

FIGURE 0.15
The Operational Diagram of DNA Ternary NOR Operation

the output based on the input sequence. The annealing temperature is more than 60°C for the DNA ternary OR operations.

0.10.5 Ternary DNA-NOR Operation

DNA ternary NOR operation is defined as $Y_{DTNOR} = \overline{\max(X, Y)}$, where X and Y are the input from $\{0, 1, 2\}$. Table 0.14 shows the truth table of DNA ternary NOR operations.

0.10.5.1 Design Procedure of DNA Ternary NOR Operation

From Table 0.14 it is found that DNA ternary NOR operation is the inverted output of the DNA ternary OR operation. The DNA ternary NOT operation is designed which inverts an input. Therefore, it is possible to construct the DNA ternary NOR operation by adding STI to the DNA ternary OR operation, which is shown in Figure 0.15.

Figure 0.15 shows that the DNA ternary NOR is constructed using one DNA ternary OR operation followed by one DNA ternary NOT operation.

TABLE 0.14
Truth Table of DNA Ternary NOR Operations

	ACCTAG	CAAGCT	TGGATC
ACCTAG	TGGATC	CAAGCT	ACCTAG
CAAGCT	CAAGCT	CAAGCT	ACCTAG
TGGATC	ACCTAG	ACCTAG	ACCTAG

TABLE 0.15

Truth Table of DNA Ternary XOR Operations

	ACCTAG	CAAGCT	TGGATC
ACCTAG	ACCTAG	CAAGCT	TGGATC
CAAGCT	CAAGCT	TGGATC	ACCTAG
TGGATC	TGGATC	ACCTAG	CAAGCT

0.10.6   Ternary DNA-XOR Operation

DNA ternary XOR operation is defined as Y_{DTXOR} = sum (X, Y), where X and Y are the input from {0, 1, 2}. Table 0.15 shows the truth table of DNA ternary XOR operations.

0.10.6.1   Design Procedure of DNA Ternary XOR Operation

Figure 0.16 shows the architecture of the DNA ternary XOR operation, where two input sequences will mix to the test tube and the fluorescence level will produce the output based on the input sequence. The annealing temperature is more than 60°C for the DNA ternary XOR operations.

0.10.7   Ternary DNA XNOR Operation

DNA ternary XNOR operation is defined as Y_{DTNOR} = $\overline{\text{sum(X, Y)}}$, where X and Y are the input from {0, 1, 2}. Table 0.16 shows the truth table of DNA ternary XNOR operations.

0.10.7.1   Design Procedure of DNA Ternary XNOR Operation

From Table 0.16, it is found that DNA ternary XNOR operation is the inverted output of the DNA ternary XOR operation. The DNA ternary NOT operation is designed which inverts an input. Therefore it is possible to construct the DNA ternary XNOR

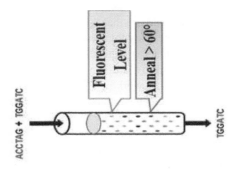

FIGURE 0.16

The Operational Diagram of DNA Ternary XOR Operation

TABLE 0.16

Truth Table of DNA Ternary XNOR Operations

	ACCTAG	CAAGCT	TGGATC
ACCTAG	TGGATC	CAAGCT	ACCTAG
CAAGCT	CAAGCT	ACCTAG	TGGATC
TGGATC	ACCTAG	TGGATC	CAAGCT

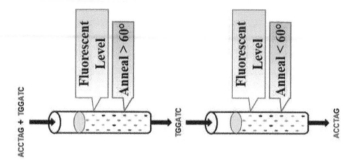

FIGURE 0.17

The Operational Diagram of DNA Ternary XNOR Operation

operation by adding STI to the DNA ternary XOR operation which is shown in Figure 0.17.

0.11 Summary

This chapter has focused on multi-valued quantum computing with some basic operations. The ternary quantum system represents one type of three-dimensional quantum system with the basis states |0>, |1>, and |2>. These basis states are called qutrit states. A ternary computer is a computer that does computations using ternary logic (three possible values) rather than binary logic (two possible values). The fundamental quantum gates that will work in ternary quantum computing are Quantum Ternary Shift Gates, Quantum Ternary Toffoli Gates, and Quantum Ternary C2 NOT Gates, where quantum ternary shift gates are working along with the Toffoli gate. This chapter has presented basic multi-valued quantum operations with their architectures and working principles. Necessary figures are also shown in their description for better understanding.

This chapter has also presented the ternary logic in the DNA computing system, that means establishing basic gates and fundamental operations in the ternary system. The circuit diagram of different DNA Computing operations in the multiple-valued system is discussed with necessary figures. As always, the design principles and working principles of the operations are shown. To be mentioned, DNA computing in multiple-valued logic is theoretically easier than in the two-valued logic system. Because no DNase Enzyme will be used to detect the DNA strands bond in the ternary system.

Bibliography

[1] Hallworth, R. P., & Heath, F. G. (1962). Semiconductor circuits for ternary logic. In Proceedings of the IEE-Part C: Monographs, 109(15), 219-225.

[2] Dhande, A. P., & Ingole, V. T. (2005, March). Design and implementation of 2 bit ternary ALU slice. In Proc. Int. Conf. IEEE-Sci. Electron., Technol. Inf. Telecommun (Vol. 17).

[3] Zheng, X., Yang, J., Zhou, C., Zhang, C., Zhang, Q., & Wei, X. (2019). Allosteric DNAzyme-based DNA logic circuit: Operations and dynamic analysis. Nucleic Acids Research, 47(3), 1097-1109.

[4] Freier, S. M., Kierzek, R., Jaeger, J. A., Sugimoto, N., Caruthers, M. H., Neilson, T., & Turner, D. H. (1986). Improved free-energy parameters for predictions of RNA duplex stability. In Proceedings of the National Academy of Sciences, 83(24), 9373-9377.

[5] Breslauer, K. J., Frank, R., Blöcker, H., & Marky, L. A. (1986). Predicting DNA duplex stability from the base sequence. In Proceedings of the National Academy of Sciences, 83(11), 3746-3750.

[6] Marella, S. T., & Parisa, H. S. K. (2022). Introduction to Quantum Computing. IntechOpen. doi: 10.5772/intechopen.94103.

[7] Miller, M. D., & Thornton, M. A. (2008). MVL concepts and algebra. In Multiple Valued Logic: Concepts and Representations (pp. 21-42). Springer, Cham.

[8] Haghparast, M., Wille, R., & Monfared, A. T. (2017). Towards quantum reversible ternary coded decimal adder. Quantum Information Processing, 16(11), 1-25.

[9] Mandal, S. B., Chakrabarti, A., & Sur-Kolay, S. (2011, May). Synthesis techniques for ternary quantum logic. In 2011 41st IEEE International Symposium on Multiple-Valued Logic (pp. 218-223). IEEE.

[10] Di, Y., Zhang, J., & Wei, H. (2008). Cartan decomposition of a two-qutrit gate. Science in China Series G: Physics, Mechanics and Astronomy, 51(11), 1668-1676.

Part I

Multi-Valued Sequential Circuits in Quantum Molecular Biology

Overview

Multi-Valued quantum computing is one of the most exciting and new scientific disciplines to emerge in the world today. Quantum computers are still a pipe dream, and they are not yet scalable. The researchers are attempting to suggest a quantum circuit that will be used in the fabrication of quantum processors, memory devices, and other devices. Multiple-valued DNA computing is also an exciting scientific field to develop in the world today, yet it is still in its early stages. In the current state of technology, multi-valued DNA computers are still a pipe dream too. The researchers hope to construct a multi-valued DNA circuit that can be used in the production of multi-valued DNA processors, memory devices, and other devices as a result of their research. The combination of these two exciting fields is a new creation that has been introduced in this book. This combination is quantum molecular biology.

The sequential circuit is a special type of circuit that has a series of inputs and outputs. The outputs of the sequential circuits depend on both the combination of present inputs and previous outputs. The previous output is treated as the present state. Sequential circuits consist of present inputs and the previous outputs used as inputs which finally produce the output result. These circuits are so much important and used in shift registers, flip flops, clocks counters, etc. In this part, sequential circuits are implemented in multi-valued quantum, DNA, quantum-DNA and DNA-quantum computing. The topics to be discussed are D flip flop, SR latch, SR flip flop, JK flip flop, T flip flop, Shift register, Ripple counter, Synchronous counter. All these sequential circuits will be described here with their general organization, circuit architecture, working principle and examples. All necessary figures will be shown also. Sequential logic is used to construct finite-state machines, a basic building block in all digital circuitry. Virtually all circuits in practical digital devices are a mixture of combinational and sequential logic.

1

Multiple-Valued Sequential Circuits in Quantum Computing

1.1 Introduction

The theory of multi-valued quantum mechanics is one of the most prominent scientific theories of the twentieth century. In addition, to give answers to some outstanding questions, it has had an impact on many modern technologies by expressing a new line of scientific thinking.

Scientists and computer scientists such as Paul A. from the Thomas J. Watson Research Center, Charles H. Bennet of IBM, Benioff of Argonne National Laboratory in Illinois, David Deutsch from the University of Oxford, and Richard P. Feynman of Caltech investigated the possibility of Multi-Valued quantum computational devices based on Multi-Valued quantum mechanics during the 1970s and early 1980s. In 1982, Feynman attempted to develop a completely new type of computer system based on notions from multi-valued quantum physics theory. The author created an abstract model to explain how a multi-valued quantum system could be used to perform computations, and also how such a machine could be used as a simulator for multi-valued quantum physics problems. Researchers in the field of quantum physics, on the other hand, have the opportunity to conduct several multi-valued quantum physics experiments within the multi-valued quantum mechanical computer. He went on to explain that multi-valued quantum computers are capable of solving some multi-valued quantum physics problems that are now intractable by conventional computing methods. Deutsch established in 1985 that multi-valued quantum computers are capable of performing tasks that are well above the capabilities of traditional computers.

A number theory difficulty known as factorization was solved in multi-valued quantum computers in 1994 by Peter Shor, who proved how the issue would be addressed in multi-valued quantum computers. He also disclosed the fact that regular computers answer issues in exponentially growing time, whereas multi-valued quantum computers solve problems in polynomial time, as opposed to traditional computers. Under Multi-Valued quantum theory and the sequence of discoveries in multi-valued quantum computation that have occurred over the century, multi-valued quantum computers may be more powerful than conventional computers. Scientists have been paying particular attention to multi-valued quantum computation in recent years as a result of the findings of this research.

DOI: 10.1201/9781003381921-1

Multi-Valued quantum computing is the use of the collective features of multi-valued quantum states, such as superposition and entanglement, to carry out computations. A Multi-Valued quantum computation system is a machine that does multi-valued quantum calculations on a large number of variables. Multi-Valued quantum computing systems include multi-valued quantum circuit models, multi-valued quantum Turing machines, adiabatic multi-valued quantum computers, one-way multi-valued quantum computers, and many types of multi-valued quantum cellular automata, to name a few examples.

Currently, the multi-valued quantum circuit, which is based in part on the multi-valued quantum bit, or "qubit," which is comparable to the bit in conventional processing, is the most widely used model of quantum computation. An individual qubit can be in either a superposition state, which is between 0 and 1 or in a multi-valued quantum state, which is either 1 or 0. With regards to measurements, however, it is always a zero or one, depending on the likelihood involved. The likelihood of each outcome is determined by the multi-valued quantum state of the qubit right before the measurement is performed.

Making a physical multi-valued quantum computer is a long-term aim that involves developing high-quality qubits using technologies such as transoms, ion traps, and topological multi-valued quantum computers. These qubits may be constructed differently depending on the computing model employed by the entire multi-valued quantum computer, such as multi-valued quantum logic gates, multi-valued quantum annealing, or adiabatic multi-valued quantum computation, among other things. There are now several significant barriers in the way to the development of viable multi-valued quantum computing systems. The fact that qubits suffer from multi-valued quantum decoherence and state integrity makes it extremely difficult to sustain their multi-valued quantum states in the long run. Because of this, multi-valued quantum computers require error correction as well as computation. According to church theory, while both multi-valued quantum computing and conventional computing are capable of solving identical tasks, multi-valued quantum computers require less time.

Although physicists have known since 1920 that the universe of subatomic particles is a world unto itself, they have been unable to prove it. The discovery that multi-valued quantum particles might be used in calculations, on the other hand, took nearly 50 years for computer scientists to make. The current state of affairs is that Google, IBM, and NASA are all working on the development of a real-world multi-valued quantum computing system, and researchers all over the world are striving to discover a way to make it a reality.

1.2 Multi-Valued Quantum D Flip Flop

In essence, a Multi-Valued Quantum D flip flop is a two-state timed flip flop. The qubit inputs of a quantum D-type flip flop are operated with a delay in a single clock cycle. The Multi-Valued Quantum D flip flop is also referred to as a delay flip flop.

One of the primary limitations of the Quantum SR NAND Gate Bistable circuit is that the indeterminate input condition of SET = "|0>" and RESET = "|0>" is prohibited. This condition overrides the feedback latching action by forcing both qubits' outputs to logic "|1>," and whichever input goes to logic "|1>" first loses control of the latch, while the other input, which is still at logic "|0>," controls the latch's ultimate state. A quantum Data Latch, quantum Delay flip flop, quantum D-type Bistable, quantum D-type Flip Flop, or simply a Multi-Valued Quantum D flip flop can be created by connecting an inverter between the "SET" and "RESET" qubit inputs.

The Multi-Valued Quantum D flip flop is by far the most important of all quantum timed flip-flops. When a quantum inverter (quantum NOT gate) is added between the Set and Reset inputs, the two inputs |S> and |R> become complements of each other, ensuring that the two inputs |S> and |R> are never equal (|0> or |1>) at the same time, allowing us to control the toggle action of the flip-flop with just one |D> (Data) input.

The Data input, designated "|D>," is then used in place of the "Set" signal, with the inverter generating the complementary "Reset" input, resulting in a level-sensitive quantum D-type flip-flop from a level-sensitive SR-latch, with |S> = |D> and |R> = |D>.

The proposed quantum D flip-flop circuit contains only one qubit input, which must be in a coherence state to perform the quantum computing function. As a result, the circuit must operate in an undefined environment. The coherence state will be disturbed if any particles arise. Heat will be generated by the quantum D flip-flop, which must be removed fast to cool down the circuit and stabilize the coherence state.

1.2.1 General Organization of Multi-Valued D Flip Flop

The ternary quantum D flip flop is one of the most crucial circuits when compared to other timed type flip flips. The ternary quantum D flip flop ensures that the ternary quantum SR flip flop's two inputs are never the same. The block diagram of multi-valued quantum D flip flop is shown in Figure 1.1.

There are two inputs to ternary quantum D flip flops: one for data and one for a clock. Multi-Valued Quantum D flip flops with multi-valued outputs have two outputs that are logically opposite. The clock input helps to synchronize the circuit with an external signal. A ternary quantum D flip flop's output can take one of two possible forms. Data input is directed to a ternary quantum NAND operation circuit in this blog diagram, while data input is reversed and routed to another ternary quantum NAND operation circuit. Both NAND procedures need the clock pulse as an input. The ternary quantum SR latch receives the result of two ternary quantum NAND operations. The ternary quantum D flip flop is constructed using the ternary quantum SR latch. This property is used to induce a delay in the circuit's data flow. The ternary quantum SR Latch is made up of two ternary quantum NAND operations. The remaining two outputs of the ternary quantum SR Latch function were uncovered. A ternary quantum D flip flop's output can be one of two types, one of which is

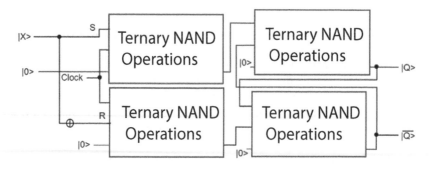

FIGURE 1.1
Block Diagram of Multi-Valued Quantum D Flip Flop

logically inverse to the other. The ternary quantum D flip flop will continue to function if the clock is enabled; else, the ternary quantum D flip flop will stop working.

1.2.2 Circuit Architecture of Multi-Valued Quantum D Flip-Flop

The D flip-flop has a single qubit input and is built with Quantum NAND operations and a Quantum SR latch. The ternary clock qubit input determines the multi-valued Quantum D flip-flop. One ternary qubit input is visible in the circuit, as shown in the diagram. The one line of this ternary qubit input will be directed into a multi-valued Quantum NAND operation, referred to as the S input in Circuit. In this multi-valued Quantum NAND operation, ternary |S> qubit input and clock qubit input are used. Figure 1.2 shows the multi-valued quantum D flip-flop circuit.

  When a Ternary |X> qubit is fed into another line, it performs a multi-valued Quantum NOT operation first. In a multi-valued Quantum NAND operation, this Quantum NOT operation was entered as R. The two-qubit inputs R and Clock are used in this ternary Quantum NAND operation. The output of the ternary Quantum NAND operations is used as an input in the multi-valued Quantum SR latch. With these two ternary inputs, the SR latch will be done. The outputs of ternary |Q> and $\overline{|Q}>$ will be the final output of a multi-valued Quantum SR latch. |Q> always produces the inverse of $\overline{|Q}>$.

1.2.3 Working Principle of Multi-Valued Quantum D Flip-Flop

There were two inputs SET and RESET in the Multi-Valued Quantum SR flip flop. However, in a D flip-flop, one ternary input and one line of input are referred to as a SET, and by attaching a Quantum NOT gate to the other line input, the D flip-flop is referred to as being RESET. Because the multi-valued SR latch is no longer available when both inputs are LOW, this complement removes the ambiguity inherent in it. One input is called a data input in multi-valued Quantum D flip-flops. If the data input is high, the flip-flop becomes SET, but if it is low, such as |0>, the flip-flop changes to state and becomes RESET.

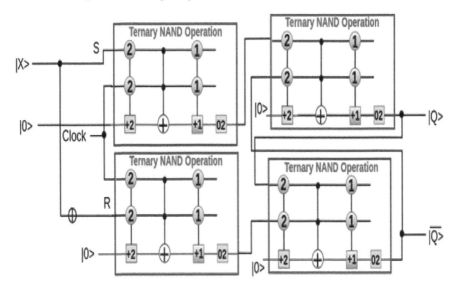

FIGURE 1.2
Multi-Valued Quantum D Flip-Flop Circuit

However, since the output of the flip flop changes with each pulse provided to this data input, this would be somewhat meaningless. To avoid this, a ternary input named "CLOCK" or "ENABLE" is used to isolate the data input from the latching circuitry of the flip flop once the necessary data is stored. When the clock input is active, the condition of the |X> input is exclusively replicated to the output |Q>. This is then used to create a D Flip Flop, which is a sequential device.

So long as the clock input is HIGH, the "multi-valued D flip flop" will store and output whatever logic level is provided to its data terminal. The "set" and "reset" inputs of the flip-flop are both kept at logic level "|1>" after the clock input becomes LOW, preventing the flip-flop from changing state and storing whatever data was present on its output before the clock transition. In other words, the output is "latched" at one of three logic levels: "|0>" or "|1> or |2>." Table 1.1 shows the truth table of multi-valued quantum D flip-flop.

TABLE 1.1
Truth Table of Multi-Valued Quantum D Flip-Flop

|Clk>	|x>	|Q>	$\overline{\|Q}$ >	Description
↓ >> |0>	X	Q	Q	Memoryno change
↑>>|1>	|0>	|0>	|1>	Reset Q>>0
↑ >>|1>	|1>	|1>	|0>	Set Q>>
↑ >> |1>	|2>	|2>	|0>	Set Q>>

1.2.4   Example

In a ternary quantum D flip-flop operational circuit, one input is qubit input |x>. Assume this input is |0>. This input will perform a ternary quantum NAND operation with the clock input |1>. Then the output will be |1>.

Then qubit input will perform ternary quantum NOT operation and then again it will perform ternary quantum NAND operation. After the ternary NOT operation the input will be |1>. The clock input is |1>. Now the output will be |0> after performing ternary quantum NAND operation.

Two ternary quantum NAND operations produce output that will be entered into an input in the ternary quantum SR latch operational circuit. This ternary quantum SR latch operational circuit output will count as a final output in the ternary quantum D flip-flop operational circuit. So, here the two outputs are |S> = |1> and |R>= |0> which will be the inputs for the ternary quantum SR latch. The final outputs are |Q>= |1> and |0> which the inverse output of |Q>.

This is the final output of ternary quantum D flip-flop operational circuit and these outputs clarify that the correct output is produced by the ternary quantum D flip-flop operational circuit.

1.3   Multi-Valued Quantum SR Latch

There are two types of memory elements, depending on the triggering that is used to operate it. A latch is one of them, and a flip-flop is the other. Flip-flops are edge sensitive, whereas latches work with an enable signal and are level sensitive.

A synchronous tool is a Multi-Valued Quantum SR Latch. It doesn't use any control signals, instead of relying on the state of the |s> and |R> inputs. A Multi-Valued Quantum SR Latch is created by combining two ternary quantum NAND operations. Nonetheless, a multi-valued Quantum SR Latch can be created by combining two ternary quantum NOR operations. Two-qubit inputs are switched and negated in the Multi-Valued Quantum SR Latch. The Multi-Valued Quantum SR Latch is also known as the SET RESET latch. Received 2 outputs from a two-qubit input in the Multi-Valued Quantum SR Latch. This output is the inverse of the previous one. This study presented a Multi-Valued Quantum SR Latch, which was constructed using two ternary Quantum NAND operation circuits. The input line is swapped between two ternary Quantum NAND operations in this proposed Multi-Valued Quantum SR Latch, but it is not negated. In quantum computers, the Multi-Valued Quantum SR Latch serves as memory, and it has a variety of uses in a quantum processor. Multi-Valued Quantum SR Latch will be utilized on this device as a memory unit if some embedded systems are created employing the quantum device.

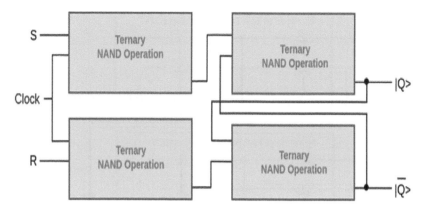

FIGURE 1.3

Block Diagram of Multi-Valued Quantum SR Latch

1.3.1 General Organization of Multi-Valued Quantum SR Latch

One of the most prevalent memory devices is the Multi-Valued Quantum SR Latch, which has an effect on the output as long as it is active. A Multi-Valued Quantum SR Latch's main qualities are that one qubit input behaves like a SET and another qubit input behaves like a RESET. Figure 1.3 shows the block diagram of multi-valued quantum SR Latch.

The Multi-Valued Quantum NAND operation is depicted in this block diagram as a basic component of the Multi-Valued Quantum SR Latch. It is a Multi-Valued Quantum SR latch made up of two basic processes. There are two input lines in the Multi-Valued Quantum SR latch, one for |s> and one for |R>. The ternary outputs |Q> and $|\overline{Q}$ > were generated from these two ternary qubit inputs. The output of the first multi-valued quantum NAND operation is used as an input for the second multi-valued quantum NAND operation, and the output of the second multi-valued quantum NAND operation is used as an input for the first multi-valued quantum NAND operation. If the input of |S> is |1>, the SR latch is activated; however, if the input of |R> is |1>, the SR latch is not having any influence on the output. In Multi-Valued Quantum SR Latch, a value of |1> cannot be used to activate two inputs.

1.3.2 Circuit Architecture of Multi-Valued Quantum SR Latch

Level-sensitive Multi-Valued Quantum SR Latches are constructed using only one fundamental operation, the ternary quantum NAND operational circuit. Figure 1.4 depicts the circuit architecture of multi-valued quantum SR Latch.

The Multi-Valued Quantum SR Latch has two ternary qubit inputs, as can be seen in the illustration above. |S> entered as an input in the first Multi-Valued Quantum NAND operation and the output of the second Multi-Valued Quantum NAND operation $|\overline{Q}$> entered as an input in the second Multi-Valued Quantum NAND operation There are two v+ gates, one v gate, and a CNOT gate present in the Multi-Valued

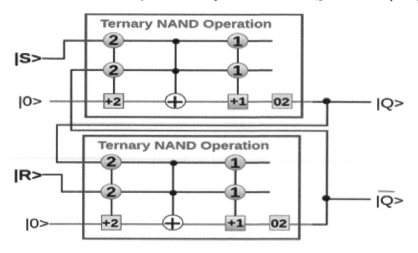

FIGURE 1.4

Circuit Architecture of Multi-Valued Quantum SR Latch

Quantum NAND operation. The architecture of Multi-Valued Quantum NAND is briefly discussed already. It is primarily obtained |Q> as a result of this Multi-Valued Quantum NAND operation.

Second, the multi-valued quantum NAND operation is performed on the ternary inputs |R> and |Q>, with the output $|\overline{Q}>$. Ancillary qubit |1> is used in each multi-valued quantum NAND operation. In quantum systems, this is a logical bit that is used to repair errors.

1.3.3   Working Principle of Multi-Valued Quantum SR Latch

In a Multi-Valued Quantum SR Latch, the state of the |S> and |R> inputs is all that matters. It is unaffected by control signals. |S> input is treated as a SET instruction, while |R> input is treated as a RESET instruction. The output |Q> will be |1> or |2>if the SET input of a Multi-Valued Quantum SR Latch is high, and the opposing output |0> will be the value of |Q> if the SET input is low. The value of |Q> is |1> or |2> when the RESET input is high, and |Q> is |0> when the RESET input is low.

The "memory" of the latch is essentially reset. The latch "latches" stay in their previously set or reset state when both inputs are low.

When both the SET and RESET inputs are high, the problem arises. As demonstrated in the circuit, the outputs |Q> and $\overline{Q}$ will have the inverse value. The circuit causes a "race condition" when the SET and RESET inputs |1> / |2> are both used together. Both gates must be identical in order for the device to be "metastable," or to remain in an undetermined state indefinitely. In actuality, only one gate will win if the proposed circuit is constructed; however, predicting which gate would win is challenging. As a result, a Multi-Valued Quantum SR Latch cannot have both the SET and RESET inputs set to high.

TABLE 1.2

Truth Table of Multi-Valued Quantum SR Latch

|S>	|R>	|Q>	$|\underline{Q}>$
|0>	|0>	Latched	
|0>	|1>	|1>	|0>
|1>	|0>	|0>	|1>
|0>	|2>	|2>	|0>
|2>	|0>	|0>	|2>
|1>	|1>	Metastable	
|2>	|2>	Metastable	

When the device is turned on, the same thing happens since both outputs, |Q> and |Q>, are low. Because of the disparities between the two gates, the device will swiftly depart the metastable state, but it's difficult to predict which of |Q> and |Q> will end up high. SR flip-flops must always be set to a known starting state before being used to avoid erroneous actions; users should not expect that they will initialize to a low state.

The truth table of multi-valued quantum SR Latch is shown in Table 1.2.

The Multi-Valued Quantum SR Latch is solved by the flip flop explained in the flip flop chapter. On the other hand, the Multi-Valued Quantum SR Latch is still an important part of a CPU or embedded device.

The quantum circuit creates a lot of heat, which makes isolating the qubit into a superposition state challenging. As a result, cooling the circuit is required to isolate the qubit into a superposition for a working quantum circuit. Any external particle can cause the qubit's coherence to be disrupted, causing it to become decoherent. The Multi-Valued Quantum SR Latch can only work if all of this is retained.

1.3.4 Example

Assume that the quantum operation contains both the ternary qubits |0> and |1> for SR Latch. SET instructions will be sent to one qubit input, while RESET instructions will be sent to the other. |0> will be used as a SET instruction in this case, and it will perform the ternary Quantum NAND operation as specified in the circuit.

The ultimate output is |0> as a result of this. A ternary quantum NAND operation's output principle states that if one of the inputs is |0>, then the output will be |0> as well.

In addition to completing the ternary quantum NAND operation, qubit input |1> now serves as a reset instruction and is placed into the circuit. If **|Q>=**|1> then **|$\overline{Q}$ >** will be |0>. As it is known, one output is always the inverse of the other one.

Multi-Valued Quantum SR with many values each ternary qubit input is handled by multi-valued Latch in the same way. However, because the computations were being done in the quantum superposition state, a lot of heat was generated during the process. The outcome will be decoherent because the computation is done in the coherence state.

1.4   Multi-Valued Quantum SR Flip Flop

Unlike Multi-Valued Quantum Combinational Logic circuits, multi-valued Quantum Sequential Logic circuits have some form of built-in "Memory" that changes state based on the real signals fed into its inputs at the time. For instance, Multi-Valued Quantum SR Flip Flops have a 1 qubit memory bistable. The SR flip flop has the same SET and RESET inputs. The SET input produces a |1>, whereas the RESET input produces a |0>.

The SET RESET flip flop is an abbreviation for the Multi-Valued Quantum SR Flip Flop. From the present state with an output, the reset input is used to return the flip flop to its initial state. The Multi-Valued NAND gate SR flip flop is a simple flip flop in which both outputs provide feedback to the input. In a memory circuit, this circuit is used to store one data bit. Set, RESET, and a found output are the three inputs. Because Multi-Valued Quantum SR Flip Flops have two inputs, most of which are from the outside, a two-qubit model will be employed. It produces more heat than quantum D flip flops at start since it uses two qubits. The basic gate in the middle determines the computation time of this Multi-Valued Quantum SR Flip Flop. In a wide range of processors and embedded devices, Multi-Valued Quantum SR Flip Flops are used. Although the proposed flip-flop produces some garbage, an error-correcting auxiliary qubit gives the desired output. The proposed quantum circuit will be more swiftly and effectively implemented in the actual world, addressing a wide range of challenges.

1.4.1   General Organization of Multi-Valued Quantum SR Latch

The Multi-Valued Quantum SR Flip Flop is also known as a SET - RESET flip-flop. The quantum SR latch clearly has a two-qubit input as a result of this. Figure 1.5 shows the block diagram of multi-valued quantum SR latch.

The Multi-Valued Quantum SR Flip Flop has two ternary inputs, |S> and |R>, as well as a clock input, |clock>. To begin, two ternary qubits |S> and |R> were combined in a Multi-Valued quantum NAND operation using clock input. The Multi-Valued Quantum NAND operation is performed via ternary input |S> and input |clock>. In addition, |R> and |clock> inputs conduct another Multi-Valued quantum NAND operation in concurrently. Because they happen in parallel, these two multi-valued quantum NAND operations take the same amount of time. Then, using the Multi-Valued Quantum NAND operations ternary output and the final ternary output $|\overline{Q}>$, then another Multi-Valued Quantum NAND operation is performed and get the result |Q>. Similar to |R> input lines Multi-Valued quantum NAND operations output and the final output |Q>, the quantum NAND operation is performed and the

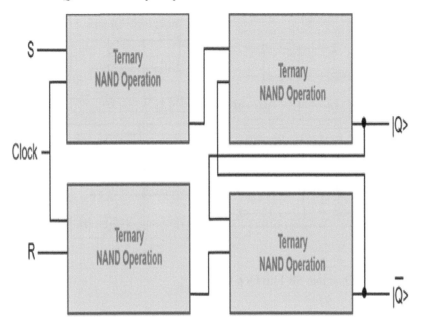

FIGURE 1.5
Block Diagram of Multi-Valued Quantum SR Latch

$|\overline{Q}>$ is produced. Actually, the created output enters Multi-Valued quantum SR latch and discovers the $|Q>$ and $|\overline{Q}>$ after the first Multi-Valued quantum NAND operations are completed. The outputs $|Q>$ and $|\overline{Q}>$ have the opposite value. The block diagram depicts all of the procedures involved in the proposed multi-valued quantum SR latch. During the procedure, all ternary qubits must be in the superposition state, which means they must be coherent. The superposition state will dissolve if any particle from the environment arrives close enough.

1.4.2 Circuit Architecture of Multi-Valued Quantum SR Flip Flop

The SR flip-flop in a quantum computer is mostly used as a SET RESET flip-flop. Figure 1.6 shows the multi-valued quantum SR Flip Flop circuits.

The proposed Multi-Valued Quantum SR Flip Flop circuit has two ternary inputs $|S>$ and $|R>$, as well as an additional clock input. Four Multi-Valued Quantum NAND operation circuits make up the Multi-Valued Quantum SR. At first, the ternary $|S>$ and clock inputs perform one Multi-Valued quantum NAND operation. Shift operation +2, C^2NOT, and CNOT are the three main operations that make up the Multi-Valued Quantum NAND operation circuit. The Multi-Valued quantum NAND operation is performed using the inputs $|R>$ and $|clk>$. Two of the outputs from the Multi-Valued Quantum NAND operations were used as inputs to the quantum SR latch. Multi-Valued quantum SR latch is performed using these two inputs.

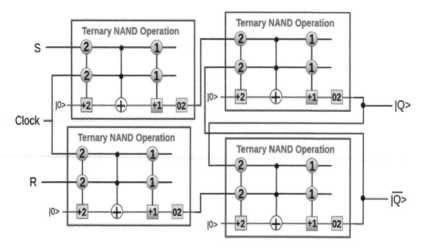

FIGURE 1.6
Multi-Valued Quantum SR Flip Flop Circuits

One output of previous Multi-Valued quantum NAND operations comes from the
|S> input line in the Multi-Valued quantum SR latch, and one of the output SR flip-
flop |$\overline{Q}$> conducts the Multi-Valued quantum NAND operation and creates the output
|Q>. The output from Multi-Valued Quantum NAND operation, which is based on
|R> input and |Q> output, enters as an input in Multi-Valued Quantum NAND oper-
ation and creates the output of Multi-Valued Quantum SR Flip Flop as |$\overline{Q}$>, just as it
did previously.

1.4.3 Working Principle of Multi-Valued Quantum SR Flip Flop

A Multi-Valued Quantum SR Flip Flop has two-qubit inputs |S> and |R>. Once the
qubit input becomes coherent and reaches a superposition state, the quantum operation is executed. Because the Quantum circuit generates a lot of heat, it needs to
transmit heat.

In a Multi-Valued Quantum SR Flip Flop, the |S> input first switches to super-
position and becomes coherent. The ternary quantum NAND operation is then con-
ducted with the |S> input and the clock qubit input |clk>. After finishing the ternary
quantum NAND operation, it generates an output. The |R> input with the |clk> input
also performs the ternary quantum NAND operation. The quantum SR latch employed the final outputs of both quantum NAND operations as inputs. One input
appears to be high, while the other appears to be low. In a quantum SR latch, which
is unaffected by control signals, the state of the |S> and |R> inputs is all that mat-
ters. The |S> input is treated as a SET command, while the |R> input is treated as
a RESET command. If the SET input of the Quantum SR latches becomes high, the
output |Q> becomes |1>, and the opposite output |0> becomes the value of |Q>. The
value of |Q> is |1> when the RESET input is high, and it is |0> when the RESET

input is low. The latch's "memory" is essentially wiped clean. When both inputs are low, the latch "latches" stay in the set or reset state.

When both the SET and RESET inputs are high, the actual issue develops. As demonstrated in the circuit, the outputs |Q> and |Q> will have polarized values. The circuit generates a "race scenario" when the SET and RESET inputs |1> are utilized. In order for the gadget to be "metastable," which implies it will be in an undetermined state for an infinite length of time, both gates must be identical. In actuality, only one gate will win if the proposed circuit is constructed; however, predicting which gate would win is challenging. As a result, with a Quantum SR latch, having both the SET and RESET inputs high is prohibited.

When the device is turned on, both ternary outputs, |Q> and |$\overline{Q}$>, are low, resulting in a condition that is identical to the one described above. The device will quickly leave the metastable state due to the differences between the two gates, but it's difficult to predict which of |Q> and |$\overline{Q}$> will end up high. To avoid spurious operations, it needs always to set SR flip-flops to a known beginning state before using them; it is nor possible to assume that they will initialize to a low state. The truth table of multi-valued quantum SR flip flop is shown in Table 1.3.

1.4.4 Example

Assume that the ternary inputs |0> and |1> are received in order for the Multi-Valued Quantum SR Flip Flop operational circuit to create the correct output. The |0> input end is wired into the |S> input end, while the |1> input end is wired into the |R> input end. The fact that the |R> input is set to |1> shows that the operation is a reset. Assume that the clock is turned on and that the input to the clock is |1>.

First, use the |clk> input to perform the quantum NAND operation. In a quantum NAND operation, the outcome is |1> if just one of the inputs is |0>; otherwise, the output will be |0>. So, **|S>= |0> and |clk>=|1>**, then the Intermediate Output is |1>.

Another |clk> and |R> input is then used to complete the ternary quantum NAND operation. The |clk> input is |1> in this example, and the |R> input is also |1>. Then the Intermediate Output is |0>

TABLE 1.3
Truth Table of Multi-Valued Quantum SR Flip Flop

|S>	|R>	|Q>	|$\overline{Q}$>
|0>	|0>	No Change	
|0>	|1>	|0>	|1>
|1>	|0>	|1>	|0>
|0>	|2>	|0>	|2>
|2>	|0>	|2>	|0>
|1>	|1>	Invalid	
|2>	|2>	Invalid	

These two ternary input qubits, |1> and |0>, will then be fed into the quantum SR latch as input. They will carry out the quantum SR latch operation. The SR latch operation circuit is made up of a series of quantum NAND operation circuits. So the final outputs are $|Q>=|0>$ and $\overline{Q}>=|1>$.

This suggested Multi-Valued Quantum SR Flip Flop circuit now has outputs |0> and |1> for the inputs |0> and |1>. This is the needed input for the specified input in a Multi-Valued Quantum SR Flip Flop. Heat is generated by Multi-Valued Quantum SR Flip Flops, however it has no influence on the output.

1.5   Multi-Valued Quantum JK Flip Flop

The most often used flip-flop in flip-flop designs will be the Multi-Valued Quantum JK flip-flop . The letters J and K are not abbreviated letters of other words, such as "S" for Set and "R" for Reset, but are distinct letters chosen by creator Jack Kilby to distinguish the flip-flop design from others. Regardless of the fact that Jack Kilby invented the digital electronics JK flip flop. The suggested Multi-Valued Quantum JK flip flop is not the same as the digital JK flip flop in terms of how it works.

The sequential operation of the Multi-Valued Quantum JK flip flop is identical to that of the preceding quantum SR flip flop, with the same "Set" and "Reset" inputs. The difference this time is that the "Multi-Valued Quantum JK flip flop" has no improper or banned quantum SR Latch input states even though S and R are both at logic "1." The Multi-Valued Quantum JK flip flop clearly does not address the issues with the quantum SR flip flop.

The Multi-Valued Quantum JK flip flop is essentially a gated quantum SR flip flop with clock qubit input circuitry to avoid the illegal or invalid output state that can occur when both inputs |S> and |R> are equal to logic level "1." Because of the extra time input, a Multi-Valued Quantum JK flip flop has four possible input combinations: "|1>," "logic |0>," "no change," and "toggle." As observed in the preceding chapter, a Multi-Valued Quantum JK flip flop has the same symbol as a quantum SR Bistable Latch. The Multi-Valued Quantum JK flip flop, like all flip flops, generates a lot of heat that must be dissipated for the process to work effectively. The Multi-Valued Quantum JK flip flop will not require as much power as other quantum circuits. Once all of the molecules are in superposition and coherence mode, the qubit may simply do the operation. In the Multi-Valued Quantum JK flip flop, a lot of trash values are found, and it is needed to investigate them further to understand what they are. This approach does not take the trash value into consideration.

1.5.1   General Organization of Multi-Valued Quantum JK Flip Flop

The Multi-Valued Quantum JK flip flop is the most commonly utilized flip flop in quantum processor construction. The two ternary qubit inputs of the Multi-Valued

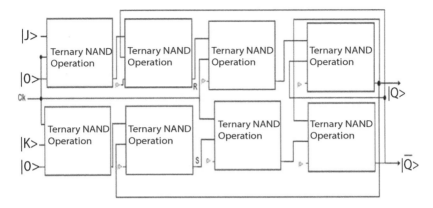

FIGURE 1.7
Block Diagram of Multi-Valued Quantum JK Flip Flop Operation

Quantum JK flip flop are designated |J> and |K>. Figure 1.7 shows the block diagram of multi-valued quantum JK flip flop operation

The ternary qubit input |J> and |k> are used in the Multi-Valued Quantum JK flip flop. Many Multi-Valued quantum NAND operations make up this Multi-Valued Quantum JK flip flop. After performing a couple of basic Multi-Valued quantum NAND operations, the ternary output of this operation is entered into the SR flip flop, yielding the ternary outputs |Q> and |$\overline{Q}$>. First, the Multi-Valued quantum NAND operation is performed using the ternary inputs |J> and |clk>. The output of the Multi-Valued quantum NAND operation and the output of the Multi-Valued Quantum JK flip flop |$\overline{Q}$> executes another Multi-Valued quantum NAND operation, resulting in the ternary output |S>. Because the ternary input is shared, both |K> and |clk> can be used. The Multi-Valued quantum NAND operation is also performed with ternary inputs. The |Q> output of the Multi-Valued Quantum JK flip flop conducts another Multi-Valued quantum NAND operation and creates the ternary output designated |R>.

The Multi-Valued quantum SR flip flop takes these |S> and |R> and produces two ternary outputs |Q> and |$\overline{Q}$>. There are four multi-valued quantum NAND operations in a multi-valued quantum SR flip flop. |S> ternary input and |clk> clock input Multi-Valued quantum NAND operation is performed using ternary input, as well as |R> ternary input and shared |clk> ternary input. The Multi-Valued quantum NAND operation is also performed via ternary input. The ternary output of these two NAND operations is used as a ternary input in the multi-valued quantum SR latch. The Multi-Valued quantum NAND operation is performed by two |Q> and one ternary input as well as |$\overline{Q}$> and another ternary input in Multi-Valued quantum SR latches. Finally, the Multi-Valued Quantum JK flip flops final ternary output |Q> and |$\overline{Q}$> is received after all of this quantum process.

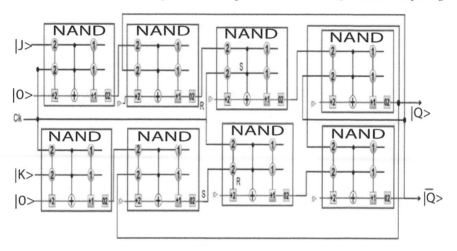

FIGURE 1.8
Multi-Valued Quantum JK Flip Flop Operation Circuit

1.5.2   Circuit Architecture of Multi-Valued Quantum JK Flip Flop

Two ternary qubit inputs and one clock shared input make up the Multi-Valued Quantum JK flip flop. The clock also affects the Multi-Valued Quantum JK flip flop. If the clock is enabled, the circuit will be enabled as well; otherwise, it will not. The multi-valued quantum JK flip flop operation circuit is shown in Figure 1.8.

|J> and |K> are ternary inputs. The Multi-Valued quantum NAND operation will be performed differently by both values with the shared |clk> input. Multi-Valued quantum NAND is performed using the ternary inputs |J> and |clk>. Multi-Valued quantum basic gates were used to perform a NAND operation. An ancillary bit was employed for error correction here. The value of an auxiliary bit in Quantum NAND operations is |1>. A ternary output qubit after performing the Multi-Valued quantum NAND operation is acquired with |J> and |clk>, and the Multi-Valued quantum NAND operation is performed with this ternary output qubit and the ternary output of the flip-flop |$\overline{Q}$> to get the ternary output |S>. The |R> ternary output is produced by the same procedure's |K>, |clk>, and |Q> ternary inputs. Multi-Valued quantum NAND operations make up the majority of these circuits. The Multi-Valued quantum SR flip-flop operation is done on this |S> and |R> input. The proposed essential component of multi-valued quantum computing is multi-valued quantum NAND operation, which is used in multi-valued quantum SR flip flop operation circuit design.

1.5.3   Working Principle of Multi-Valued Quantum JK Flip Flop

The two-qubit circuit is the multi-valued quantum JK flip flop. The JK flip flop is the most often utilized flip flop in multi-valued quantum computing. The ternary input

|J> and |K> of a Multi-Valued Quantum JK flip flop performs two Multi-Valued quantum processes in tandem.

The Multi-Valued quantum NAND operation is performed with |J> and a shared ternary input |clk>. The Multi-Valued Quantum NAND operation is then performed on one ternary output of the Multi-Valued Quantum JK flip flop, yielding |S>.The |K> and |clk> ternary inputs are processed first in the Multi-Valued quantum NAND operation. Then, using the ternary output of the Multi-Valued Quantum JK flip flop |Q> as well as the result of the Multi-Valued quantum first NAND operation, another Multi-Valued quantum NAND operation is performed, yielding |R>. The steps for creating |S> and |R> are carried out simultaneously. Multi-Valued quantum operations are known for doing many operations simultaneously, and this is exactly what is happening here. The Multi-Valued Quantum SR flip flop operation is then conducted on these |S> and |R> ternary inputs. The Multi-Valued Quantum SR flip flop is likewise based on the Multi-Valued Quantum NAND operation, which is used in this case. Two ternary outputs are discovered after doing the multi-valued quantum SR flip-flop operation. The opposite of one ternary output is the other. This Multi-Valued Quantum JK flip flop solves the Multi-Valued Quantum SR flip flop problem. In the fifth chapter, the multi-valued quantum SR flip-flop is briefly described.

Initially, molecules entered a state of superposition and formed qubits. When a qubit is in a superposition state, it is coherent. The flip flop process will be continued until the qubits are coherent. The operation will be disturbed if an outside particle breaks the coherence and causes the qubit to become decoherent. If the heat in the circuit is not reduced, the operation in this circuit will likewise come to a halt. Table 1.4 shows the truth table of multi-valued quantum JK flip flop.

Assume that the Multi-Valued Quantum JK flip flop's truth table is always enabled by the clock input. It works like a quantum SR latch circuit when any of the inputs is |0>, but when both inputs are |1> or |2>, it toggles to create the output, according to the truth table.

TABLE 1.4
Truth Table of Multi-Valued Quantum JK Flip Flop

|J>	|K>	|Q>	$|\overline{Q}>$
|0>	|0>	No Change	
|0>	|1>	|0>	|1>
|1>	|0>	|1>	|0>
|0>	|2>	|2>	|0>
|2>	|2>	|0>	|2>
|1>	|1>	|1>	|0>
|1>	|1>	|0>	|1>
|2>	|2>	|2>	|0>

The JK flip flop is a better-performing timed SR flip flop. The "race" issue, however, persists. This problem happens when the state of the output |Q> is changed before the clock input's timing pulse has chance to go "Off."

1.5.4   Example

Assume that the ternary qubit input is |0> and |1> for the purposes of testing the proposed Multi-Valued Quantum JK flip flop circuit. The Multi-Valued Quantum JK flip flop will work if and only if the clock input is high or |1>. Qubits |0> and |clk> will conduct the quantum NAND operation if the clock input is high. Then Intermediate Output will be |1>

The input |1> and the |clk> inputs are working in parallel in addition to complete the quantum NAND operation and creating the appropriate Intermediate Output |0>

Two independent multi-valued quantum NAND operations are now carried out using these two intermediate qubit outputs. The Intermediate Output ternary qubit |1> and the final output of the Multi-Valued Quantum JK flip flop $|\overline{Q}>$ perform the quantum NAND operation. Assume that |Q> is the most recent state |1>.
The truth table's output is |S>=|0>. Now, the quantum NAND operation is then conducted on the Multi-Valued Quantum JK flip flop's intermediate output |0> and final output |Q>= |0>. The result is |R>=|1>.

These qubits named |S> and |R> will now execute the quantum SR flip-flop operation, according to the Multi-Valued Quantum JK flip flop circuit. So the final output will be |0> and |1>

A quantum suggestion was eventually made. The JK flip-flop supplies the required qubit inputs |0> and |1>, indicating that the suggested circuit yields the ideal output in theory.

1.6   Multi-Valued Quantum T Flip Flop

Multi-Valued quantum T flip–flop is also known as "Multi-Valued Quantum Toggle Flip–flop." In a Multi-Valued Quantum SR flip–flop, just one input, called the Trigger qubit input or Toggle input, should be sent to the flip–flop to avoid the intermediate state. Toggling is defined as "changing the next state output to complement the current state output." The quantum T flip–flop is built by making minor adjustments to the quantum JK flip–flop. Because the quantum T flip–flop has just one qubit input, a quantum JK flip–flop can be converted to a quantum T flip–flop by joining the |J> and |K> inputs and giving them the label |T>.

T flip flops, like any other ternary quantum operation circuit, struggle to generate more heat. This Multi-Valued Quantum T flip flop will learn if and only if the heat is reduced to near-zero levels. The ternary quantum AND and NOR operations are the most basic components of a Multi-Valued Quantum T flip flop. Basic ternary quantum gates make up this fundamental component. This ternary quantum process is extraordinarily fast when compared to the operation of classical computers.

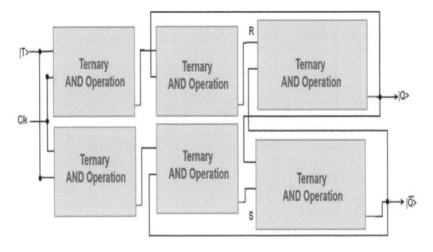

FIGURE 1.9
Block Diagram of Multi-Valued Quantum T Flip Flop

1.6.1 General Organization of Multi-Valued Quantum T Flip Flop

The problem of multi-valued quantum JK and multi-valued quantum SR flip-flops is solved with the Multi-Valued Quantum T flip-flop. Multi-Valued Quantum T flip flops have a single ternary input |T>. Figure 1.9 shows the block diagram of multi-valued quantum T flip flop.

Two quantum AND operations share a single ternary qubit input |T>. Ternary |clk> input and ternary |T> output produce two ternary qubit outputs by performing two multi-valued quantum AND operations in tandem. As inputs, these ternary outputs are fed into multi-valued quantum SR flip-flops. Two multi-valued quantum AND operations and two multi-valued quantum NOR operations are used to create these multi-valued quantum SR flip-flops. Two ternary qubit inputs are used in this multi-valued quantum SR flip-flop.

First, the multi-valued quantum AND operation is performed on the output of the previous multi-valued quantum AND operation and the output of the multi-valued quantum SR flip-flop $|\overline{Q}>$. |S> is the name of the ternary output. Then |Q> executes the multi-valued quantum AND operation on another output of the previous operation. Ternary qubit |R> is the ternary output of these multi-valued quanta AND operations. The multi-valued quantum NOR gate constructs a multi-valued quantum SR latch with ternary input |S> and ternary input |R>.

The multi-valued quantum NOR operation is performed by the ternary output of the quantum SR latch |Q> and |S>. The multi-valued quantum NOR operation is performed in parallel by |R> and $|\overline{Q}>$. Finally, from the Multi-Valued Quantum T flip flop, the ternary output of |Q> and $|\overline{Q}>$ are received. These multi-valued Quantum T flip flops produce two ternary outputs that are opposite to one another. Depending on the ternary input qubit, these ternary outputs are truly toggled.

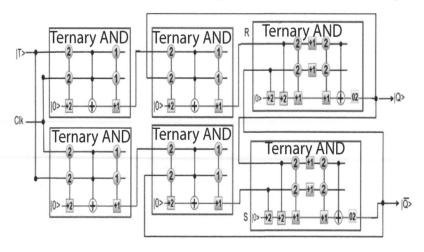

FIGURE 1.10
Multi-Valued Quantum T Flip Flop Operation Circuit

1.6.2 Circuit Architecture of Multi-Valued Quantum T Flip Flop

The multi-valued quantum JK flip flop problem is solved by the multi-valued Quantum T flip flop, which has only one input labeled "T." This quantum flip flop action's main application is toggling. Figure 1.10 shows the multi-valued quantum T flip flop operation circuit.

In the Multi-Valued quantum T flip flop operation circuit, design some ternary quantum AND operation circuits and some ternary quantum NOR operation circuits. A closer look reveals that the Multi-Valued Quantum T flip flop design is very similar to the quantum JK flip flop. Quantum JK flip-flop operation has a number of challenges that Multi-Valued Quantum T flip-flop operation solves. Multi-Valued Quantum T flip-flops have a shared input, whereas Quantum JK flip-flops have a two-qubit input. Between two Multi-Valued Quantum T flip flops is a quantum SR flip-flop. This flip-flop was made using two concurrent quantum techniques. The quantum SR flip-flop is slightly different from the SR flip-flop explained in the previous chapter. The quantum SR latch in this quantum SR flip-flops concept was created using two quantum AND operations and a quantum NOR SR latch. However, the input and output will be the same as in the quantum SR flip-flop working circuit described earlier. The Multi-Valued Quantum T flip flop has two parallel ternary quantum processes built-in. Then, in the multi-valued quantum SR flip-flop section, two quantum, AND operations are connected in parallel, and two quantum NOR operations are set up in parallel in the multi-valued quantum SR latches. Although it's just a theoretical circuit, the Multi-Valued Quantum T flip flop design retains quantum parallelism's features. This circuit architecture can be altered depending on the qubit origin, heat, and temperature.

1.6.3 Working Principle of Multi-Valued Quantum T Flip Flop

A few features distinguish the Multi-Valued Quantum T flip flop from the quantum JK flip flop. Multi-Valued Quantum T flip flops are used to toggle when it is needed. Only one input, |T>, and one clock, |clk>, are available in this operational circuit. If the clock input is |1>, the circuit will be turned on; otherwise, it will be turned off. As a result, the Multi-Valued Quantum T flip flop operation was enabled, and the basic operation started to work. This circuit, like any other circuit, will function if the qubit is in a coherent state. This circuit must be kept at a temperature close to 0 degrees Fahrenheit to prevent any particles from disrupting coherence and making it decoherent.

The ternary quantum AND operation, which is the most basic operation in ternary quantum computing, is performed first in the Multi-Valued Quantum T flip flop operational circuit. Two ternary quantum AND operations execute simultaneously, requiring the same amount of time because the |T> and |clk> inputs are shared. This output is then used as an input for a quantum SR flip-flop. In this quantum, the SR flip-flop performs two ternary quantum AND operations in tandem. The ternary quantum SR flip-flop, whose output is passed into the ternary quantum SR latch as input, is the first of two ternary quantum AND operations that operate simultaneously. In this case, a ternary quantum NOR SR latch is employed. The ternary quantum NOR SR latch is built using two ternary quantum NOR operations. The Quantum SR latch performs two NOR operations simultaneously, each taking the same amount of time.

After completing all of the operations, it is possible to retrieve the output of the Multi-Valued Quantum T flip flop operation circuit. This technique produces two outputs, one of which is diametrically opposed to the other. Table 1.5 shows the truth table of multi-valued quantum T flip flop

In a Multi-Valued Quantum T flip flop, an auxiliary bit will be employed for error correction, and this flip flop may produce trash, however, this is not the subject of the study.

1.6.4 Example

Assume the clock input is set to |1> and the value of input |T> is set to |0>. To begin, two simultaneous ternary quantum AND operations will be executed. Two outputs are created after two ternary quantum AND operations, both of which are |0>.

TABLE 1.5

Truth Table of Multi-Valued Quantum T Flip Flop

|T>	|Q>	|$\overline{Q}$ >
|0>	|0>	|1>
|1>	|0>	|1>
|0>	|1>	|0>
|2>	|1>	|0>
|1>	|1>	|0>

These two |0> were sent into a multi-valued quantum SR flip-flop as an input. Two ternary quantum AND operations are also running in parallel here. Following these processes, each ternary quantum AND operation generates an output |0>.

Because, as it is known that if the input to any ternary quantum AND operation is |0>, the output will be |0> as well. The ternary quantum NOR SR latch is then fed two outputs as inputs. These two inputs then perform the ternary quantum NOR operation in parallel. The result of each of these procedures is |0> as it is known from NOR operation characteristics.

1.7   Multi-Valued Quantum Shift Register

In a Multi-Valued Quantum flip flop, a single bit of two-valued qubit data (|1>or |0>) is stored. To store many qubits of data, however, additional Multi-Valued Quantum flip-flops are necessary. N Multi-Valued Quantum flip flops must be linked in a specific order to store n qubits of data. This data is stored in a Multi-Valued Quantum Register, which is a device. It is made up of a series of quantum flip flops that are used to store data in multiple qubits.

Information contained in these multi-valued quantum registers can be communicated using multi-valued quantum shift registers. A Multi-Valued Quantum Shift Register is a set of flip flops that can store multiple bits of data. The qubits contained in such multi-valued quantum registers can be made to travel inside and out of them by applying clock pulses to them. An n-qubit Multi-Valued Quantum Shift register can be constructed by connecting n quantum flip-flops, each of which stores a single qubit of data. Quantum registers called "Multi-Valued Quantum Shift left registers" shift the qubits to the left. Quantum registers that shift the qubits to the right are known as "Multi-Valued Quantum Shift Right Registers."

Multi-Valued Quantum Shift registers are basically of four types. These are:

1. Multi-Valued Quantum Serial In Serial Out shift register

2. Multi-Valued Quantum Serial In parallel Out shift register

3. Multi-Valued Quantum Parallel In Serial Out shift register

4. Multi-Valued Quantum Parallel In parallel Out shift register

A multi-valued quantum shift register is created in this work that converts serial data into quantum data using a Quantum D flip-flop operating circuit. The quantum Serial-In Serial-Out shift register is a form of Multi-Valued Quantum Shift Register that allows one qubit of serial input at a time over a single data line and generates a serial output. Due to the fact that there is only one qubit output, data exits the Multi-Valued Quantum Shift register one qubit at a time in a serial pattern, hence the moniker quantum Serial-In Serial-Out Shift Register. In this circuit, four Quantum D flip-flops are connected in a serial method. Because each multi-valued Quantum flip flop receives the same clock signal, they are all synced with one another. A Multi-Valued Quantum Shift right register accepts serial data from the quantum flip flop's

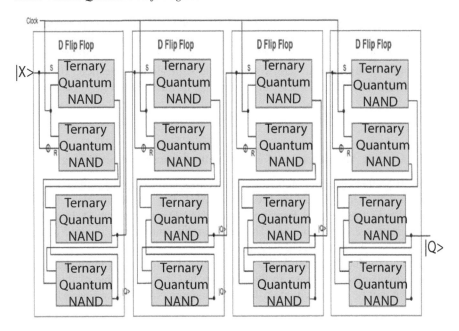

FIGURE 1.11
Block Diagram of Multi-Valued Quantum Shift Register

left side, as seen in the circuit above. The primary function of a QSISO is to act as a delay element.

1.7.1 General Organization of Multi-Valued Quantum Shift Register

A Multi-Valued Quantum Shift register is created using four quantum D flip-flop operational circuits. A Multi-Valued Quantum Shift register is employed as a basic component for data shift, and a quantum D flip-flop is used to make it happen. Figure 1.11 shows the block diagram of multi-valued quantum shift register.

Multi-Valued Quantum D flip flops have two inputs: one for data and the other for the clock. Multi-Valued Quantum D flip flips have logically opposing results. The clock input aids in the circuit's synchronization with an external signal. The output of a Quantum D flip flop can have two different values. In this block diagram, data input is sent to a Ternary Quantum NAND operation circuit, while data input reverse is sent to another Ternary Quantum NAND operation circuit. The clock pulse input is used by both NAND operations. The result of two Ternary Quantum NAND operations is fed into the Quantum SR Latch. The SR latch is used to make the D flip flop. This characteristic is used to create a delay in the data flow of the circuit. Two Ternary Quantum NAND Operations make up the multi-valued Quantum SR Latch. The multi-valued Quantum SR Latch function's final two outputs were discovered. A Quantum D flip flop can output two different forms of data, one of which is logically inverse to the other. If the clock is enabled, the quantum D flip flop will continue to function; otherwise, the quantum D flip flop will stop operating.

In the Multi-Valued Quantum Shift register block diagram, operational circuits for quantum D flip-flops are also connected through serial connection. The Multi-Valued Quantum Shift register is built around the quantum D flip-flop operating circuit.

1.7.2   Circuit Architecture of Multi-Valued Quantum Shift Register

The Multi-Valued Quantum Shift Register uses four D flip-flops as well as a Ternary Quantum AND Gate. These are used by the Shift Register to produce four qubit outputs. The D flip-flop uses Ternary Quantum NAND and Quantum SR latch operations to create a single qubit input. The circuit architecture of multi-valued quantum shift register is shown in Figure 1.12.

The multi-valued quantum D flip-flop requires the clock qubit input. The circuit has one qubit input, as shown in the diagram. In-Circuit, one line of this ternary qubit input will be channeled into a Ternary Quantum NAND operation, which will

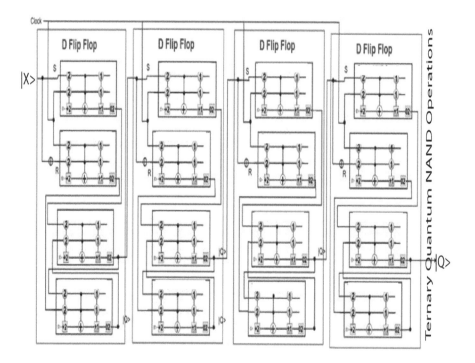

FIGURE 1.12
Circuit Architecture of Multi-Valued Quantum Shift Register

be referred to as |S> input. Ternary |S> qubit input and Clock qubit input is used to perform Ternary Quantum NAND.

Another line receives the |X> qubit, which executes a Ternary Quantum NOT operation first. The R designation was given to this Ternary Quantum NOT operation, which was included in the Ternary Quantum NAND operation. In this Ternary Quantum NAND operation, the two-qubit inputs R and Clock are employed. As an input, the Ternary Quantum NAND operations' output is passed to the Quantum SR latch. An SR latch will be created using these two inputs. The final output of |Q> and $\overline{|Q}$ > will be the Quantum SR latch's output. |Q> will always result in the inverse of $\overline{|Q}$ >. Earlier in the book, in the chapter titled Quantum SR Latch, the Quantum SR latch is explored. After processing the inputs in a multi-valued Quantum D flip-flop, it outputs one output. The multi-valued Quantum D flip-output flop is used as a clock input for the next multi-valued Quantum D flip-flop. As a result, the output of a Multi-Valued Quantum Shift register is a single qubit.

1.7.3   Working Principle of Multi-Valued Quantum Shift Register

Multi-Valued Quantum Shift registers are a type of register in which both qubit data loading and data retrieval from the Multi-Valued Quantum Shift register happen in serial mode. The positive edge of the clock pulse is sensitive in this research synchronous quantum SISO shift register. At the qubit input of the first quantum flip-flop, the data word to be stored is fed bit by bit. Furthermore, the outputs of the preceding ones influence the qubit inputs of all subsequent flip-flops, for example, the input of quantum D flip-flop number-2 is driven by the output of quantum D flip-flop number-1. Finally, the data saved in the quantum register is serially retrieved from the output pin of the nth quantum D flip-flop.

All of the quantum flip-flops in the quantum register are cleared at first by setting their clear pins to high. Following that, the incoming data word is serially fed into quantum D Flip-flop number-1. As soon as the first leading edge of the clock occurs, the qubit arriving at the first pin is placed into quantum D flip-flop number-1. B1 is placed in quantum D flip-flop number-2 at the second clock tick, while a new bit enters quantum flip-flop number-2.

Every rising edge of the clock pulse results in a similar shift in data qubits. This means that the data in the quantum register shifts a single bit to the right with every clock pulse. Following the qubit data transmission, the first qubit of an input word appears at the output of the nth flip-flop during the nth clock tick, as discussed previously. The serial output of the next succeeding qubits of the qubit input data word can be obtained by applying additional clock cycles. Table 1.6 shows the truth table of multi-valued quantum shift register.

It creates one output after processing the inputs in a Quantum D flip-flop. The output is fed into the following Quantum D flip-flop as a clock input. A Multi-Valued Quantum Shift Register generates one final qubit output as a result.

TABLE 1.6

Truth Table of Multi-Valued Quantum Shift Register

|Clk>	|x>	|Q>	$\overline{|Q}$ >	Description
↓ >> |0>	X	Q	Q	Memoryno change
↑ >> |1>	|0>	|0>	|1>	Reset Q >> 0
↑ >> |1>	|1>	|1>	|0>	Set Q >>

1.7.4   Example

There is only one ternary qubit input required to conduct the required operation in the Multi-Valued Quantum Shift register. If and only if the clock input is enabled, the Multi-Valued Quantum Shift register will activate. Multi-Valued Quantum Shift registers were presented in this study to shift one qubit data into the correct location.

Assume that the Multi-Valued Quantum Shift register operation will be performed by |1>|1>|1>|1>|0> data. To begin, |0> will conduct a multi-valued quantum D flip flop number-1 operation, with the result being |0>. Then this |0>, as well as like every other qubit, will shift to the right side for the first time.

Then |1> will be placed into the Multi-Valued Quantum Shift register, which will conduct the midway quantum D flip-flop operation, and then the data will be shifted once on the right side, like the previous one.

After that, in the same way as these previous two steps, |1> performs the midway quantum D flip-flop operation in the Multi-Valued Quantum Shift register in the necessary time. Then each piece of data will be shifted once more. It shows that every time qubit data shifts to the right, it shifts once. The operational circuit of a Multi-Valued Quantum Shift Register will do n+1 operations, and the final data table will be like the table below. Table 1.7 shows data shifting process in multi-valued quantum shift register.

Because no data is accessible at the n+1 operation time, |0> appears in the truth table, and one data qubit is shifted once on the right side according to the Multi-Valued Quantum Shift register concept. The final data after the n+1 operation is |0>|1>|1>|1>, these results only demonstrated that the Multi-Valued Quantum Shift register produces the right output while adhering to the quantum principle. The

TABLE 1.7

Data Shifting Process in Multi-Valued Quantum Shift Register

|clk>	|x>	|Q>	|Q1>	|Q2>	|Q3>
|0>	|0>	|0>	|0>	|0>	|0>
|1>	|0>	|0>	|0>	|0>	|0>
|1>	|1>	|1>	|0>	|0>	|0>
|1>	|1>	|1>	|1>	|0>	|0>
|1>	|1>	|1>	|1>	|1>	|0>
|1>	|1>	|0>	|1>	|1>	|1>

Multi-Valued Quantum Shift Register can produce some junk values, however this is not the focus of this research.

1.8 Multi-Valued Quantum Ripple Counter

A Multi-Valued quantum counter is basically used to count the number of clock pulses applied to a multi-valued quantum flip-flop. It can also be used for Multi-Valued quantum Frequency divider, Multi-Valued quantum time measurement, Multi-Valued quantum frequency measurement, Multi-Valued quantum distance measurement and also for generating square waveforms. In this, the multi-valued quantum flip-flops are Multi-Valued quantum asynchronous counters and are supplied with different clock signals, there may be a delay in producing output. Also, a few numbers of multi-valued quantum logic gates are needed to design asynchronous counters. So they are elementary in design and also are less expensive.

An n-qubit ripple counter can count up to $2n$ states. It is also known as MOD n counter. It is known as a ripple counter because of the way the clock pulse ripples its way through the flip-flops. It is an asynchronous counter. Different flip-flops are used with a different clock pulse. All the flip-flops are used in toggle mode. Only one flip-flop is applied with an external clock pulse and another flip-flop clock is obtained from the output of the previous flip-flop. The flip-flop applied with an external clock pulse acts as LSB (Least Significant qubit) in the counting sequence. A counter may be an up counter that counts upwards or can be a down counter that counts downwards or can do both, i.e., count up as well as count downwards depending on the input control. The sequence of counting usually gets repeated after a limit

Multi-Valued quantum Ripple Counter is made out of four JK flip flops. Using these multi-valued quantum JK Flip flops, the multi-valued quantum Ripple Counter creates four qubit outputs. Here in JK flip flop, J and K are not shortened abbreviated letters of other words, such as "S" for Set and "R" for Reset, but are autonomous letters chosen by its inventor Jack Kilby to distinguish the flip-flop design from other types. Though Jack Kilby invented the digital electronics JK flip flop. Multi-Valued quantum Ripple Counter is an asynchronous counter. It is created using multi-valued quantum JK flip flops and these flip flops are only controlled by clock pulse input.

Multi-Valued quantum ripple counter produces much heat to produce the molecule's superposition state and also produces some garbage value.

1.8.1 General Organization of Multi-Valued Quantum Ripple Counter

Multi-Valued quantum Ripple Counter uses four multi-valued quantum JK flip flop to create four qubit outputs. Multi-Valued quantum JK flip flop has the two-qubit input named as |J> and |K>. Figure 1.13 shows the block diagram of multi-valued quantum ripple counter.

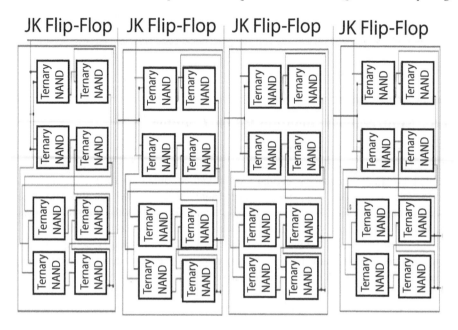

FIGURE 1.13

Block Diagram of Multi-Valued Quantum Ripple Counter

Multi-Valued quantum JK flip-flop has the qubit input |J> and |k>. This Multi-Valued quantum JK flip flop consists of many Ternary quantum NAND operations. At first basic Ternary quantum NAND operation performs a couple of then this operation output is entered into the SR flip flop as well as got the output |Q> and $|\overline{Q}>$ First of all |J> and |clk> input perform the Ternary quantum NAND operation. The output of the Ternary quantum NAND operation and output of the Multi-Valued quantum JK flip flop $|\underline{Q}>$ performs another Ternary quantum NAND operation and produces the output named as |S>. |clk> input is shared so |K> and |clk> input also performs the Ternary quantum NAND operation. This Ternary quantum NAND operations output and |Q> output of the Multi-Valued quantum JK flip flop performs another Ternary quantum NAND operation as well as produces the output named as |R>.

These |S> and |R> entered into the Multi-Valued quantum SR flip flop and produce two outputs |Q> and $|\overline{Q}>$. In Multi-Valued quantum SR flip flop, it has four Ternary quantum NAND operations. |S> input and |clk> input performs Ternary quantum NAND operation as well as |R> input and shared |clk> input also perform the Ternary quantum NAND operation. These two NAND operations output entered the Multi-Valued quantum SR latch as input. In Multi-Valued quantum, SR latches two |Q> and one input as well as $|\overline{Q}>$ and other inputs perform the Ternary quantum NAND operation. Finally, after all of this Multi-Valued quantum operation, got the Multi-Valued quantum JK flip flop's final output |Q> and $|\overline{Q}>$.

After processing the inputs in a Multi-Valued quantum JK flip flop, the output of the flip flop is going to be stored as an output of the multi-valued quantum Ripple

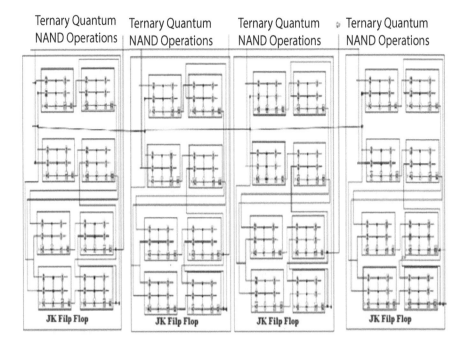

FIGURE 1.14
Circuit Architecture of Multi-Valued Quantum Ripple Counter

Counter. A multi-valued quantum ripple counter can be performed as an up counter and also a down counter. Here Multi-Valued quantum every Multi-Valued quantum JK flip-flops output |Q> enters another Multi-Valued quantum JK flip-flop as a clock pulse. This |clk> will decide thus the Multi-Valued quantum JK flip-flop operational circuit will perform or not. Every Multi-Valued quantum JK flip-flop operational circuit will produce the final output.

1.8.2   Circuit Architecture of Multi-Valued Quantum Ripple Counter

Multi-Valued quantum Ripple Counter uses four multi-valued quantum JK flip flops to create four qubit outputs. Multi-Valued quantum JK flip flop has a two-qubit input and one clock shared input. Multi-Valued quantum JK flip flops also depend on the clock. If the clock is enabled then the circuit will enable, otherwise not. Figure 1.14 shows the circuit architecture of multi-valued quantum ripple-counter.

In Multi-Valued quantum ripple counter there's one clock input and one logic input which is shared in both |J> and |K> input port. Input |J> and input |K> both the value will perform the Ternary quantum NAND operation with the shared |clk> input differently. |J> and |clk> input performs the Ternary quantum NAND operation. NAND operation made by using multi-valued quantum basic gates. Basic gates in mainly multi-valued quantum computing are V, V+, and CNOT gates. For error correction here used an ancillary qubit. In Quant NAND operation the value of an ancillary qubit in |1>. After |J> and |clk> perform the Ternary quantum NAND operation, produced an output qubit, and this output qubit and the output of the flip flop $|\overline{Q}>$ perform the Ternary quantum NAND operation and produce the output |S>. Like in the same procedure |K>, |clk>, and |Q> input produces the |R> output. These operation circuits are mainly Ternary quantum NAND operations. This |S> and |R> input is performed the Multi-Valued quantum SR flip-flop operation. In Multi-Valued quantum SR flip flop operational circuit architecture, the proposed basic component of multi-valued quantum computing is Ternary quantum NAND operation.

After processing the inputs in a Multi-Valued quantum JK flip flop, the output of the flip flop is going to be stored as an output of the multi-valued quantum Ripple Counter. Multi-Valued quantum ripple counter intermediate architectures four multi-valued quantum JK flip-flops connected in serial connection using the logical qubit input. Every clock input as |clk> enters every qubit and from the 2nd Multi-Valued quantum JK flip-flop |clk> input is previous Multi-Valued quantum Jk flip-flops first output |Q>. With the same architecture, Multi-Valued quantum ripple counter can be performed as an up counter, or as a down counter, but clock pulse as |clk> need to sometimes have a positive edge triggered and sometimes a negative edge triggered.

1.8.3   Working Principle of Multi-Valued Quantum Ripple Counter

In the Multi-Valued quantum ripple counter, there are four multi-valued quantum JK flip-flop operational circuits. These Multi-Valued quantum JK flip-flop operational circuits are connected in serial connection. In this Multi-Valued quantum ripple counter, there is one clock input and a logic input which are shared into the port |J> and |K> of Multi-Valued quantum JK flip-flop operational circuit. The inputs |J> and |K> of a Multi-Valued quantum Jk flip-flop conduct two Multi-Valued quantum processes in parallel. The Ternary quantum NAND operation is performed using |J> and shared input |clk>. The Ternary quantum NAND operations are then completed, and one of the Multi-Valued quantum JK flip flop's outputs executes the Ternary quantum NAND operation, producing |S>. The |K> and |clk > inputs are performed first in the Ternary quantum NAND operation. The output of the Multi-Valued quantum JK flip-flop |Q>, as well as the result of the Multi-Valued quantum first NAND operation, are then used to perform another Ternary quantum NAND operation, yielding |R>. The steps for creating |S> and |R> are carried out simultaneously. It is known that one of the distinctive properties of multi-valued quantum operations is that they may do several operations at the same time, and this is exactly what is happening. The Multi-Valued quantum SR flip flop operation is then conducted on these |S> and |R> inputs. The Ternary quantum NAND operation, which is employed here, is also used to make multi-valued quantum SR flip flops. Two outputs are discovered after conducting the multi-valued quantum SR flip-flop operation. The opposite of one output is the opposite of the other.

TABLE 1.8

Truth Table of Multi-Valued Quantum Ripple Counter

|clk>	$\|Q_0 >$	$\|Q_1 >$	$\|Q_2 >$	$\|Q_3 >$
|0>	|0>	|0>	|0>	|0>
|1>	|1>	|1>	|1>	|1>
|1>	|0>	|0>	|0>	|0>

Here if the |clk> is |0> then the Multi-Valued quantum JK flip-flop will not be triggered but if |clk> is |1> or |2> the Multi-Valued quantum JK flip-flop will be triggered and it will toggle the output. First of all Multi-Valued quantum JK flip-flop operational circuit gets |clk> value is |1> or |2> and toggles the output value from the previous state value. Then the output of the initial Multi-Valued quantum JK flip-flop will be |clk> input of the next Multi-Valued quantum JK flip-flop. If the |clk> value is |1> or |2> then the output value will be toggled, otherwise, the output will be the previous state output. Maintaining the same procedure every multi-valued quantum JK flip-flop operated in the multi-valued quantum ripple counter. Multi-Valued quantum JK flip-flop is toggled very much, that's why in the Multi-Valued quantum ripple counter Multi-Valued quantum JK flip-flop operational circuit here is used as a basic component. Table 1.8 shows the truth table of multi-valued quantum ripple counter.

Ternary value |2> will work as like high input such as |1> in multi-valued quantum ripple counter.

1.8.4   Example

To check the Multi-Valued quantum ripple counter proposed circuit, assume a clock signal is |1> and the logical qubit is high. So, if initially the clock signal |1> is delivered, then for the first Multi-Valued quantum JK flip-flop operational circuit, it will be |1>.

According to the working principle of Multi-Valued quantum ripple counter, if clock input value is |1>, then the previous state value will be toggled. Then, from the first multi-valued quantum JK flip-flop, we get one output, which is |1>. This |1> will be the clock input for the second Multi-Valued quantum JK flip-flop circuit according to the architecture of the Multi-Valued quantum ripple counter.

As with the previous multi-valued quantum JK flip-flop, for the third and fourth multi-valued quantum flip-flops, the principle is the same. For both of the flip-flops, the clock value will be the previous multi-valued quantum JK flip-flop's output.

Hence, from the above truth table (Table 1.8), we can see that the proposed multi-valued quantum ripple counter is correct theoretically. Multi-Valued quantum JK flip-flop used here for toggling. This Multi-Valued quantum ripple counter produces much heat when the molecule's normal state becomes a superposition state. Multi-Valued quantum ripple-counter's full operation needs to be in an environment where other particles are totally prohibited to maintain the coherence state.

1.9   Multi-Valued Quantum Synchronous Counter

A ternary Quantum counter is a ternary Quantum gadget that can count any specific event based on how many times the event(s) has occurred. This ternary Quantum counter, which is based on a ternary Quantum clock signal, can count and store the number of times any given event or process has occurred in a ternary Quantum logic system or computer. A sequential ternary Quantum logic circuit with a single clock input and many qubit outputs is the most frequent type of ternary Quantum counter. The outputs of the qubits are two-valued decimal numbers. Each clock pulse either adds to or subtracts from the total.

A multi-valued Quantum Synchronous circuit is anything that is time-synchronized with others. Multi-Valued Quantum Synchronous signals have the same clock rate as single-valued Quantum Synchronous signals, and all clocks are synchronized with the same reference clock. The qubit output of the ternary Quantum counter is directly coupled to the input of the next following counter, forming a chain system. As a result of this chain system propagation delay, counting delays develop throughout the counting stage. The clock qubit input across all ternary Quantum flip-flops in a Multi-Valued Quantum Synchronous counter uses the same source and generates the same clock signal at the same time in a Multi-Valued Quantum Synchronous counter. As a result, a Multi-Valued Quantum Synchronous counter is a ternary Quantum counter that uses the same clock signal from the same source at the same time.

Four JK flip flips and two ternary Quantum AND Gates make up a Multi-Valued Quantum Synchronous Counter. The Multi-Valued Quantum Synchronous Counter generates four qubit outputs using these ternary Quantum JK Flip flops and ternary Quantum AND Gates. A Multi-Valued Quantum Synchronous counter generates a lot of heat, and this circuit operation must take place in a ternary Quantum computing environment.

1.9.1   General Organization of Multi-Valued Quantum Synchronous Counter

Four ternary Quantum JK flip flops are used to create four qubit outputs in the Multi-Valued Quantum Synchronous Counter. The two-qubit inputs of the ternary Quantum JK flip flop are designated $|J>$ and $|K>$. Figure 1.15 shows the block diagram of multi-valued quantum synchronous counter.

The qubit inputs $|J>$ and $|k>$ are used in the Ternary Quantum JK flip-flop. Many ternary Quantum NAND operations make up this ternary Quantum JK flip flop. After performing a couple of simple ternary Quantum NAND operations, the output of this operation is entered into the SR flip flop, yielding the output $|Q>$ and $|\overline{Q}>$. First, the ternary Quantum NAND operation is performed using the $|J>$ and $|clk>$ inputs. The output of the ternary Quantum NAND operation and the output of the ternary Quantum JK flip flop $|\overline{Q}>$ is used to execute another ternary Quantum NAND operation, yielding the output $|S>$. Because the $|clk>$ input is shared, the ternary Quantum NAND operation is performed on both the $|K>$ and $|clk>$ inputs. The ternary

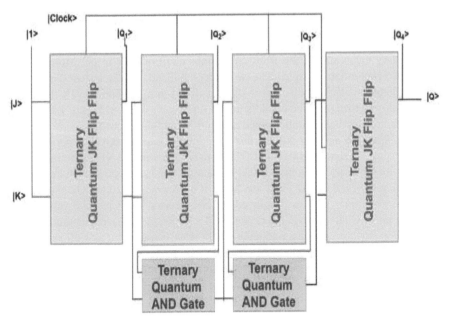

FIGURE 1.15

Block Diagram of Multi-Valued Quantum Synchronous Counter

Quantum JK flip flop's |Q> output conducts another ternary Quantum NAND operation and yields the |R> output.

These |S> and |R> are fed into the ternary Quantum SR flip flop, which generates two |Q> and $|\overline{Q}>$. There are four ternary Quantum NAND operations in a ternary Quantum SR flip flop. The ternary Quantum NAND operation is performed by the |S> input and shared |clk> input, as well as the |R> input and shared |clk> input. The output of these two NAND operations was used as input to the ternary Quantum SR latch. The ternary Quantum NAND operation is performed by two |Q> and one input as well as $|\overline{Q}>$ and other inputs in ternary Quantum SR latches. Finally, the ternary Quantum JK flip flops final output |Q> and $|\overline{Q}>$ are received after all of this ternary Quantum operation. The output of a ternary Quantum JK flip flop will be stored as an output of the Multi-Valued Quantum Synchronous Counter after it has processed the inputs. Using four ternary Quantum JK flip flops, the Multi-Valued Quantum Synchronous Counter generates four qubit outputs. All four ternary Quantum JK flip flops get their clock inputs from the same source. All of the flip flops work ternary synchronously for this. The second and third flip flops each have one output that goes through ternary Quantum AND Gates.

1.9.2 Circuit Architecture of Multi-Valued Quantum Synchronous Counter

Four ternary Quantum JK flip flops create four qubit outputs in the Multi-Valued Quantum Synchronous Counter. The two-qubit input and one clock shared input of a ternary Quantum JK flip flop. The clock also affects ternary Quantum JK flip-flops. The circuit will turn on if the clock is turned on; else, it will not. Figure 1.16 shows the circuit architecture of multi-valued quantum synchronous counter.

Both inputs |J> and |K> will conduct the ternary Quantum NAND operation differently using the shared |clk> input. The ternary Quantum NAND operation is performed with inputs |J> and |clk>. Ternary Quantum basic gates are used in the NAND operation. v, V+, and CNOT gates are the most basic gates in ternary Quantum computing. An auxiliary bit was employed for error correction here. The value of an auxiliary bit in Quant NAND operations is |1>. After the ternary Quantum NAND operation is performed by |J> and |clk>, there is an output qubit as well as the output of the flip flop |$\overline{Q}$>. Produce the output |S> by performing the ternary Quantum NAND operation. The |R> output is produced by the same technique that creates the

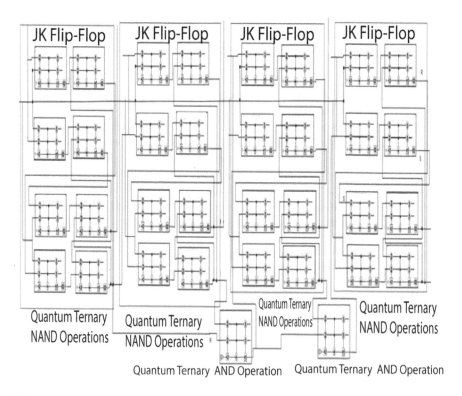

FIGURE 1.16
Circuit Architecture of Multi-Valued Quantum Synchronous Counter

|K>, |clk>, and |Q> inputs. The majority of these operating circuits are ternary Quantum NAND operations. The ternary Quantum SR flip-flop operation is done on this |S> and |R> input. The proposed basic component of ternary Quantum computing is ternary Quantum NAND operation, which is also used in ternary Quantum SR flip flop operation circuit architecture.

The output of a ternary Quantum JK flip flop will be stored as an output of the Multi-Valued Quantum Synchronous Counter after the inputs have been processed in the flip flop. Using four ternary Quantum JK flip flops, the Multi-Valued Quantum Synchronous Counter produces four qubit outputs. All four ternary Quantum JK flip flops share a clock input. All of the flip flops operate in tandem to do this. The second and third flip flops each have one output that goes via a ternary Quantum AND Gate.

1.9.3 Working Principle of Multi-Valued Quantum Synchronous Counter

One logical qubit input and one clock input are shared in a Multi-Valued Quantum Synchronous counter. The main component of a multi-valued Quantum Synchronous counter is a ternary Quantum JK flip-flop operational circuit. The two-qubit circuit is a ternary Quantum JK flip flop. In ternary quantum computing, the JK flip flop is the most commonly utilized flip flop. The inputs |J> and |K> of the ternary Quantum JK flip-flop perform two ternary Quantum operations in simultaneously. The ternary Quantum NAND operation is performed with |J> and shared input |clk>. Then, one output of the ternary Quantum JK flip flop conducts the ternary Quantum NAND operation and produces |S>. The |K> and |clk > inputs are processed first, just like in ternary Quantum NAND. The output of the ternary Quantum JK flip-flop |Q> is then combined with the result of the ternary Quantum first NAND operation to generate |R>. The steps for creating |S> and |R> run in parallel. It is known that ternary Quantum operations have the iconic feature of doing many operations simultaneously, and this is exactly what is happening here. The ternary Quantum SR flip flop operation is then conducted on the |S> and |R> inputs. The ternary Quantum SR flip flop is also created using the ternary Quantum NAND operation. Two outputs are found after conducting the ternary Quantum SR flip-flop operation. One output is diametrically opposed to another. The ternary Quantum JK flip flop solves the ternary Quantum SR flip flop problem. The 5th chapter briefly covers the ternary Quantum SR flip-flop.

When the clock value is |1>, the Ternary Quantum JK flip-flop is triggered. As a result, the counter clock in Multi-Valued Quantum Synchronous must always be high. Counters must be toggled according to the principle, which is why the ternary Quantum JK flip-flop is ideal for Multi-Valued Quantum Synchronous counters. When the ternary Quantum JK flip-flop operational circuit is triggered, its output is shared with the second ternary Quantum JK flip-flop operational circuit. The outputs of the first and second ternary Quantum JK flip-flops are then sent into the ternary Quantum AND operation circuit, which conducts the ternary Quantum AND operation. The ternary Quantum JK flip-flop operation is performed with the output from the ternary Quantum AND

TABLE 1.9

Truth Table of Multi-Valued Quantum
Synchronous Counter

clk	$|Q_0>$	$|Q_1>$	$|Q_2>$	$|Q_3>$
$|0>$	$|0>$	$|0>$	$|0>$	$|0>$
$|1>$	$|0>$	$|0>$	$|0>$	$|1>$
$|1>$	$|0>$	$|0>$	$|1>$	$|0>$
$|1>$	$|0>$	$|0>$	$|1>$	$|1>$
$|1>$	$|0>$	$|1>$	$|0>$	$|0>$
$|1>$	$|0>$	$|1>$	$|0>$	$|1>$
$|1>$	$|0>$	$|1>$	$|1>$	$|0>$
$|1>$	$|0>$	$|1>$	$|1>$	$|1>$
$|1>$	$|1>$	$|0>$	$|0>$	$|0>$
$|1>$	$|1>$	$|0>$	$|0>$	$|1>$
$|1>$	$|1>$	$|0>$	$|1>$	$|0>$
$|1>$	$|1>$	$|0>$	$|1>$	$|1>$
$|1>$	$|1>$	$|1>$	$|0>$	$|0>$
$|1>$	$|1>$	$|1>$	$|0>$	$|1>$
$|1>$	$|1>$	$|1>$	$|1>$	$|0>$
$|1>$	$|1>$	$|1>$	$|1>$	$|1>$
$|1>$	$|0>$	$|0>$	$|0>$	$|0>$

operation. The first ternary Quantum AND operation's output is then used to execute another ternary Quantum AND operation, as is the third ternary Quantum JK flip-flop operation's result. As a result, the preceding ternary Quantum AND operations output is shared as two inputs into the ternary Quantum JK flip-flops, and the ternary Quantum JK flip-flop operational circuit is completed. The Multi-Valued Quantum Synchronous counter circuit is mostly used as a finite counter.

When the clock value is $|2>$, the Ternary Quantum JK flip-flop is triggered. As a result, the counter clock in Multi-Valued Quantum Synchronous must always be high. Counters must be toggled according to the principle, which is why the ternary Quantum JK flip-flop is ideal for Multi-Valued Quantum Synchronous counters. As usual, all the operation performance is the same for $|2>$ as like ternary input $|1>$. Table 1.9 shows the truth table of multi-valued quantum synchronous counter.

1.9.4 Example

Assume that the input qubit enters into the Multi-Valued Quantum Synchronous counter with the value $|1>$. The core component of Multi-Valued Quantum Synchronous counters is a ternary Quantum JK flip-flop operational circuit, thus the circuit clock value is also $|1>$ when triggered. Both $|J>$ and $|K>$ inputs are now $|1>$ in the first ternary Quantum JK flip-flop. The flip-flop is activated and the value toggles if two values are $|1>$. Now $|Q_0>$ is $|1>$ in comparison to the prior state $|0>$.

However, following the initial ternary Quantum JK flip-flop, the value of $|\overline{Q_0}>$ became $|0>$. Then, in the second ternary Quantum JK flip-flop operation, $|\overline{Q_0}>$ is inserted as an input, and the circuit is not triggered. Then again $|\overline{Q_1}>$ and $|\overline{Q_0}>$ performs the ternary Quantum AND operation. As shown by the ternary Quantum AND operation, if any input is $|0>$, the resulting output will be $|0>$. The third ternary Quantum JK flip-flop operational circuit will then receive $|0>$ as an input. If two inputs are both $|0>$, the ternary Quantum JK flip-flop will not be triggered, as seen in this circuit. Again third ternary Quantum flip-flop's output $|\overline{Q_2}>$ and previous ternary Quantum AND operations produced output performs ternary Quantum AND operation again. Because the previous ternary Quantum AND operation output is $|0>$ in this proposed circuit, the output of this ternary Quantum AND operation will also be $|0>$. As a result, $|0>$ will be the fourth ternary Quantum JK flip-flop operation's input, and it will also produce $|0>$.

1.10 Summary

Sequential circuits are a useful section in quantum computing. Quantum D flip-flop is used to build a quantum asynchronous counter. Multi-Valued quantum D flip-flop can be used in the register, multiplexer, frequency divider, etc. quantum circuits. This Quantum D flip-flop is very essential to prepare the quantum computing processor. Multi-Valued quantum SR latch operational circuits are utilized as a memory device in computers and, in certain cases, in IoT devices. Latches are used in the construction of memory devices like flip-flops. Quantum SR flip-flops are made utilizing the quantum SR latch operational circuit in quantum computing. Multi-Valued quantum sequential circuits are an essential part of quantum electronics.

Bibliography

[1] Muthukrishnan, A., & Stroud Jr, C. R. (2000). Multi-Valued logic gates for quantum computation. Physical Review A, 62(5), 052309.

[2] Biswas, R., Jiang, Z., Kechezhi, K., Knysh, S., Mandra, S., O'Gorman, B., & Wang, Z. (2017). A NASA perspective on quantum computing: Opportunities and challenges. Parallel Computing, 64, 81-98.

[3] Greenberger, D., Hentschel, K., & Weinert, F. Experimental Observation of Decoherence. https://citeseerx.ist.psu.edu/document?repid=rep1&type=pdf&doi=8d2bcbbf9d856b33c770ec7bc3dbf419e03efba3.

[4] Diósi, L. (2011). A Short Course in Quantum Information Theory: An Approach from Theoretical Physics (Vol. 827). Springer.

2

Multi-Valued Sequential Circuits in DNA Computing

2.1 Introduction

Multi-Valued DNA   Computing is one of the most exciting and challenging research topics in the present world. Multi-Valued DNA computing is a branch of bio-molecular computing where Multi-Valued DNA   molecule sequence is used to perform a logical operation or arithmetical operation.

It is already known that the first theory of Multi-Valued DNA computation was proposed by Leonard Adleman in 1994. He put his experimental theory to the test with a seven-point Hamiltonian path problem or also called the traveling salesman problem. The salesman in this problem needs to find the shortest path between seven cities whose distances are known in such a way that he crosses no city twice and returns to the original city.

The most stable form of nucleic acid is called deoxyribonucleic acid (Multi-Valued DNA). Each of the Multi-Valued DNA strands forms helical structures that are long polymers of millions of linked nucleotides. These nucleotides consist of one of four nitrogen bases, a five-carbon sugar, and a phosphate group. The nitrogen bases – A (Adenine), T (Thymine), G (Guanine), and C (Cytosine) encode the genetic information while the others provide structural stability. The strands are linked to each other by the base-pairing rule, T with A and C with G. The arrangement of these bases is important as they decide the functionality of different genes.

Although the practice has inspired a great deal of interest among science and technology circles, Multi-Valued DNA computing has never been implemented on any broad scale, as its processing is considerably less efficient than that of standard hardware configuration. The Adleman Multi-Valued DNA computer's achievement is evidence that Multi-Valued DNA could be used to analyze complex math equations. This initial Multi-Valued DNA machine is very far from intimidating computers built on silicone in terms of efficiency, nevertheless. Computer scientists at Davis and Caltech University of California have formulated Multi-Valued DNA molecules that can be self-assembled into frameworks by using six-bit inputs to effectively run their programs. Microsoft also has a programming language for Multi-Valued DNA

DOI: 10.1201/9781003381921-2

computing which will help make Multi-Valued DNA computing functional once bio-processor technology is progressing to the stage where it can operate more sophisticated algorithms.

2.2   Multi-Valued DNA D Flip Flop

A Multi-Valued DNA D flip flop is essentially a two-state timed flip flop. In one clock cycle, the qubit molecular input sequences of a ternary DNA D-type flip flop are actuated with a delay. A delay flip flop is another term for the Multi-Valued DNA D flip flop.

The indeterminate molecular input sequence condition of SET = "TGGATC" and RESET = "TGGATC" is banned in the basic ternary DNA SR NAND Gate Bistable circuit, which is one of its fundamental drawbacks. This condition forces both qubit output sequences to logic "ACCTAG," overriding the feedback latching action, and whichever molecular input sequence goes to logic "ACCTAG" first loses control, while the other molecular input sequence, which is still at logic "TGGATC," controls the latch's final state. However, an inverter may be connected between the "SET" and "RESET" qubit molecular input sequences to create a ternary DNA Data Latch, ternary DNA Delay flip flop, ternary DNA D-type Bistable, ternary DNA D-type Flip Flop, or simply a Multi-Valued DNA D flip flop as it is most often known.

By far the most essential of all the ternary DNA timed flip-flops is the Multi-Valued DNA D flip flop. The |S> and |R> molecular input sequences become complements of each other when a ternary DNA inverter (ternary DNA NOT gate) is added between the Set and Reset molecular input sequences, ensuring that the two molecular input sequences |S> and |R> are never equal (TGGATC or ACCTAG) to each other at the same time, allowing us to control the toggle action of the flip-flop with just one |D> (Data) molecular input sequence.

The Data molecular input sequence, labeled "|D>," is then utilized in place of the "Set" signal, and the inverter is used to create the complementary "Reset" molecular input sequence, resulting in a level-sensitive ternary DNA D-type flip-flop from a level-sensitive SR-latch, with |S> = |D> and |R> = |D>.

The proposed Multi-Valued DNA D flip flop circuit has just one qubit molecular input sequence, and the qubit molecular input sequence must be in a coherence state in order to conduct the ternary DNA computational function. As a result, the circuit must exist in an environment that does not exist. If any particle emerges, the coherence state will be disrupted. The Multi-Valued DNA D flip flop will generate heat, which must be removed quickly in order to cool down the circuit and stabilize the coherence state.

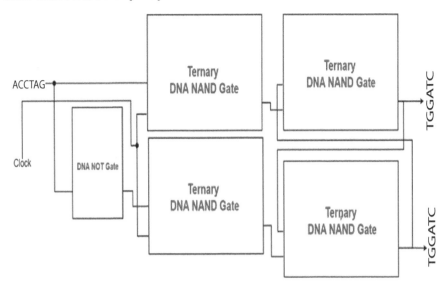

FIGURE 2.1

Block Diagram of Multi-Valued DNA D Flip Flop

2.2.1   General Organization of Multi-Valued DNA D Flip Flop

When compared different types of flip flops, the D flip flop is the most significant filp flop among them. D flip flop verifies that the two molecular input sequences of the SR flip flop are never the same. Figure 2.1 shows the block diagram of multi-valued DNA D Flip Flop.

D flip flops have two molecular input sequences: one for data and the other for clock. D flip flops have two output sequences that are logically opposite to one another. The clock molecular input sequence aids in the circuit's synchronization with an external signal. The output sequence of a D flip flop can have two potential values. This blog diagram shows that data molecular input sequence is sent to a Ternary DNA NAND operation circuit, while the reversal of data molecular input sequence is routed to another Ternary DNA NAND operation circuit. The clock pulse molecular input sequence is used by both NAND processes. The result of two Ternary DNA NAND operations is fed into the SR Latch. The SR latch is used to build the D flip flop. This attribute is utilized to create a delay in the data flow in the circuit. Two Ternary DNA NAND Operations create the SR Latch. The remaining two output sequences are discovered of the ternary DNA SR Latch function. The output sequence of a D flip flop can be of two sorts, one of which is logically inverse to the other. If the clock is enabled, the multi-valued DNA D flip flop will continue to function; otherwise, the multi-valued DNA D flip flop will cease to function.

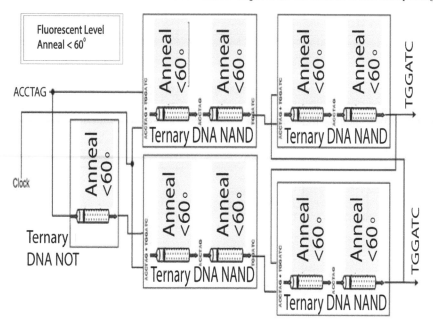

FIGURE 2.2
Multi-Valued DNA D Flip Flop Circuit

2.2.2　Circuit Architecture of Multi-Valued DNA D Flip Flop

The D flip-flop has a single qubit molecular input sequence and is developed using Ternary DNA NAND operations and a ternary DNA SR latch. Figure 2.2 shows the multi-valued DNA D Flip Flop circuit.

The clock ternary molecular input sequence affects the DNA D flip-flop. Seeing the diagram that the circuit has one molecular input sequence. One line of this molecular input sequence will be directed into the ternary DNA NAND operation known as S molecular input sequence in Circuit. In this case, the ternary S molecular input sequence and the Clock molecular input sequence are used in a ternary DNA NAND operation.

When an ACCCTAG molecular sequence traverses another line, it first undertakes a DNA not operation. This multi-valued DNA NOT operation was dubbed R when it was entered into the multi-valued DNA NAND operation. The R and Clock molecular input sequences are used in this multi-valued DNA NAND operation. The output sequence of the multi-valued DNA NAND operations is sent into the multi-valued DNA SR latch as a molecular input sequence. With these two ternary molecular input sequences, a multi-valued SR latch will be done. The multi-valued DNA SR latch will be used, and the output sequence of the multi-valued DNA SR latch will be the final output sequence of Q and $\overline{Q}$. The conclusion of Q will always be the inverse of Q.

2.2.3 Working Principle of Multi-Valued DNA D Flip Flop

There are two molecular input sequences in the ternary DNA SR flip flop: SET and RESET. Alternatively, in a D flip-flop, one molecular input sequence and the molecular input sequence's one line are referred to as a SET, and by coupling a ternary DNA NOT gate towards the other line molecular input sequence, the D flip-flop is set as a RESET. This complement resolves the contradiction inherent in the SR latch when both molecular input sequences are LOW because that circumstance is no longer feasible. Multi-Valued DNA D flip flops have a single molecular input sequence, which is sometimes alluded to as a data molecular input sequence. If this data molecular input sequence is high, the flip-flop becomes SET; if the data molecular input sequence is low, such as TGGATC, the flip-flop changes its state and becomes RESET.

However, this would be pretty futile because the output sequence of the flip flop will always vary with each pulse delivered to this data molecular input sequence. To circumvent this, an extra molecular input sequence known as the "CLOCK" or "ENABLE" molecular input sequence is used to separate the data molecular input sequence from the latching circuitry of the flip flop after the appropriate data has been stored. The result is that the |X> molecular input sequence condition is only replicated to the output sequence |Q> while the clock molecular input sequence is active. This then serves as the foundation for yet another sequential gadget known as a D Flip Flop.

As long as the clock molecular input sequence is HIGH, the "D flip flop" will store and output sequence any logic level that is applied to its data terminal. Once the clock molecular input sequence is changed to LOW, the flip-"set" flop's and "reset" molecular input sequences are both kept at logic level "ACCTAG," preventing the flip-flop from altering the underlying and preserving whatever statistics are available on its output sequence prior to the clock transition. In other words, either logic "TGGATC" or logic "ACCTAG" latches the output sequence. Table 2.1 shows the truth table of multi-valued DNA D Flip Flop.

2.2.4 Example

In ternary DNA D flip-flop operational circuit one input is molecule input x sequence. Assume this input is TGGATC. This input will perform ternary DNA NAND operation with the clock input ACCTAG. Then the output will be TGGATC.

TABLE 2.1
Truth Table of Multi-Valued DNA D Flip Flop

|Clk>	|x>	|Q>	$\overline{\|Q>}$	Description
↓ >> TGGATC	X	Q	Q	Memory no change
↑ >> ACCTAG	TGGATC	TGGATC	ACCTAG	Reset Q >> 0
↑ >> ACCTAG	ACCTAG	ACCTAG	TGGATC	Set Q >>
↑ >> ACCTAG	CGGATC	CGGATC	TGGATC	Set Q >>

Then molecule sequence input will perform ternary DNA NOT operation and then again it will perform ternary DNA NAND operation. So, x = ACCTAG and clk = ACCTAG will produce output TGGATC.

Two ternary DNA NAND operations produced output will be entered into an input in the ternary DNA SR latch operational circuit. This ternary DNA SR latch operational circuit output will count as a final output in ternary DNA D flip-flop operational circuit. Here, S = ACCTAG and R = TGGATC will produce output AC-CTAG and TGGATC.

This is the final output of ternary DNA D flip-flop operational circuit and these outputs clarify that correct output is produced by ternary DNA D flip-flop operational circuit.

2.3   Multi-Valued DNA SR Latch

Based on the triggering that is suited to operate it, there are two types of memory elements. One of them is a latch, and the other is a flip-flop. Latches operate with enable signal, level-sensitive, whereas flip-flops are edge sensitive.

A Multi-Valued DNA SR Latch is an asynchronous tool. It operates without the use of control signals, relying solely on the state of the |s> and |R> molecular ternary input sequences. Two ternary DNA NAND operations can make a Multi-Valued DNA Latch. Nevertheless, two ternary DNA NOR operations can also make a Multi-Valued DNA SR Latch. In Multi-Valued DNA SR Latch, two-qubit molecular ternary input sequences are swapped and negated. Multi-Valued DNA SR Latch can be said as SET RESET latch. In Multi-Valued DNA SR Latch from two-qubit molecular ternary input sequence, got two ternary output sequences. This ternary output sequence is reversed to one another. This research proposed a Multi-Valued DNA SR Latch, and two ternary DNA NAND operation circuits designed for this ternary SR Latch. This proposed Multi-Valued DNA SR Latch has the molecular ternary input sequence line swapped between two ternary DNA NAND operations, but it is not negated. Multi-Valued DNA SR Latch works as memory stuff in ternary DNA computers, and it has several applications in a ternary DNA processor. If it is possible to design some embedded system using the ternary DNA device, then Multi-Valued DNA SR Latch will be used on this device as a memory unit. Multi-Valued DNA SR Latch is level sensitive and has few disadvantages, and this will be recovered by the ternary DNA flip flop, which is described earlier.

2.3.1   General Organization Multi-Valued DNA SR Latch

The Multi-Valued DNA SR Latch is one of the most common memory devices, and it has an effect on the ternary output sequence as long as it is active. The essential properties of a Multi-Valued DNA SR Latch are that one qubit molecular ternary

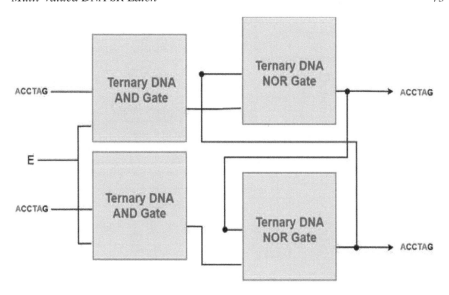

FIGURE 2.3
Block Diagram of Multi-Valued DNA SR Latch

input sequence behaves like a SET and another qubit molecular ternary input sequence behaves like a RESET. Figure 2.3 depicts the block diagram of multi-valued DNA SR Latch .

The Multi-Valued DNA SR Latch is made up of two fundamental processes, which are depicted in this block diagram of the Multi-Valued DNA SR Latch. There are two molecular ternary input sequence lines in the Multi-Valued DNA SR Latch, one for |s> and the other for |R>. Two ternary output sequences are obtained from this two-qubit molecular ternary input sequence: |Q> and $\overline{Q}$. The ternary output sequence of the first ternary DNA NAND operation is used as molecular ternary input sequence in the second ternary DNA NAND operation, and the ternary output sequence of the second ternary DNA NAND operation is used as a molecular ternary input sequence in the first ternary DNA NAND operation. If the molecular ternary input sequence of |S> is ACCTAG, the ternary SR Latch is activated; however, if the molecular ternary input sequence of |R> is ACCTAG, the ternary SR Latch has no influence on the ternary output sequence. In a Multi-Valued DNA SR Latch, a value of ACCTAG cannot be used to activate two molecular ternary input sequences.

2.3.2 Circuit Architecture of Multi-Valued DNA SR Latch

Multi-Valued DNA SR Latches are level sensitive and are built using only one fundamental operation, the ternary DNA NAND function. Figure 2.4 shows the circuit architecture of multi-valued DNA SR Latch.

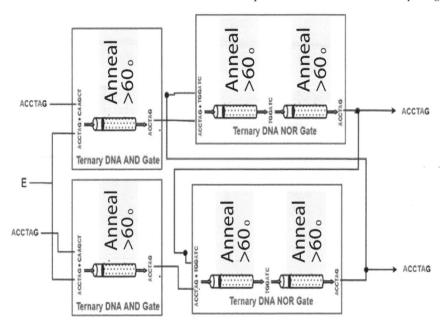

FIGURE 2.4
Circuit Architecture of Multi-Valued DNA SR Latch

As seen in the diagram above, the Multi-Valued DNA SR Latch has two-qubit molecular ternary input sequences. In the first ternary DNA NAND operation, the molecular ternary input sequence S and the ternary output sequence of the second ternary DNA NAND operation Q are both entered as molecular ternary input sequences. The ternary output sequence of this ternary DNA NAND operation is mostly Q.

Second, the ternary DNA NAND operation is performed on the molecular ternary input sequences R and Q, yielding Q as the ternary output sequence. The base sequence ACCTAG is employed in every ternary DNA NAND operation.

2.3.3   Working Principle of Multi-Valued DNA SR Latch

The state of the S and R molecular ternary input sequences is all that matters in a Multi-Valued DNA SR Latch. It acts independently of control signals. S molecular ternary input sequence behaves as if it were a SET instruction, and R molecular ternary input sequence behaves as if it were a RESET instruction. If the SET molecular ternary input sequence of a Multi-Valued DNA SR Latch is high, the ternary output sequence Q will be ACCTAG or CGGATC and the opposing ternary output sequence TGGATC will be the value of $\overline{Q}$. When the RESET molecular ternary input sequence is high, the value of |Q> is ACCTAG or CGGTAC, and when the RESET molecular ternary input sequence is low, the value of Q is TGGATC. The latch's "memory" is basically reset. When both molecular ternary input sequences are low, the latch "latches" stay in their previously set or reset state.

The actual problem comes when both the molecular ternary input sequences SET and RESET go high. The ternary output sequences Q and $\overline{Q}$ will have the opposite value as shown in the circuit. When the SET and RESET molecular ternary input sequences ACCTAG or CGGATC are used together, the circuit creates a "race situation." In order for the device to be "metastable," which implies it will remain in an indeterminate state indefinitely, both gates must be identical. In reality, if the suggested circuit is to be manufactured, one gate will win; however, determining which gate won is difficult. Because of this, having both the SET and RESET molecular ternary input sequences high is forbidden in a Multi-Valued DNA SR Latch.

The same thing happens when the device is switched on, since both ternary output sequences, Q and $\overline{Q}$, are low. The device will quickly leave the metastable state due to the differences between the two gates, but it's difficult to forecast which of Q and $\overline{Q}$ will end up high. To avoid erroneous actions, multi-valued DNA SR flip-flops must always be set to a known starting state before being used; users should not assume that they would initialize to a low state. Table 2.2 shows the truth table of multi-valued DNA SR Latch.

The flip flop discussed in the flip flop chapter solves the difficulty of the Multi-Valued DNA SR Latch. The Multi-Valued DNA SR Latch, on the other hand, is still a vital component of a CPU or embedded device.

The ternary DNA circuit generates a lot of heat, making it difficult to isolate the qubit into a superposition state. As a result, the circuit is needed to be cooled to isolate the qubit into a superposition for an able ternary DNA circuit. Any type of external particle can disrupt the qubit's coherence and cause it to become decoherent. If all of this is preserved, the Multi-Valued DNA SR Latch can truly function.

2.3.4 Example

Presume that the qubits TGGATC and ACCTAG are both present in the ternary DNA SR Latch. One qubit molecular ternary input sequence will be used for SET instructions, while the other will be used for RESET instructions. Here, TGGATC will be

TABLE 2.2

Truth Table of Multi-Valued DNA SR Latch

|S>	|R>	|Q>	$|\overline{Q}>$
TGGATC	TGGATC	Latched	
TGGATC	ACCTAG	ACCTAG	TGGATC
ACCTAG	TGGATC	TGGATC	ACCTAG
TGGTAC	CGGTAC	CGGATC	TGGATC
CGGTAC	TGGATC	TGGATC	CGFGATC
ACCTAG	ACCTAG	Metastable	
CGGATC	CGGTAC	Metastable	

used as a SET instruction, and it will conduct the ternary DNA NAND operation according to the suggested circuit idea.

As a result, the final ternary output sequence is TGGATC. The ternary output sequence principle of a ternary DNA NAND operation is that if one of the molecular ternary input sequences is TGGATC, the ternary output sequence will also be TGGATC. As a corollary, whether Q is TGGATC or ACCTAG, the ternary output sequence will be ACCTAG.

Qubit molecular ternary input sequence ACCTAG now functions as a reset instruction and is inserted into the circuit, as well as performing the ternary DNA NAND operation.

Multi-Valued DNA SR Latch operates in the same mechanism with each qubit molecular ternary input sequence. However, because all of the computations took place in the ternary DNA superposition state, a lot of heat was generated throughout the process. All computation takes place in the coherence state, and the result will be decoherent.

2.4   Multi-Valued DNA SR Flip Flop

Ternary DNA Sequential Logic circuits, unlike ternary DNA Combinational Logic circuits, include some type of built-in "Memory" that changes state depending on the real signals supplied to its molecular ternary input sequences at the moment. Multi-Valued DNA SR flip-flops, for example, have a 1 molecule sequence memory bistable. The SET and RESET molecular ternary input sequences of the ternary SR flip flop are the same. The ternary output sequence of the SET molecular ternary input sequence is a ACCTAG or CGGTAC, whereas the ternary output sequence of the RESET molecular ternary input sequence is a TGGATC.

The Multi-Valued DNA SR flip flop is often referred to as the SET RESET flip flop. The reset molecular ternary input sequence is used to restore the flip flop to its starting state from the current state with a ternary output sequence. The NAND gate ternary SR flip flop is a basic flip flop with both ternary output sequences providing feedback to its opposite molecular ternary input sequence. This circuit is used to store a single data bit in a memory circuit. The three molecular ternary input sequences are SET, RESET, and a found ternary output sequence. A two-qubit model will be used since Multi-Valued DNA SR flip flops have two molecular ternary input sequences that are mostly from the outside. Because using two-qubits, it generates more heat at first than ternary DNA D flip flops. The computation time of this Multi-Valued DNA SR flip-flop is determined by the fundamental gate in its middle. Multi-Valued DNA SR flip flops may be found in a wide range of processors and embedded systems. Although the suggested flip-flop can generate some trash, an error-correcting auxiliary qubit provides the desired ternary output sequence. The real-world

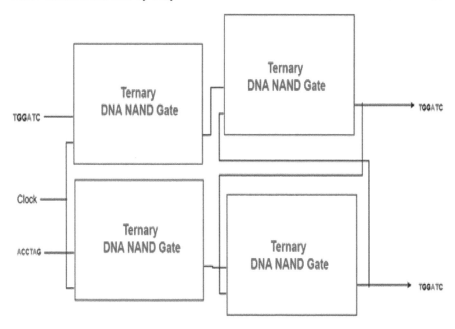

FIGURE 2.5
Block Diagram **of** Multi-Valued DNA SR Flip-Flop Circuit

implementation of the suggested ternary DNA circuit will address a wide range of issues more quickly and effectively.

2.4.1   General Organization of Multi-Valued DNA SR Flip Flop

A SET - RESET flip-flop is a common name for a Multi-Valued DNA SR flip-flop. As a result, it is evident that the Multi-Valued DNA SR latch has a two-qubit molecular ternary input sequence. Figure 2.5 shows the block diagram of multi-valued DNA SR Flip-Flop Circuit.

The major molecular ternary input sequences of a Multi-Valued DNA SR flip flop are S and R, as well as one clock molecular ternary input sequence called clock. First, a ternary DNA NAND operation with clock molecular ternary input sequence was performed the S and R. The ternary DNA NAND operation is performed on molecular ternary input sequence S and molecular ternary input sequence clock. A ternary DNA NAND operation is also performed in parallel by the R and clock molecular ternary input sequences. Because the two ternary DNA NAND operations are performed in parallel, they take the same amount of time. Then, using ternary DNA NAND operations ternary output sequence from the S molecular ternary input sequence lines and the final ternary output sequence Q, another ternary DNA NAND operation will be executed and generate the ternary output sequence Q. Similarly,

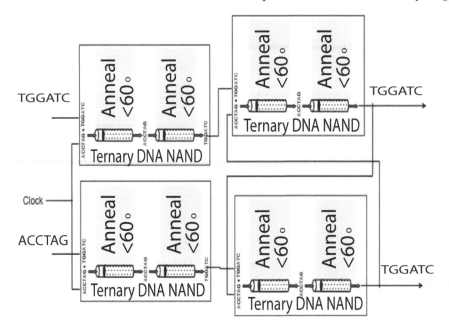

FIGURE 2.6

Circuit Architecture of Multi-Valued DNA SR Flip-Flop

ternary DNA NAND operations are ternary output sequence from the R molecular ternary input sequence lines, and the final ternary output sequence Q conducts the ternary DNA NAND operation and generates the Q.

Actually, the created ternary output sequence enters the Multi-Valued DNA SR latch and finds the Q and $\overline{Q}$ after the first ternary DNA NAND operations are completed. The values of ternary output sequence Q and $\overline{Q}$ are diametrically opposed. The block diagram depicts all of the procedures involved in the proposed Multi-Valued DNA SR latch. During the operation, all qubits must be in the superposition state, which implies they must be coherent. If any particle from the environment approaches close enough, the superposition state will evaporate.

2.4.2 Circuit Architecture of Multi-Valued DNA SR Flip-Flop

In ternary DNA, ternary SR flip-flop mainly works as a SET RESET flip-flop in a ternary DNA computer. Figure 2.6 depicts the circuit architecture of multi-valued DNA SR Flip-Flop.

This Multi-Valued DNA SR flip-flop circuit contains two molecular ternary input sequences S and R, as well as a clock molecular ternary input sequence. Four ternary DNA NAND operation circuits make up the Multi-Valued DNA SR. To begin, one ternary DNA NAND operation is performed on the S and clock molecular ternary input sequences. The ternary DNA NAND operation is carried out by another molecular ternary input sequence R and clk. Two of the ternary DNA NAND operations' ternary output sequences were used as molecular ternary input sequences into the Multi-Valued DNA SR latch. The Multi-Valued DNA SR latch is controlled by these two molecular ternary input sequences. One ternary output sequence from earlier ternary DNA NAND operations comes from the S molecular ternary input sequence line in a Multi-Valued DNA SR latch, and one of the ternary output sequence ternary SR flip-flops Q conducts the ternary DNA NAND operation and provides the ternary output sequence Q. The ternary output sequence of the ternary DNA NAND operation, which is based on the R molecular ternary input sequence and Q ternary output sequence, enters as a molecular ternary input sequence in the ternary DNA NAND operation and creates the ternary output sequence of the Multi-Valued DNA SR flip flop as Q, just as it did previously.

2.4.3   Working Principle of Multi-Valued DNA SR Flip-Flop

There are two input molecular ternary input sequences S and R in a Multi-Valued DNA SR flip flop. The ternary DNA operation is performed once the qubit molecular ternary input sequence becomes coherent and enters a superposition state. It is needed to transport the heat from the ternary DNA circuit since it creates a lot of it.

First, the S molecular ternary input sequence in a Multi-Valued DNA SR flip-flop changes its state to superposition and becomes coherent. The ternary DNA NAND operation is then performed using the clock qubit molecular ternary input sequence clk using the S molecular ternary input sequence. It generates a ternary output sequence after completing the ternary DNA NAND operation. The ternary DNA NAND operation is also performed by the R molecular ternary input sequence with the clk molecular ternary input sequence. The final ternary output sequences of both ternary DNA NAND operations were used as molecular ternary input sequences in the Multi-Valued DNA SR latch. One molecular ternary input sequence acts as if it is high, while the other acts as if it is low. The state of the S and R molecular ternary input sequences is all that matters in a Multi-Valued DNA SR latch, which is independent of control signals. S molecular ternary input sequence behaves as if it were a SET command, whereas R behaves as if it were a RESET command. The ternary output sequence Q will be ACCTAG if the SET molecular ternary input sequence of the Multi-Valued DNA SR latches becomes high, and the opposing ternary output sequence TGGATC will be the value of Q. When the RESET molecular ternary input sequence is high, the value of Q is ACCTAG or CGGTAC and when the RESET molecular ternary input sequence is low, the value of Q is TGGATC. The "memory" of the latch is basically reset. The latch "latches" stay in their previously set or reset state when both molecular ternary input sequences are low.

The real issue arises when both the molecular ternary input sequences SET and RESET go high. The ternary output sequences Q and $\overline{Q}$ will have the opposite values as shown in the circuit. When the SET and RESET molecular ternary input sequences ACCTAG are used, the circuit creates a "race situation." Both gates should be identical in order for the device to be "metastable," which means it will be in an

TABLE 2.3

Truth Table of Multi-Valued DNA SR Flip-Flop

S	R	Q	$\overline{Q}$
TGGATC	TGGATC	No Change	
TGGATC	ACCTAG	TGGATC	ACCTAG
ACCTAG	TGGATC	ACCTAG	TGGATC
TGGATC	CGGTAC	TGGATC	CGGTAC
ACCTAG	ACCTAG	Invalid	

indeterminate state for an endless amount of time. In reality, if the suggested circuit is manufactured, one gate will win; however, determining which gate won is difficult. Because of this, having both the SET and RESET molecular ternary input sequences high in a Multi-Valued DNA SR latch is unlawful.

When the device is turned on, both ternary output sequences, Q and $\overline{Q}$ are low, resulting in a similar situation. Because of the disparities between the two gates, the device will swiftly depart the metastable state, but it's hard to anticipate which of Q and $\overline{Q}$ will end up high. Table 2.3 shows the truth table of multi-valued DNA SR Flip-Flop.

2.4.4 Example

Assume that this study effort receives the molecular ternary input sequences TG-GATC and ACCTAG in order to ensure that the Multi-Valued DNA SR flip-flop operational circuit produces the right ternary output sequence. The TGGATC molecular ternary input sequence end is connected to the S molecular ternary input sequence end, while the ACCTAG molecular ternary input sequence end is connected to the R molecular ternary input sequence end. The fact that the R molecular ternary input sequence is ACCTAG indicates that it is for a reset operation. Assume that the clock is activated and that the clock's molecular ternary input sequence is ACCTAG.

First, do the ternary DNA NAND operation using the clk molecular ternary input sequence. If just one of the molecular ternary input sequences is TGGATC in a ternary DNA NAND operation, the result is ACCTAG; otherwise, the ternary output sequence is TGGATC.

The ternary DNA NAND operation is then performed using another clk and R molecular ternary input sequence. In this case, the clk molecular ternary input sequence is ACCTAG, and the R molecular ternary input sequence is also ACCTAG. Then Intermediate ternary output sequence is TGGATC.

Then, as the molecular ternary input sequence to the Multi-Valued DNA SR latch, these two molecular ternary input sequence qubits, ACCTAG and TGGATC, will be molecular ternary input sequenced. The Multi-Valued DNA SR latch operation will be performed by them. A collection of ternary DNA NAND operation circuits makes up the ternary SR latch operation circuit.

For the molecular ternary input sequences TGGATC and ACCTAG, this proposed Multi-Valued DNA SR flip-flop circuit now has the ternary output sequences

TGGATC and ACCTAG. In a Multi-Valued DNA SR flip-flop, this is the needed molecular ternary input sequence for the provided molecular ternary input sequence. Multi-Valued DNA SR flip-flops generate a lot of heat, yet this has no effect on the ternary output sequence.

2.5 Multi-Valued DNA JK Flip-Flop

In flip-flop designs, the Multi-Valued DNA JK flip-flop will be the most extensively utilized flip-flop. J and K are not abbreviated letters of other words, such as "S" for Set and "R" for Reset, but are independent letters chosen by the inventor Jack Kilby to identify the flip-flop design from others. Despite the fact that the digital electronics ternary JK flip flop was created by Jack Kilby. The functioning concept of the proposed Multi-Valued DNA JK flip flop differs from that of the digital ternary JK flip flop.

The Multi-Valued DNA JK flip flop's sequential operation is identical to that of the prior ternary DNA SR flip flop, with the same "Set" and "Reset" molecular ternary input sequences. The distinction this time is that even though S and R are both at logic "1," the "Multi-Valued DNA JK flip flop" has no incorrect or prohibited ternary DNA SR Latch molecular ternary input sequence states. It is evident that the Multi-Valued DNA JK flip flop does not solve the disadvantages of the ternary DNA SR flip flop.

The Multi-Valued DNA JK flip flop is essentially a gated ternary DNA SR flip flop with the addition of clock bit molecular ternary input sequence circuitry to avoid the unlawful or invalid ternary output sequence state that can arise when both molecular ternary input sequences S and R are equal to logic level "1." A Multi-Valued DNA JK flip-flop has four potential molecular ternary input sequence combinations due to the extra timed molecular ternary input sequence: "ACCTAG," "logic TG-GATC," "no change," and "toggle." A Multi-Valued DNA JK flip flop has the same symbol as a ternary DNA SR Bistable Latch, as seen in the preceding chapter. The Multi-Valued DNA JK flip flop, like other flip flops, generates a lot of heat, which must be dissipated in order for the operation to run properly. As compared to other ternary DNA circuits, the Multi-Valued DNA JK flip flop will not require as much power. The bit may simply conduct the operation once all of the molecules are in superposition state and coherence mode. A lot of junk values are received in the Multi-Valued DNA JK flip flop, and it is needed to do more investigation to figure out what they are. The trash value is not taken into account in this procedure.

2.5.1 General Organization of Multi-Valued DNA JK Flip Flop

In the construction of ternary DNA computers, the Multi-Valued DNA JK flip-flop is the most often utilized flip-flop. J and k are two-bit molecular ternary input sequences in a Multi-Valued DNA JK flip flop. Figure 2.7 displays the block diagram of multi-valued DNA JK Flip Flop Operation Circuit.

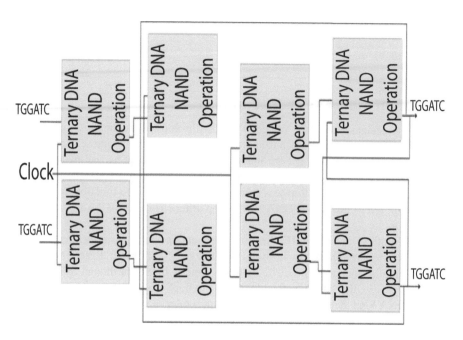

FIGURE 2.7
Block Diagram of Multi-Valued DNA JK Flip Flop Operation Circuit

The bit molecular ternary input sequences J and k are used in the Multi-Valued DNA JK flip-flop. Many ternary DNA NAND gate operations make up this Multi-Valued DNA JK flip flop. After performing a pair of fundamental ternary DNA NAND gate operations, the result of this operation is inserted into the SR flip flop, yielding the ternary output sequence Q and $\overline{Q}$. First, the ternary DNA NAND gate operation is performed using the J and clk molecular ternary input sequences. The ternary output sequence of the ternary DNA NAND gate operation and the ternary output sequence of the Multi-Valued DNA JK flip flop Q conducts another ternary DNA NAND gate operation and creates the ternary output sequence designated S. Because the k and clk molecular ternary input sequences are shared, the ternary DNA NAND gate operation is performed on both of them. The Multi-Valued DNA JK flip flop's Q ternary output sequence executes another ternary DNA NAND gate operation and generates the R ternary output sequence.

The ternary DNA SR flip flop accepts these S and R and produces two ternary output sequences, Q and $\overline{Q}$. There are four ternary DNA NAND gate operations in the ternary DNA SR flip flop. The ternary DNA NAND gate operation is performed by the S molecular ternary input sequence and clk molecular ternary input sequence. The ternary DNA NAND gate operation is also performed by the R molecular ternary input sequence and shared clk molecular ternary input sequence. The result of these two NAND operations is used as molecular ternary input sequence in the ternary DNA SR latch. The ternary DNA NAND gate operation is performed in ternary DNA SR latches using two Q and one molecular ternary input sequence, as well as Q and another molecular ternary input sequence. Finally, the Multi-Valued DNA JK flip flops end ternary output sequences Q and $\overline{Q}$ were obtained after all of this ternary DNA operation. In Chapters 5 and 6, the ternary DNA SR flip flop and the SR latch are discussed in general.

2.5.2 Circuit Architecture of Multi-Valued DNA JK Flip-Flop

The two-bit input molecular ternary input sequence and one clock shared molecular ternary input sequence of the Multi-Valued DNA JK flip flop are shared. The clock affects the Multi-Valued DNA JK flip flop as well. The circuit will turn on if the clock is turned on, otherwise it will not. Figure 2.8 shows the multi-valued DNA JK Flip-Flop operation circuit.

Both molecular ternary input sequence J and molecular ternary input sequence k will conduct the ternary DNA NAND gate operation differently using the shared clk molecular ternary input sequence. The ternary DNA NAND gate operation is performed using the J and clk molecular ternary input sequences. Ternary DNA fundamental gates were used to perform the DNA NAND gate operation. Acquiring a ternary output sequence bit after the ternary DNA NAND gate operation is

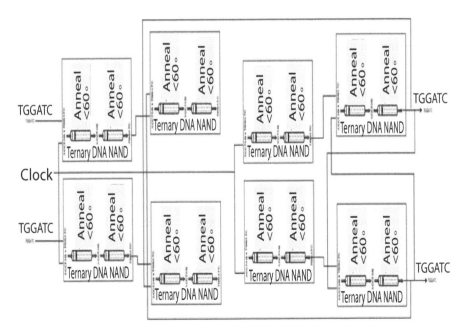

FIGURE 2.8
Multi-Valued DNA JK Flip-Flop Operation Circuit

performed by J and clk, and this ternary output sequence bit, together with the ternary output sequence of the flip flop Q, is used to execute the ternary DNA NAND gate operation and generate the ternary output sequence S. The R ternary output sequence is produced by the same technique that creates the k, clk, and Q molecular ternary input sequences. The majority of these circuits are ternary DNA NAND gate operations. The ternary DNA SR flip-flop operation is conducted on the S and R molecular ternary input sequences. The suggested essential component of ternary DNA computing is ternary DNA NAND gate operation, which is also used in ternary DNA SR flip flop operation circuit design.

2.5.3   Working Principle of Multi-Valued DNA JK Flip Flop

The two-bit circuit is the Multi-Valued DNA JK flip flop. In ternary DNA computing, the Multi-Valued DNA JK flip flop is the most often utilized flip flop. The molecular ternary input sequences J and k of a Multi-Valued DNA JK flip-flop conduct two ternary DNA processes in parallel. The ternary DNA NAND gate operation is performed using J and shared molecular ternary input sequence clk. The ternary DNA NAND gate operations are then completed, and one of the Multi-Valued DNA JK flip flop's ternary output sequences executes the ternary DNA NAND gate operation, producing S. The k and clk molecular ternary input sequences are performed first in the ternary DNA NAND gate operation. The ternary output sequence of the Multi-Valued DNA JK flip-flop Q, as well as the result of the ternary DNA first NAND operation, are then used to perform another ternary DNA NAND gate operation, yielding R. The steps for creating S and R are carried out in simultaneously. It is known that one of the distinctive properties of ternary DNA operations is that they may do several operations at the same time, and this is exactly what is happening. The ternary DNA SR flip flop operation is then conducted on these S and R molecular ternary input sequences. The ternary DNA NAND gate operation, which is employed here, is also used to make ternary DNA SR flip flops. Two ternary output sequences are discovered after conducting the ternary DNA SR flip-flop operation. The opposite of one ternary output sequence is the opposite of the other. The Multi-Valued DNA JK flip flop truly solves the ternary DNA SR flip flop problem. In the fifth chapter, the ternary DNA SR flip-flop is briefly described.

Initially, molecules entered a state of superposition and formed bits. When a bit is in a superposition state, it is being coherent. The flip flop process will be maintained until the bits are coherent. The process will be disturbed if a particle from outside breaks the coherence and causes the bit to become decoherent. If the heat in the circuit is not reduced, the operation in this circuit will likewise come to a halt. Table 2.4 shows the truth table of multi-valued DNA JK Flip-Flop.

Assume that the clock molecular ternary input sequence always enables the truth table of the Multi-Valued DNA JK flip-flop. When any molecular ternary input sequence is TGGATC, it acts like a ternary DNA SR latch circuit, but when both molecular ternary input sequences are ACCTAG or CGGTAC, it toggles to create the ternary output sequence, according to the truth table.

TABLE 2.4

Truth Table of Multi-Valued DNA JK Flip-Flop

J	k	Q	$|\overline{Q}>$
TGGATC	TGGATC	No Change	
TGGATC	ACCTAG	TGGATC	ACCTAG
ACCTAG	TGGATC	ACCTAG	TGGATC
CGGTAC	TGGATC	TGGATC	CGGTAC
ACCTAG	ACCTAG	TGGATC	ACCTAG
ACCTAG	ACCTAG	ACCTAG	TGGATC

The ternary JK flip flop is a timed SR flip flop with better performance. However, the "race" issue still exists. When the state of the ternary output sequence Q is altered before the timing pulse of the clock molecular ternary input sequence has time to go "Off," this issue occurs.

2.5.4 Example

For the purpose of testing the proposed Multi-Valued DNA JK flip-flop circuit, assume that the molecular ternary input sequence is TGGATC and ACCTAG. If and only if the clock molecular ternary input sequence is high or ACCTAG, the Multi-Valued DNA JK flip flop will function perfectly and produce ACCTAG and TGGATC as Intermediate ternary output sequence.

These two intermediate ternary output sequences are now used to conduct two independent ternary DNA NAND gate operations. Assume $|\overline{Q}>$ = ACCTAG, so the first Intermediate ternary output sequence ACCTAG will produce S = TGGATC.

Then Q = TGGATC, so the second intermediate ternary output sequence TGGATC will produce R= ACCTAG.

According to the suggested circuit of Multi-Valued DNA JK flip-flop, these bits labeled S and R will now conduct the ternary DNA SR flip-flop operation and the final output will be Q= TGGATC and $\overline{Q}$ = ACCTAG

Finally, a ternary DNA proposal was made. The ternary JK flip-flop provides the necessary molecular ternary input sequences TGGATC and ACCTAG, demonstrating that the proposed circuit theoretically produces the ideal ternary output sequence.

2.6 Multi-Valued DNA T Flip-Flop

The "Multi-Valued DNA Toggle Flip–flop" is another name for the Multi-Valued DNA T flip–flop. To avoid the occurrence of the intermediate state in a ternary DNA SR flip–flop, just one molecular ternary input sequence, called the Trigger qubit molecular ternary input sequence or Toggle molecular ternary input sequence, should be sent to the flip–flop. 'Changing the next state ternary output sequence to complement the current state ternary output sequence' is referred to as toggling. By making modest changes to the ternary DNA JK flip–flop, it is possible to create the Multi-Valued DNA T flip–flop. Because the Multi-Valued DNA T flip–flop is a single qubit molecular ternary input sequence device, a ternary DNA JK flip–flop can be transformed into a Multi-Valued DNA T flip–flop by linking the J and K molecular ternary input sequences together and giving them a single molecular ternary input sequence named T.

Ternary T flip flops, like any ternary DNA operation circuits, face the difficulty of producing additional heat. If and only if the heat is lowered, and the temperature is near to zero, this Multi-Valued DNA T flip flop will learn. The most fundamental component of a Multi-Valued DNA T flip flop is the ternary DNA AND and NOR operations. This fundamental component is made up of basic ternary DNA gates. When compared to the functioning of classical computers, this ternary DNA process is extremely quick. Only this flip flop operation will be performed if the qubits are in a state of coherence; otherwise, this operation will not be performed. Any particle in the surroundings can disrupt the state of coherence, causing it to become decoherent.

2.6.1   General Organization of Multi-Valued DNA T Flip-Flop

The Multi-Valued DNA T flip-flop solves the problems of the ternary DNA JK and SR flip-flops. One molecular ternary input sequence T is used in Multi-Valued DNA T flip-flops. Figure 2.9 depicts the block diagram of multi-valued DNA T Flip-Flop.

Two ternary DNA AND computations share one molecular ternary input sequence T. The combination of the clk molecular ternary input sequence and the T ternary output sequence yields two-qubit ternary output sequences by performing two ternary DNA AND operations in tandem. As molecular ternary input sequences, their ternary output sequences are fed into ternary DNA SR flip-flops. Two ternary DNA AND operations and two ternary DNA NOR operations are used to create these ternary DNA SR flip-flops. There are two molecular ternary input sequences to this ternary DNA SR flip-flop. First, the ternary DNA AND operation is performed using the ternary output sequence of the previous ternary DNA AND operation and the ternary output sequence of the ternary DNA SR flip-flop Q. S is the name given to the ternary output sequence. The ternary DNA AND operation is then performed on another ternary output sequence of the prior operation by using Q. R is the moniker given to the result of these ternary DNA AND operations. The ternary DNA NOR gate uses molecular ternary input sequence S and molecular ternary input sequence R to build a ternary DNA SR latch. The ternary DNA NOR operation is performed on the ternary output sequence of ternary DNA SR latch Q and S. The ternary DNA NOR operation is carried out in parallel by both R and Q. Finally, the Multi-Valued DNA T flip-flop gives us the ternary output sequences Q and $\overline{Q}$. The ternary output sequences of these Multi-Valued DNA T flip-flops are diametrically opposed.

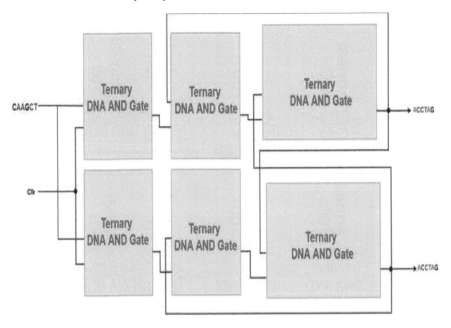

FIGURE 2.9
Block Diagram of Multi-Valued DNA T Flip-Flop

2.6.2 Circuit Architecture of Multi-Valued DNA T Flip-Flop

Multi-Valued DNA T flip flop is an operation with one molecular ternary input sequence named "T" that alleviates the JK flip flop issue. Toggling is the primary use of this ternary DNA flip flop action. Figure 2.10 displays the multi-valued DNA T Flip-Flop operation circuit.

Basically build some ternary DNA AND operation circuits and some ternary DNA NOR operation circuits in the Multi-Valued DNA T flip-flop operation circuit. However, a detailed examination reveals that the Multi-Valued DNA T flip-flop design is similar to the ternary DNA JK flip-flop. Ternary DNA JK flip-flop operation has various issues that are addressed by Multi-Valued DNA T flip-flop operation. Ternary DNA JK flip-flops have a two-qubit molecular ternary input sequence, whereas Multi-Valued DNA T flip-flops have a shared molecular ternary input sequence. A ternary DNA SR flip-flop is sandwiched between two Multi-Valued DNA T flip-flops. Two concurrent ternary DNA procedures were used to create this flip-flop. When compared to the SR flip-flop discussed in the preceding chapter, the ternary DNA SR flip-flop is slightly different. Two ternary DNA AND operations and a ternary DNA NOR SR latch were used to create the ternary DNA SR latch in this ternary DNA SR flip-flops design. First, two parallel ternary DNA operations are built in the Multi-Valued DNA T flip-flop. Then two ternary DNA AND operations are connected in parallel in the SR flip-flop section, and two ternary DNA NOR operations are set up in parallel in the SR latches. The Multi-Valued DNA T flip-flop

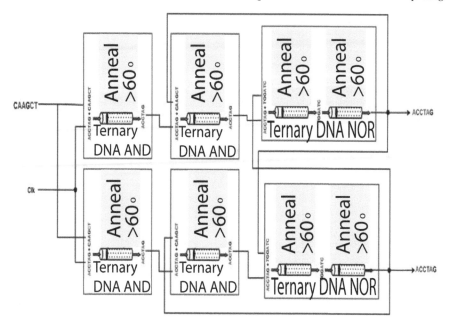

FIGURE 2.10
Multi-Valued DNA T Flip-Flop Operation Circuit

design keeps ternary DNA parallelism's properties, although it's simply a theoretical circuit. Depending on the qubit origin, heat, and temperature, this circuit design can be modified.

2.6.3   Working Principle of Multi-Valued DNA T Flip-Flop

The Multi-Valued DNA T flip-flop differs from the ternary DNA JK flip-flop in a few ways. When it is needed to be toggled, Multi-Valued DNA T flip-flop is employed. This operating circuit only has one molecular ternary input sequence, T, and one clock, clk. The circuit will be activated if the clock molecular ternary input sequence is ACCTAG; else, the circuit will be disabled. As a result, the Multi-Valued DNA T flip-flop operation was enabled, and the fundamental operation began to function. This circuit will operate if the qubit is in a coherent mood, just like any other circuit. Because any environment particle entering the circuit would disrupt coherence and make it decoherent, this circuit must be kept at a temperature close to 0 degrees Fahrenheit.

The basic operation of ternary DNA computing, known as the ternary DNA AND operation, is performed initially in the Multi-Valued DNA T flip-flop operating circuit. Because the T and clk molecular ternary input sequences are shared, two ternary DNA AND operations run in parallel, requiring the same amount of time. After the process, this ternary output sequence acts as a molecular ternary input sequence for a ternary DNA SR flip-flop. The SR flip-flop conducts two ternary DNA AND

TABLE 2.5

Truth Table of Multi-Valued DNA T
Flip-Flop

T	Q	$|\overline{Q}>$
TGGATC	TGGATC	ACCTAG
ACCTAG	TGGATC	ACCTAG
TGGATC	ACCTAG	TGGATC
ACCTAG	ACCTAG	TGGATC
TGGATC	TGGATC	CGGTAC
CGGTAC	TGGATC	CGGTAC

operations in parallel in this ternary DNA. The working processes for ternary DNA AND operation are briefly detailed in the earlier section. Two ternary DNA AND operations run in parallel, the first of which is the ternary DNA SR flip-flop, and the ternary output sequence of which is sent into the ternary DNA SR latch as the molecular ternary input sequence. Here a ternary DNA NOR SR latch is used. Two ternary DNA NOR procedures are used to build the ternary DNA NOR SR latch. The two NOR operations of the ternary DNA SR latch run in parallel and take the same amount of time.

The ternary output sequence is obtained of the Multi-Valued DNA T flip-flop operation circuit after finishing all of the operations. This procedure generates two ternary output sequences, one of which is the polar opposite of the other. Table 2.5 shows the truth table of multi-valued DNA T Flip-Flop.

2.6.4   Example

Presume that the clock molecular ternary input sequence is ACCTAG and that the value of molecular ternary input sequence T is entered as TGGATC. To begin, two ternary DNA AND operations will be performed simultaneously.

After conducting two ternary DNA AND operations, two ternary output sequences are produced, both of which are TGGATC.

As a molecular ternary input sequence, these two TGGATC were fed into a ternary DNA SR flip-flop. Here, too, two ternary DNA AND processes run in parallel. Each ternary DNA AND operation produces a ternary output sequence TGGATC after conducting these procedures. Two ternary output sequences are then fed as molecular ternary input sequences into the ternary DNA NOR SR latch. The ternary DNA NOR operation is then carried out in parallel by these two molecular ternary input sequences. Each of these processes produces the result TGGATC.

2.7   Multi-Valued DNA Shift Register

A single bit of multi valued-valued data (ACCTAG or TGGATC or CGGTAC) can be stored in a ternary DNA flip flop. However, many ternary DNA flip-flops are required to store multiple qubits of data. To store n qubits of data, N ternary DNA flip flops must be coupled in a certain order. A ternary DNA register is a gadget that stores this type of data. It consists of a sequence of ternary DNA flip flops used to store multiple qubits of data.

Multi-Valued DNA Shift Registers enable the information stored in these ternary DNA registers to be transmitted. A Multi-Valued DNA Shift Register is a collection of flip flops that stores several bits of information. By applying clock pulses to the bits contained in such ternary DNA registers, they may be made to move inside them and in and out of them. By linking n ternary DNA flip-flops, each of which stores a single qubit of data, an n-qubit Multi-Valued DNA Shift Register may be built. "Multi-Valued DNA Shift left registers" are ternary DNA registers that will shift the qubits to the left. "Multi-Valued DNA Shift right registers" are ternary DNA registers that will shift the qubits to the right.

Multi-Valued DNA Shift Registers are basically of 4 types. These are:

1. ternary DNA Serial In Serial Out ternary Shift Register

2. ternary DNA Serial In parallel Out ternary Shift Register

3. ternary DNA Parallel In Serial Out ternary Shift Register

4. ternary DNA Parallel In parallel Out ternary Shift Register

In this study, a ternary Shift Register is built utilizing a ternary DNA D flip-flop operational circuit to convert serial data into ternary DNA data. The ternary DNA Serial-In Serial-Out ternary Shift Register is a type of Multi-Valued DNA Shift Register that permits serial molecular ternary input sequence one qubit at a time over a single data line and ternary output sequences a serial ternary output sequence. The data exits the Multi-Valued DNA Shift Register one qubit at a time in a serial pattern since there is only one qubit ternary output sequence, thus the term ternary DNA Serial-In Serial-Out ternary Shift Register.

Four ternary DNA D flip-flops are linked in a serial fashion in this circuit. Because the same clock signal is supplied to each ternary DNA flip flop, they are all synchronized with one another. The circuit above is an example of a Multi-Valued DNA Shift right register, which accepts serial data from the ternary DNA flip flop's left side. A QSISO's principal function is to operate as a delay element.

2.7.1   General Organization of Multi-Valued DNA Shift Register

Four ternary DNA D flip-flop operational circuits are used to make a Multi-Valued DNA Shift Register. As a fundamental component, a Multi-Valued DNA Shift Register is utilized for data shift, and a ternary DNA D flip-flop is used to make it happen. Figure 2.11 shows the block diagram of multi-valued DNA Shift Register.

Two molecular ternary input sequences are used in ternary DNA D flip flops one for data and the other for the clock. The ternary output sequences of ternary DNA D flip flips are logically opposite to one another. The circuit's synchronization with an external signal is aided by the clock molecular ternary input sequence. A ternary DNA D flip flop's ternary output sequence can have two possible values.

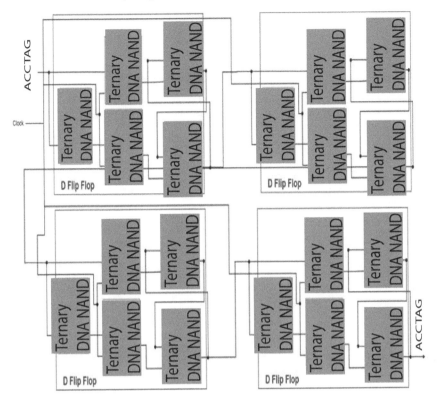

FIGURE 2.11
Block Diagram of Multi-Valued DNA Shift Register

Data molecular ternary input sequence is directed to a ternary DNA NAND operation circuit in this block diagram, while data molecular ternary input sequence reverse is routed to another ternary DNA NAND operation circuit. Both NAND procedures use the clock pulse molecular ternary input sequence. Ternary DNA SR Latch receives the result of two ternary DNA NAND operations. The D flip flop is constructed using the SR latch. This property is used to induce a delay in the circuit's data flow. Ternary DNA SR Latch is made up of two ternary DNA Nand Operations. The final two ternary output sequences of the ternary DNA SR Latch function were uncovered. A ternary DNA D flip flop may provide two types of the ternary output sequence, one of which is logically inverse to the other. The ternary DNA D flip flop will continue to function if the clock is enabled; otherwise, the ternary DNA D flip flop will stop working.

Ternary DNA D flip-flop operational circuits are also coupled through a serial connection in the Multi-Valued DNA Shift Register block diagram. The ternary DNA D flip-flop operational circuit is the fundamental component of the Multi-Valued DNA Shift Register.

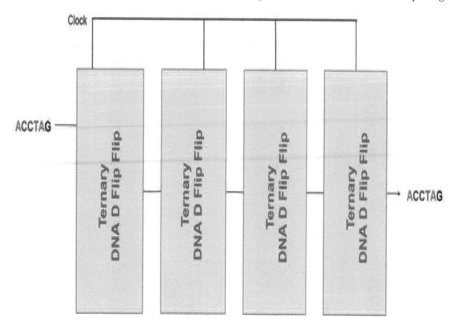

FIGURE 2.12
Simplified Block Diagram of Multi-Valued DNA Shift Register

2.7.2   Circuit Architecture of Multi-Valued DNA Shift Register

Four D flip-flops and a ternary DNA AND Gate are used in the Multi-Valued DNA Shift Register. Ternary Shift Register generates four qubit ternary output sequences using these. The D flip-flop has a single qubit molecular ternary input sequence and is developed using ternary DNA NAND and ternary DNA SR latch operations. Figure 2.12 shows the simplified block diagram of multi-valued DNA Shift Register.

The clock molecular ternary input sequence is required for the ternary DNA D flip-flop. It can be seen from the diagram that the circuit has one molecular ternary input sequence. One line of this qubit molecular ternary input sequence will be directed into a ternary DNA NAND operation termed S molecular ternary input sequence in Circuit. Ternary DNA NAND is performed here using S qubit molecular ternary input sequence and Clock qubit molecular ternary input sequence.

X molecular ternary input sequences another line, which performs a ternary DNA not operation first. This ternary DNA NOT operation was included in the ternary DNA NAND operation with the designation R. The two-qubit molecular ternary input sequences R and Clock are used in this ternary DNA NAND operation. The ternary output sequence of the ternary DNA NAND operations is sent into the ternary DNA SR latch as a molecular ternary input sequence. These two molecular ternary input sequences will be used to produce an SR latch. The ternary output sequence of the ternary DNA SR latch will be the final ternary output sequence of Q and $\overline{Q}$. Q will always produce the opposite of $\overline{Q}$. The ternary DNA SR latch is discussed in the chapter titled ternary DNA SR Latch earlier in the book.

It produces one ternary output sequence after processing the molecular ternary input sequences in a ternary DNA D flip-flop. The ternary output sequence of the ternary DNA D flip-flop is utilized as a clock molecular ternary input sequence for the following ternary DNA D flip-flop. As a result, a Multi-Valued DNA Shift Register generates a single qubit ternary output sequence.

2.7.3   Working Principle of Multi-Valued DNA Shift Register

Multi-Valued DNA Shift Registers are a kind of registers where both qubit data loading, as well as data retrieval to/from the Multi-Valued DNA Shift Register, occurs in serial mode sometimes. This synchronous ternary DNA SISO Shift Register is sensitive to the positive edge of the clock pulse. Here the data word which is to be stored is fed bit-by-bit at the qubit molecular ternary input sequence of the first ternary DNA flip-flop. Further, it is seen that the qubit molecular ternary input sequences of all other flip-flops are driven by the ternary output sequences of the preceding ones says, for example, the molecular ternary input sequence of ternary DNA D Flip-flop number-2 is driven by the ternary output sequence of ternary DNA D flip-flop number-1. At last, the data stored within the ternary DNA register is obtained at the ternary output sequence pin of the nth ternary DNA D flip-flop in a serial fashion.

Initially, all the ternary DNA flip-flops in the ternary DNA register are cleared by applying high on their clear pins. Next, the molecular ternary input sequence data word is fed serially to ternary DNA D Flip-flop number-1.

This causes the molecule sequence appearing at the first pin to be stored into ternary DNA D flip-flop number-1 as soon as the first leading edge of the clock appears. Further at the second clock tick, B1 gets stored into ternary DNA D flip-flop number-2 while a new bit enters into ternary DNA flip-flop number-2.

This kind of shift in data qubits continues for every rising edge of the clock pulse. This indicates that for every single clock pulse the data within the ternary DNA register moves towards the right by a single bit. Following the qubit data transmission, as explained, one can note that the first qubit of a molecular ternary input sequence word appears at the ternary output sequence of nth flip-flop for the nth clock tick. On applying further clock cycles, one gets the next successive qubits of the qubit molecular ternary input sequence data word as the serial ternary output sequence. Table 2.6 shows the truth table of multi-valued DNA Shift Register.

TABLE 2.6

Truth Table of Multi-Valued DNA Shift Register

clk	x	Q	$\overline{Q}$	Description
↓ >> TGGATC	X	Q	Q	Memory no change
↑ >> ACCTAG	TGGATC	TGGATC	ACCTAG	Reset Q >> 0
↑ >> ACCTAG	ACCTAG	ACCTAG	TGGATC	Set Q >>

CGGTAC and ACCTAG both the DNA sequences are high input in the multi-valued quantum shift register circuit. After processing the molecular ternary input sequences in a ternary DNA D flip-flop, it creates one ternary output sequence. The ternary output sequence is used as a clock molecular ternary input sequence for the next ternary DNA D flip-flop. Thus, a Multi-Valued DNA Shift Register creates one final qubit ternary output sequence.

2.7.4 Example

In the Multi-Valued DNA Shift Register, there is only one qubit molecular ternary input sequence needed to perform the required operation. Multi-Valued DNA Shift Register will enable if and only when the clock molecular ternary input sequence is enabled. This research proposed Multi-Valued DNA Shift Register shift one qubit data into the right position.

Assume ACCTAG data is going to perform the Multi-Valued DNA Shift Register operation. First of all 0 will perform the ternary DNA D flip-flop number-1 operation and produce the ternary output sequence is TGGATC. Then this TGGATC will shift to the right side as well as every qubit will shift to the right side once. Then ACC-TAG will be entered into the Multi-Valued DNA Shift register and perform the midst ternary DNA D flip-flop operation and then again like the previous one the data will be shifted once on the right side.

Then like these two step by step ACCTAG, ACCTAG also entered into the Multi-Valued DNA Shift Register in the relevant period and perform the midst ternary DNA D flip-flop operation. Then again each data will be shifted once. Every time qubit data shift once on the right side. In Multi-Valued DNA Shift Register operational circuit will perform n+1 operation then truth final data will be like the table below (Table 2.7).

Here in n+1 operation time, no data is available that's why TGGATC appears in the truth table and according to the Multi-Valued DNA Shift Register principle, one data qubit shifted once on the right side. After n+1 operation, final data is TG-GATC. These data purely proved that the Multi-Valued DNA Shift Register gives the correct required ternary output sequence by maintaining the ternary DNA principle.

TABLE 2.7

Data Shifting Process in Multi-Valued DNA Shift Register

clk	x	Q	Q1	Q2	Q3
TGGATC	TGGATC	TGGATC	TGGATC	TGGATC	TGGATC
ACCTAG	TGGATC	TGGATC	TGGATC	TGGATC	TGGATC
ACCTAG	ACCTAG	ACCTAG	TGGATC	TGGATC	TGGATC
ACCTAG	ACCTAG	ACCTAG	ACCTAG	TGGATC	TGGATC
ACCTAG	ACCTAG	ACCTAG	ACCTAG	ACCTAG	TGGATC
ACCTAG		TGGATC	ACCTAG	ACCTAG	ACCTAG

Multi-Valued DNA Shift Register can produce some garbage value but in this research, this is not the part of the study.

2.8 Multi-Valued DNA Ripple Counter

A ternary DNA counter is basically used to count the number of clock pulses applied to a ternary DNA flip-flop. It can also be used for ternary DNA Frequency divider, ternary DNA time measurement, ternary DNA frequency measurement, and ternary DNA distance measurement and also for generating square waveforms. In this, the ternary DNA flip-flops are ternary DNA asynchronous counters and are supplied with different clock signals, there may be a delay in producing ternary output sequence. Also, a few numbers of ternary DNA logic gates are needed to design asynchronous counters. So they are elementary in design and also are less expensive.

An n-bit ternary Ripple Counter can count up to 2n states. It is also known as MOD n counter. It is known as a ternary Ripple Counter because of the way the clock pulse ripples its way through the flip-flops. It is an asynchronous counter. Different flip-flops are used with a different clock pulse. All the flip-flops are used in toggle mode. Only one flip-flop is applied with an external clock pulse and another flip-flop clock is obtained from the ternary output sequence of the previous flip-flop. The flip-flop applied with an external clock pulse acts as LSB (Least Significant Bit) in the counting sequence. A counter may be an up counter that counts upwards or can be a down counter that counts downwards or can do both i.e. count up as well as count downwards depending on the molecular ternary input sequence control. The sequence of counting usually gets repeated after a limit

Multi-Valued DNA Ripple Counter is made out of four JK flip flops. Using these ternary DNA JK Flip flops, the Multi-Valued DNA Ripple Counter creates four qubit ternary output sequences. Here in JK flip flop, J and K are not shortened abbreviated letters of other words, such as "S" for Set and "R" for Reset, but are autonomous letters chosen by its inventor Jack Kilby to distinguish the flip-flop design from other types. Though Jack Kilby invented the digital electronics JK flip flop. Multi-Valued DNA Ripple Counter is an asynchronous counter. It is created using ternary DNA JK flip flops and these flip flops are only controlled by clock pulse molecular ternary input sequence.

Multi-Valued DNA Ripple Counter produces much heat to produce the molecule's superposition state and also produces some garbage value.

2.8.1 General Organization of Multi-Valued DNA Ripple Counter

Multi-Valued DNA Ripple Counter uses four ternary DNA JK flip flop to create four qubit ternary output sequences. Ternary DNA JK flip flop has the two-qubit molecular ternary input sequence named as J and K. Figure 2.13 shows the block diagram of multi-valued DNA Ripple Counter.

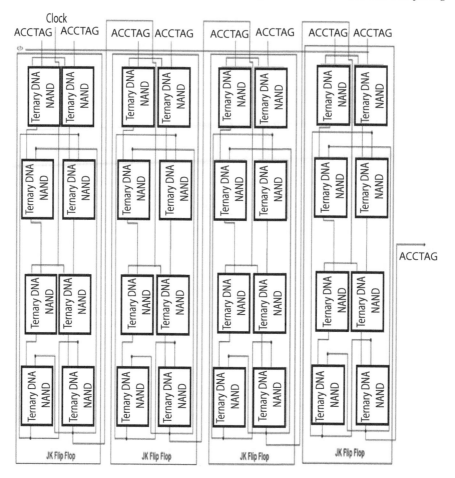

FIGURE 2.13
Block Diagram of Multi-Valued DNA Ripple Counter

Ternary DNA JK flip-flop has the qubit molecular ternary input sequence J and K. This ternary DNA JK flip flop consists of many ternary DNA NAND operations. At first basic ternary DNA NAND operation performs a couple of then this operation ternary output sequence is entered into the SR flip flop as well as got the ternary output sequence Q and $\overline{Q}$. First of all J and clk molecular ternary input sequence perform the ternary DNA NAND operation. The ternary output sequence of the ternary DNA NAND operation and ternary output sequence of the ternary DNA JK flip flop $\overline{Q}$ performs another ternary DNA NAND operation and produces the ternary output sequence named as S. clk molecular ternary input sequence is shared, so K and clk molecular ternary input sequence also perform the ternary DNA NAND operation. This ternary DNA NAND operations ternary output sequence and ternary output sequence of the ternary DNA JK flip flop performs another ternary DNA NAND operation as well as produces the ternary output sequence named as R.

These S and R entered into the ternary DNA SR flip flop and produce two ternary output sequences DNA $\overline{Q}$. In ternary DNA SR flip flop, it has four ternary DNA NAND operations. S molecular ternary input sequence and clk molecular ternary input sequence performs ternary DNA NAND operation as well as R molecular ternary input sequence and shared clk molecular ternary input sequence also performs the ternary DNA NAND operation. These two NAND operations ternary output sequences entered the ternary DNA SR latch as molecular ternary input sequence. In ternary DNA SR latches two Q and one molecular ternary input sequence as well as $\overline{Q}$ and other molecular ternary input sequences perform the ternary DNA NAND operation. Finally, after all of this ternary DNA operation, got the ternary DNA JK flip flops final ternary output sequence $\overline{Q}$.

After processing the molecular ternary input sequences in a ternary DNA JK flip flop, the ternary output sequence of the flip flop is going to be stored as a ternary output sequence of the Multi-Valued DNA Ripple Counter. Multi-Valued DNA Ripple Counter can be performed as an up counter and also a down counter. This clk will decide thus the ternary DNA JK flip-flop operational circuit will perform or not. Every ternary DNA JK flip-flop operational circuit will produce the final ternary output sequence.

2.8.2   Circuit Architecture of Multi-Valued DNA Ripple Counter

Multi-Valued DNA Ripple Counter uses four ternary DNA JK flip flops to create four qubit ternary output sequences. Ternary DNA JK flip flop has a two-qubit molecular ternary input sequence and one clock shared molecular ternary input sequence. Ternary DNA JK flip flops also depend on the clock. If the clock is enabled then the circuit will enable, otherwise not. Figure 2.14 depicts the simplified block diagram of multi-valued DNA Ripple Counter.

In Multi-Valued DNA Ripple Counter, there's a one clock molecular ternary input sequence and one logic molecular ternary input sequence which is shared in both J and K molecular ternary input sequence port. molecular ternary input sequence J and molecular ternary input sequence K both the value will perform the ternary DNA NAND operation with the shared clk molecular ternary input sequence differently. J and clk molecular ternary input sequence perform the ternary DNA NAND operation. NAND operation made by using ternary DNA basic gates. In DNA NAND operation the value of an ancillary bit in ACCTAG. After J and clk perform the ternary DNA NAND operation produced a ternary output sequence qubit and this ternary output sequence qubit and the ternary output sequence of the flip flop $\overline{Q}$ perform the ternary DNA NAND operation and produce the ternary output sequence S. Like in the same procedure K, clk and Q molecular ternary input sequence produces the R ternary output sequence. These operation circuits are mainly ternary DNA NAND operations. This S and R molecular ternary input sequence is performed the ternary DNA SR flip-flop operation. In ternary DNA SR flip flop operational circuit

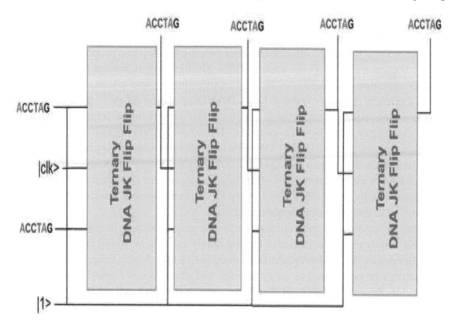

FIGURE 2.14
Simplified Block Diagram of Multi-Valued DNA Ripple Counter

architecture also made by the proposed basic component of ternary DNA computing is ternary DNA NAND operation.

After processing the molecular ternary input sequences in a ternary DNA JK flip flop, the ternary output sequence of the flip flop is going to be stored as a ternary output sequence of the Multi-Valued DNA Ripple Counter. Multi-Valued DNA Ripple Counter intermediate architectures four ternary DNA JK flip-flops connected in serial connection using the logical qubit molecular ternary input sequence. Every clock molecular ternary input sequence as clk enters in every qubit and from 2nd ternary DNA JK flip-flop, clk molecular ternary input sequence is previous ternary DNA Jk flip-flops first ternary output sequence Q. With the same architecture, Multi-Valued DNA Ripple Counter can be performed as an up counter or as a down counter but clock pulse as clk need to sometimes have a positive edge-triggered and sometimes a negative edge triggered.

2.8.3   Working Principle of Multi-Valued DNA Ripple Counter

In the Multi-Valued DNA Ripple Counter, there are four ternary DNA JK flip-flop operational circuits. These ternary DNA JK flip-flop operational circuits are connected in serial connection. In this Multi-Valued DNA Ripple Counter there is one clock molecular ternary input sequence and a logic molecular ternary input sequence which are shared in the port J and K of ternary DNA JK flip-flop operational circuit. The molecular ternary input sequences J and K of a ternary DNA Jk flip-flop conduct two ternary DNA processes in parallel. The ternary DNA NAND operation is performed using J and shared molecular ternary input sequence clk The ternary DNA NAND operations are then completed, and one of the ternary DNA JK flip flop's ternary output sequences executes the ternary DNA NAND operation, producing S. The K and clk molecular ternary input sequences are performed first in the ternary DNA NAND operation. The ternary output sequence of the ternary DNA JK flip-flop Q, as well as the result of the ternary DNA first NAND operation, are then used to perform another ternary DNA NAND operation, yielding R. The steps for creating S and R are carried out simultaneously. It is known that one of the distinctive properties of ternary DNA operations is that they may do several operations at the same time, and this is exactly what is happening. The ternary DNA SR flip flop operation is then conducted on these S and R molecular ternary input sequences. The ternary DNA NAND operation, which is employed here, is also used to make ternary DNA SR flip flops. Two ternary output sequences are discovered after conducting the ternary DNA SR flip-flop operation. The opposite of one ternary output sequence is the opposite of the other.

Here if the clk is TGGATC then the ternary DNA JK flip-flop will not be triggered but if clk is ACCTAG or CGGTAC the ternary DNA JK flip-flop will be triggered and it will toggle the ternary output sequence. First of all ternary DNA JK flip-flop operational circuit getting clk value is ACCTAG or CGGTAC and toggles the ternary output sequence value from the previous state value. Then the ternary output sequence of the initial ternary DNA JK flip-flop will be clk molecular ternary input sequence of next ternary DNA JK flip-flop. If the clk value is ACCTAG or CG-GTAC then the ternary output sequence value will be toggled, otherwise the ternary output sequence will be the previous state ternary output sequence. Maintaining the same procedure every ternary DNA JK flip-flop operated in the Multi-Valued DNA Ripple Counter. Ternary DNA JK flip-flop is toggled very much, that's why in the Multi-Valued DNA Ripple Counter ternary DNA JK flip-flop operational circuit here is used as a basic component. Table 2.8 shows the truth table of multi-valued DNA Ripple Counter.

As like ACCTAG, CGGTAC is also a high input molecule sequence and performs in multi-valued DNA shift register.

2.8.4 Example

To check the Multi-Valued DNA Ripple Counter proposed circuit assumed a clock signal is ACCTAG and the logical qubit is high. So, if initially clock signal ACCTAG

TABLE 2.8

Truth Table of Multi-Valued DNA Ripple Counter

clk	Q_0	Q_1	Q_2	Q_3
TGGATC	TGGATC	TGGATC	TGGATC	TGGATC
ACCTAG	ACCTAG	ACCTAG	ACCTAG	ACCTAG
ACCTAG	TGGATC	TGGATC	TGGATC	TGGATC

is delivered then for the first ternary DNA JK flip-flop operational circuit produce ACCTAG.

According to the working principle of Multi-Valued DNA Ripple Counter if clock molecular ternary input sequence value is ACCTAG then the previous state value will be toggled. Then now from the first ternary DNA JK flip-flop got one ternary output sequence which is ACCTAG. This ACCTAG will be the clock molecular ternary input sequence for the second ternary DNA JK flip-flop circuit according to the architecture of the Multi-Valued DNA Ripple Counter.

As like the previous ternary DNA JK flip-flop for third and fourth ternary DNA flip-flops the principle is the same. For both of the flip-flops the clock value will be the previous ternary DNA JK flip-flop's ternary output sequence.

Hence from the above discussion, it is seen that the proposed Multi-Valued DNA Ripple Counter is correct theoretically. Ternary DNA JK flip-flop is used here for toggling.

2.9   Multi-Valued DNA Synchronous Counter

A ternary DNA counter is a ternary DNA device that can count any particular event on the basis of how many times the particular event(s) has occurred. In a ternary DNA logic system or computer, this ternary DNA counter can count and store the number of times any particular event or process has occurred, depending on a ternary DNA clock signal. The most common type of ternary DNA counter is a sequential ternary DNA logic circuit with a single clock molecular ternary input sequence and multiple ternary output sequences. The ternary output sequences represent two-valued decimal numbers. Each clock pulse either increases the number or decreases the number.

Multi-Valued DNA Synchronous circuit generally refers to something which is coordinated with others based on time. Multi-Valued DNA Synchronous signals occur at the same clock rate and all the clocks follow the same reference clock. Ternary DNA asynchronous Counter has shown that the ternary output sequence of that ternary DNA counter is directly connected to the molecular ternary input sequence of the next subsequent counter and making a chain system, and due to this chain system propagation delay appears during the counting stage and create counting delays. In a Multi-Valued DNA Synchronous counter, the clock molecular ternary input sequence across all the ternary DNA flip-flops uses the same source and creates the same clock signal at the same time. So, a ternary DNA counter which is using the same clock signal from the same source at the same time is called a Multi-Valued DNA Synchronous counter.

Multi-Valued DNA Synchronous Counter is made out of four JK flip flops and two ternary DNA AND Gates. Using these ternary DNA JK Flip flops and ternary DNA AND Gates, the Multi-Valued DNA Synchronous Counter creates four ternary output sequences. A Multi-Valued DNA Synchronous counter produces much heat and this circuit operation needs to happen in the required environment of ternary DNA computing.

2.9.1   General Organization of Multi-Valued DNA Synchronous Counter

Multi-Valued DNA Synchronous Counter uses four ternary DNA JK flip flops to create four ternary output sequences. Ternary DNA JK flip flop has the two-molecular ternary input sequence named NAND K.

Ternary DNA JK flip-flop has the molecular ternary input sequence K. This ternary DNA JK flip flop consists of many ternary DNA NAND operations. At first basic ternary DNA NAND operation performs a couple of then this operation ternary output sequence is entered into the SR flip flop as well as got the ternary output sequence Q and $\overline{Q}$. First of all, the NAND clk molecular ternary input sequence performs the ternary DNA NAND operation. The ternary output sequence of the ternary DNA NAND operation and ternary output sequence of the ternary DNA JK flip flop $\overline{Q}$ performs another ternary DNA NAND operation and produces the ternary output sequence named as S. clk molecular ternary input sequence is shared, so K and clk molecular ternary input sequence also performs the ternary DNA NAND operation. This ternary DNA NAND operations produces ternary output sequence, and the Q ternary output sequence of the ternary DNA JK flip flop performs another ternary DNA NAND operation as well as produces the ternary output sequence named as R.

These S and R entered into the ternary DNA SR flip flop and produce two ternary output sequences Q and $\overline{Q}$. In ternary DNA SR flip flop, it has four ternary DNA NAND operations. S molecular ternary input sequence and clk molecular ternary input sequence perform ternary DNA NAND operation as well as R molecular ternary input sequence and shared clk molecular ternary input sequence also performs the ternary DNA NAND operation. These two NAND operations ternary output sequences entered the ternary DNA SR latch as the molecular ternary input sequences. In ternary DNA SR latches two Q and one molecular ternary input sequence as well as $\overline{Q}$ and other molecular ternary input sequences perform the ternary DNA NAND operation. Finally, after all of this ternary DNA operation, got the ternary DNA JK flip flops final ternary output sequence Q and $\overline{Q}$.

After processing the molecular ternary input sequences in a ternary DNA JK flip flop, the ternary output sequence of the flip flop is going to be stored as a ternary output sequence of the Multi-Valued DNA Synchronous Counter. Thus, Multi-Valued DNA Synchronous Counter creates four ternary output sequences using four ternary DNA JK flip flops. The clock molecular ternary input sequences for all of the four ternary DNA Jk flip flops come from the same source. For this, all of the flip flops work synchronously. One ternary output sequence of each of the second and third flip flops go through ternary DNA AND Gates.

2.9.2   Circuit Architecture of Multi-Valued DNA Synchronous Counter

Multi-Valued DNA Synchronous Counter uses four ternary DNA JK flip flops to create four ternary output sequences. Ternary DNA JK flip flop has a two-molecular

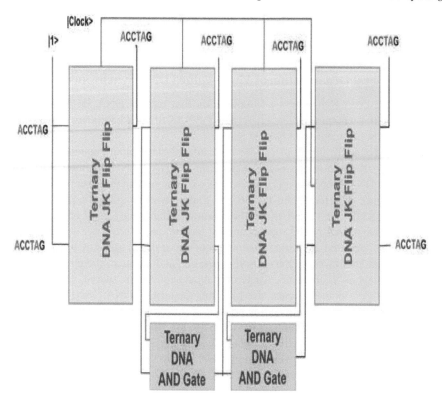

FIGURE 2.15
Simplified Block Diagram of Multi-Valued DNA Synchronous Counter

ternary input sequence and one clock shared molecular ternary input sequence. Ternary DNA JK flip flops also depend on the clock. If the clock is enabled then the circuit will enable, otherwise not. Figure 2.15 shows simplified block diagram of multi-valued DNA Synchronous Counter.

Molecular ternary input sequence NAND molecular ternary input sequence K both the value will perform the ternary DNA NAND operation with the shared clk molecular ternary input sequence differently. NAND clk molecular ternary input sequence performs the ternary DNA NAND operation. NAND operation made by using ternary DNA basic gates. Basic gates in mainly ternary DNA computing are v, V+, and CNOT gates. For error correction here used an ancillary bit. In DNA NAND operation the value of an ancillary bit in ACCTAG. After NAND clk perform the ternary DNA NAND operation, a ternary output sequence is achieved and this ternary output sequence and the ternary output sequence of the flip flop $|\overline{Q}>$ perform the ternary DNA NAND operation and produce the ternary output sequence S. Like in the same procedure K, clk and Q molecular ternary input sequence produces the R ternary output sequence. These operational circuits are mainly ternary DNA NAND operations. This S and R molecular ternary input sequence is performed the ternary DNA SR flip-flop operation. In ternary DNA SR flip flop operation circuit architecture also made by the proposed basic component of ternary DNA computing is ternary DNA NAND operation.

After processing the molecular ternary input sequences in a ternary DNA JK flip flop, the ternary output sequence of the flip flop is going to be stored as a ternary output sequence of the Multi-Valued DNA Synchronous Counter. Thus, Multi-Valued DNA Synchronous Counter creates four ternary output sequences using four ternary DNA JK flip flops. The clock molecular ternary input sequences for all of the four ternary DNA JK flip flops come from the same source. For this, all of the flip flops work synchronously. One ternary output sequence of each of the second and third flip flops go through ternary DNA AND operations.

2.9.3 Working Principle of Multi-Valued DNA Synchronous Counter

Multi-Valued DNA Synchronous counter has one logical molecular ternary input sequence which is shared and one clock molecular ternary input sequence. Multi-Valued DNA Synchronous counter is constructed by the basic component as a ternary DNA JK flip-flop operational circuit. Ternary DNA JK flip flop is the two- circuit. Ternary DNA JK flip flop is mostly used flip-flop in ternary DNA computing. Ternary DNA Jk flip-flop's molecular ternary input sequence NAND K perform two ternary DNA operations parallelly. NAND shared molecular ternary input sequence clk performs the ternary DNA NAND operation. Then this ternary DNA NAND operations result and one ternary output sequence of the ternary DNA JK flip flop performs the ternary DNA NAND operation and produce S. Like ternary DNA NAND operation performs K and |clk > molecular ternary input sequence first. Then the ternary output sequence of ternary DNA JK flip-flop Q and the result of ternary DNA first NAND operation executes again another ternary DNA NAND operation and produce R. The procedure of producing S and R are executed in parallel. It is known that ternary DNA operations have one of the iconic characteristics that perform multiple operations parallelly and here it is happening. These S and R molecular ternary input sequences are performed then ternary DNA SR flip flop operation. Ternary DNA SR flip flop is also basically made by the ternary DNA NAND operation that is used here. After performing the ternary DNA SR flip-flop operation, two ternary output sequences are found. One ternary output sequence is the opposite of another. This ternary DNA JK flip flop actually removes the problem of ternary DNA SR flip flop. Ternary DNA SR flip-flop describes briefly in the 5th chapter.

Ternary DNA JK flip-flop is triggered when the value of clock is ACCTAG. So, in Multi-Valued DNA Synchronous, the counter clock needs to be always high. According to the principle, counters need to be toggled, that's why ternary DNA JK flip-flop is perfect for Multi-Valued DNA Synchronous counters. Ternary DNA JK flip-flop operational circuit triggered then ternary output sequence of ternary DNA JK flip-flop will be shared molecular ternary input sequence of second ternary DNA JK flip-flop operational circuit. Then first ternary DNA JK flip-flops ternary output sequence and second ternary DNA JK flip-flops ternary output sequence enters in the ternary DNA AND operation circuit and performs the ternary DNA AND operation.

TABLE 2.9

Truth Table of Multi-Valued DNA Synchronous Counter

clk	Q_3	Q_2	Q_1	Q_0
TGGATC	TGGATC	TGGATC	TGGATC	TGGATC
ACCTAG	TGGATC	TGGATC	TGGATC	ACCTAG
ACCTAG	TGGATC	TGGATC	ACCTAG	TGGATC
ACCTAG	TGGATC	TGGATC	ACCTAG	ACCTAG
ACCTAG	TGGATC	ACCTAG	TGGATC	TGGATC
ACCTAG	TGGATC	ACCTAG	TGGATC	ACCTAG
ACCTAG	TGGATC	ACCTAG	ACCTAG	TGGATC
ACCTAG	TGGATC	ACCTAG	ACCTAG	ACCTAG
ACCTAG	ACCTAG	TGGATC	TGGATC	TGGATC
ACCTAG	ACCTAG	TGGATC	TGGATC	ACCTAG
ACCTAG	ACCTAG	TGGATC	ACCTAG	TGGATC
ACCTAG	ACCTAG	TGGATC	ACCTAG	ACCTAG
ACCTAG	ACCTAG	ACCTAG	TGGATC	TGGATC
ACCTAG	ACCTAG	ACCTAG	TGGATC	ACCTAG
ACCTAG	ACCTAG	ACCTAG	ACCTAG	TGGATC
ACCTAG	ACCTAG	ACCTAG	ACCTAG	ACCTAG
ACCTAG	TGGATC	TGGATC	TGGATC	TGGATC

The produced ternary output sequence from ternary DNA AND operation performs the ternary DNA JK flip-flop operation. Then again produced ternary output sequence from the first ternary DNA AND operation and the third ternary DNA JK flip-flop operation's ternary output sequence performs another ternary DNA AND operation. Hence the previous ternary DNA AND operations ternary output sequence is shared into the ternary DNA JK flip-flops as two molecular ternary input sequences as well as performs the ternary DNA JK flip-flop operational circuit. Multi-Valued DNA Synchronous counter circuit mainly performs as a finite counter. Table 2.9 shows the truth table of multi-valued DNA Synchronous Counter.

Like ACCTAG molecule sequence in multi-valued synchronous counter, CGGTAC is assumed as also a high input for this research. So all the output and performance will be the same form CGGTAC and ACCTAG molecule sequences as an input.

2.9.4   Example

As suppose the molecular ternary input sequence is ACCTAG entered into the Multi-Valued DNA Synchronous counter. Multi-Valued DNA Synchronous counters basic component is ternary DNA Jk flip-flop operational circuit so for triggered the circuit clock value is also ACCTAG. Now first ternary DNA JK flip-flop both Molecular ternary input sequence and K molecular ternary input sequence is ACCTAG. If two value is ACCTAG the flip-flop triggered and value toggles. Now Q_0 is ACCTAG from previous state TGGATC.

But, after the first ternary DNA JK flip-flop operation $\overline{Q_0}$'s value became TG-GATC. Then $\overline{Q_0}$ entered as a molecular ternary input sequence in the second ternary DNA JK flip-flop operation and this circuit won't be triggered. Then again $\overline{Q_1}$ and $\overline{Q_0}$ performs the ternary DNA AND operation. As ternary DNA AND operation depicts that if any molecular ternary input sequence is TGGATC then the produced ternary output sequence will TGGATC. Then TGGATC will be molecular ternary input sequence for the third ternary DNA JK flip-flop operational circuit. As this circuit ternary DNA JK flip-flop depicts that if two molecular ternary input sequence is TGGATC then ternary DNA Jk flip-flop won't be triggered. Again third ternary DNA flip-flop's ternary output sequence $\overline{Q_2}$ and previous ternary DNA AND operations produced ternary output sequence performs ternary DNA AND operation again. As this proposed circuit depicts previous ternary DNA AND operation ternary output sequence is TGGATC so, this ternary DNA AND operations ternary output sequence will be TGGATC. Hence TGGATC will be molecular ternary input sequence of NAND K for the fourth ternary DNA JK flip-flop operation as well as finally, it produces TGGATC.

So, this proposed circuit illustrates the exact required ternary output sequence of a Multi-Valued DNA Synchronous counter theoretically.

2.10   Summary

Multi-Valued DNA Shift register is a very much useful circuit in DNA computing. It can be used as counter, data format convertor, data processor, etc. The multi-valued DNA synchronous counter contains flip-flops which are all in sync with each other, i.e., their clock molecules sequence inputs are connected together and are triggered by the same external clock signal. This implies that all the DNA flip-flops update their values at the same time. As the name suggests, DNA Synchronous counters perform "counting" such as time and electronic pulses (external source like infrared light). They are widely used in lots of other designs as well such as DNA computing processors, DNA calculators, real time clocks, etc. All other multi-valued DNA sequential circuits are useful too in modern science.

Bibliography

[1] Adleman, L. M. (1994). Molecular computation of solutions to combinatorial problems. Science, 266(5187), 1021-1024.

[2] Smith, L. M., Sanders, J. Z., Kaiser, R. J., Hughes, P., Dodd, C., & Connell, C. R. (1986). C. Heiner, S. B. Kent, and L. E. Hood. Nature, 321(674), 7679.

[3] Zheng, X., Yang, J., Zhou, C., Zhang, C., Zhang, Q., & Wei, X. (2019). Allosteric DNAzyme-based DNA logic circuit: Operations and dynamic analysis. Nucleic Acids Research, 47(3), 1097-1109.

3

Multi-Valued Sequential Circuits in Quantum-DNA Computing

3.1 Introduction

It is widely acknowledged that quantum mechanics is one of the most significant theories in the history of science, particularly during the twentieth century. Beyond finding answers to some unsolved questions, it has had an impact on many modern technologies by expressing an alternative line of scientific thinking. Physicists and computer scientists such as Paul Benioff from the Thomas J. Watson Research Center, Charles H. Bennett from IBM, Benioff from Argonne National Laboratory in Illinois, David Deutsch from the University of Oxford, and Richard P. Feynman from Caltech first proposed quantum computing devices in the 1970s and 1980s, and they were met with a great deal of opposition at the time. When Richard Feynman attempted to develop an entirely new sort of computer system by incorporating quantum physics notions, it was in 1982 that he was successful. Specifically, the abstract model that he constructed was meant for demonstration of how a quantum system could be used to conduct computations and also for demonstration of how a machine of this type could be used as a simulator for quantum physics problems. Stanford University awarded him a doctorate in philosophy. In 1994, Peter Shor demonstrated how a number theory problem known as factorization will be addressed in quantum computers, and it was a game-changer in the field of computing. He also discovered that traditional computers answer issues in exponentially growing time, whereas quantum computers, in contrast to traditional computers, solve problems in polynomial time, according to his research. It is possible that quantum computers will be more powerful than conventional computers in the future, based on quantum theory and the series of discoveries in quantum processing that have occurred over the course of the last century. As a result of the findings of the research, scientists have been paying particular attention to quantum computation in recent years.

Multi-Valued DNA Computing is one of the most fascinating and difficult study issues to be investigated in the modern world, and it is one of the most promising and difficult research topics to be examined. When performing a logical operation or an arithmetical operation on a computer, the sequence of DNA molecules is employed to conduct a DNA computing operation, which is a subset of bio molecular computing. Adleman proposed the initial theory of DNA computation in 1994, which was subsequently amended] and it was later altered again. He conducted an experiment

DOI: 10.1201/9781003381921-3

to put his theoretical predictions to the test using the seven-point Hamiltonian path problem, also known as the traveling salesman's problem, which he developed. Assume that a salesperson must determine the shortest journey between seven cities whose distances are known. He must do it without passing through any of the cities more than once and without returning to the beginning city, which is not possible.

The quantum mechanical nature of all matter, whether it is alive or inanimate, is the fundamental nature of all things. It is made up of ions, atoms, and/or molecules, and the equilibrium properties of these particles are exactly specified by the principles of quantum theory. As a result, it is possible to declare that quantum mechanical principles regulate the entire field of biological science. A combination of multi-valued quantum computing and multi-valued DNA computing, according to the researchers, has resulted in quantum molecular biology being declared a fact. In the future decades, quantum molecular biology will illustrate how far bioinspired quantum gadgets may beat their classical counterparts in terms of overall performance. Yet another more fundamental concern is how quantum-dynamical processes at the nanoscale might be used to give a complete organism a selection advantage in a competitive environment. It is a fact of quantum biology that the building of quantum circuits from DNA occurs.

3.2   Multi-Valued Quantum-DNA D Flip Flop

A Multi-Valued Quantum-DNA D flip flop is essentially a two-state timed flip flop. In one clock cycle, the qubit inputs of a Multi-Valued Quantum D-type flip flop are actuated with a delay. A delay flip flop is another term for the Multi-Valued Quantum-DNA D flip flop.

The indeterminate input condition of SET = "|0>" and RESET = "|0>" is banned in the basic Multi-Valued Quantum SR NAND Gate Bistable circuit, which is one of its fundamental drawbacks. This condition forces both Multi-Valued DNA molecule sequence outputs to logic "ACCTAG," overriding the feedback latching action, and whichever input goes to logic "ACCTAG" first loses control, while the other input, which is still at logic "TGGATC," controls the latch's final state. However, an inverter may be connected between the "SET" and "RESET" qubit inputs to create a Multi-Valued Quantum-DNA Data Latch, Multi-Valued Quantum-DNA Delay flip flop, Multi-Valued Quantum-DNA D-type Bistable, Multi-Valued Quantum-DNA D-type Flip Flop, or simply a Multi-Valued Quantum-DNA D Flip Flop as it is most often known.

By far the most essential of all the Multi-Valued Quantum-DNA timed flip-flops is the Multi-Valued Quantum-DNA D Flip Flop. The |S> and |R> inputs become complements of each other when a Multi-Valued Quantum inverter (ternary Quantum NOT gate) is added between the Set and Reset inputs, ensuring that the two inputs |S> and |R> are never equal (|0> or |1>) to each other at the same time, allowing us to control the toggle action of the flip-flop with just one |D> (Data) input.

The Data input, labeled "|D>," is then utilized in place of the "Set" signal, and the inverter is used to create the complementary "Reset" input, resulting in a level-sensitive Multi-Valued Quantum D-type flip-flop from a level-sensitive SR-latch, with |S> = |D> and |R> = $\overline{|D>}$.

The proposed Multi-Valued Quantum-DNA D flip-flop circuit has just one qubit input, and the qubit input must be in a coherence state in order to conduct the Multi-Valued Quantum computational function. As a result, the circuit must exist in an environment that does not exist. If any particle emerges, the coherence state will be disrupted. The Multi-Valued Quantum D flip-flop will generate heat, which must be removed quickly in order to cool down the circuit and stabilize the coherence state.

Hence Multi-Valued Quantum-DNA flip-flops have two portions based on the principle of the Multi-Valued Quantum-DNA circuit. Multi-Valued Quantum -Multi-Valued DNA D flip-flop operational circuit's first portion was constructed using Multi-Valued Quantum principle and second portion was constructed using Multi-Valued DNA computing principle. Multi-Valued Quantum circuit's portion produced qubit and stored into Multi-Valued Quantum cache memory. These qubits perform NMR relaxation process operation and make them normal molecules. Then these molecules perform Multi-Valued DNA computing operations. Multi-Valued DNA computing is very good for memory storage and Multi-Valued Quantum computing has a super-fast computation speed.

3.2.1 General Organization of Multi-Valued Quantum-DNA D Flip-Flop

When compared different types of flip flops, the Multi-Valued Quantum-DNA D flip flop is treated as the most significant flip flop among them. Multi-Valued Quantum-DNA D flip-flop verifies that the two inputs of the Multi-Valued Quantum-DNA SR flip flop are never the same. Figure 3.1 shows the block diagram of multi-valued Quantum-DNA D Flip Flop.

In Multi-Valued Quantum-DNA D flip-flop operational circuit there is one input which is |x>. There is another |clk> input is also there. Multi-Valued Quantum-DNA D flip flops have two output molecule sequences that are logically opposite to one another. The clock qubit input aids in the circuit's synchronization with an external signal. The output of a Multi-Valued Quantum-DNA D flip flop can have two potential values. This block diagram of Multi-Valued Quantum-DNA D flip-flop operational circuit shows that data input is sent to a ternary Quantum NAND operation circuit, while the reversal of data input is routed to another ternary Quantum NAND operation circuit. The clock pulse input is used by both ternary Quantum NAND operation processes. The result of two ternary Quantum NAND operations is fed into Multi-Valued Quantum cache memory. Multi-Valued Quantum cache memory is made by using some Multi-Valued Quantum register and it saves the Multi-Valued Quantum qubit data. When the accurate time appears then these qubit servers into the NMR machine and it performs NMR relaxation process as well as making molecules from the qubit. The Multi-Valued DNA SR latch is used to build the Multi-Valued Quantum-DNA D flip flop . This attribute is utilized to create a delay in the data flow

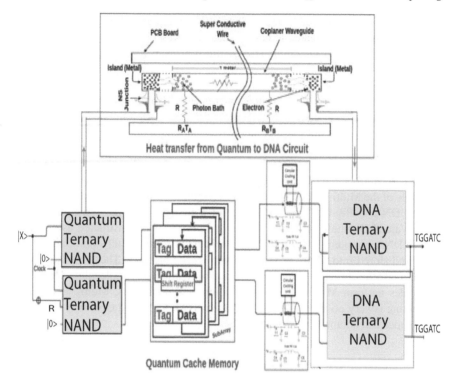

FIGURE 3.1

Block Diagram of Multi-Valued Quantum-DNA D Flip Flop

in the circuit. Two ternary DNA NAND Operations create the Multi-Valued DNA SR Latch. The proposed circuit discovered the remaining two output molecules of the Multi-Valued DNA SR Latch function. The output of a Multi-Valued Quantum-DNA D flip flop can be of two sorts, one of which is logically inverse to the other. If the clock is enabled, the Multi-Valued Quantum-DNA D flip flop will continue to function; otherwise, the Multi-Valued Quantum-DNA D flip flop will cease to function.

3.2.2   Circuit Architecture of Multi-Valued Quantum-DNA D Flip-Flop

The Multi-Valued Quantum-DNA D flip-flop has a single qubit input and is developed using ternary Quantum NAND operations, ternary DNA NAND gate and a Multi-Valued DNA SR latch. Figure 3.2 depicts the multi-valued Quantum-DNA D Flip-Flop operation circuit

The clock qubit input affects the Multi-Valued Quantum-DNA D flip-flop. The diagram has one qubit input. One line of this qubit input will be directed into the Multi-Valued Quantum-DNA NAND operation known as |S> input in Circuit. In this

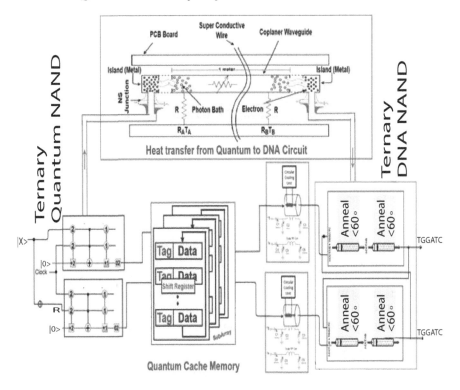

FIGURE 3.2
Multi-Valued Quantum-DNA D Flip-Flop Operation Circuit

case, the |S> qubit input and the Clock qubit input are used in a ternary Quantum NAND operation.

When a |X> qubit traverses another line, it first undertakes a ternary Quantum NOT operation. This ternary Quantum NOT operation was dubbed R when it was entered into the ternary Quantum NAND operation. The R and Clock qubit inputs are used in this ternary Quantum NAND operation. The output of these ternary Quantum NAND operations are getting stored into Multi-Valued Quantum cache memory by using the line. Multi-Valued Quantum cache memory made by shift register where qubit data stored in some sub-array. When the actual time appears the Multi-Valued Quantum cache memory serves the qubit into the NMR relaxation process using line. Qubit first of all performs the NMR relaxation process where EMR emission is totally prohibited. Hence qubit is converted into molecule sequence and then performs the Multi-Valued DNA SR latch operation. In Multi-Valued DNA SR latch operation two ternary DNA NAND operations perform parallely and produce the final molecules sequence.

Multi-Valued Quantum Circuit produces more heat and the Multi-Valued DNA circuit needs to process the input. For that reason, it would be best to transfer the heat into Multi-Valued DNA circuits. Multi-Valued Quantum portion Heat transfers using the heat transfer circuit into Multi-Valued DNA circuit's portion. In the heat transfer circuit, two junctions are connected to Multi-Valued Quantum Circuits and Multi-Valued DNA circuits. A maximum one meter can transfer heat to this circuit. This heat transfer circuit has superconductive ware, photon batch, PCB board to make the architecture. This circuit mainly transfers the heat using the photon bath. With this circuit proposed Multi-Valued Quantum-DNA T flip-flop circuit is constructed fully.

3.2.3   Working Principle of Multi-Valued Quantum-DNA D Flip-Flop

There are two inputs in the Multi-Valued Quantum-DNA SR flip flop: SET and RESET. Alternatively, in a Multi-Valued Quantum-DNA D flip-flop, one input and the input's one line are referred to as a SET, and by coupling a Multi-Valued Quantum-DNA NOT gate towards the other line input, proposed circuit may designate the Multi-Valued Quantum-DNA D flip-flop as a RESET. This complement resolves the contradiction inherent in the Multi-Valued Quantum-DNA SR latch when both inputs are LOW because that circumstance is no longer feasible. Multi-Valued Quantum-DNA D flip-flops have a single qubit input, which is sometimes alluded to as a data qubit input. If this qubit data input is high, the Multi-Valued Quantum-DNA flip-flop becomes SET; if the data input is low, such as |0>, the flip-flop changes state and becomes RESET.

However, this would be pretty futile because the output of the flip flop will always vary with each pulse delivered to this data input. To circumvent this, an extra input known as the "CLOCK" or "ENABLE" input is used to separate the data input from the latching circuitry of the flip flop after the appropriate data has been stored. The result is that the |X> input condition is only replicated to the output |Q> while the clock input is active. This then serves as the foundation for yet another sequential gadget known as a Multi-Valued Quantum-DNA D Flip Flop.

As long as the clock input is HIGH, the "Multi-Valued Quantum-DNA D flip flop" will store and output any logic level that is applied to its data terminal. Once the clock input is changed to LOW, the flip-"set" flop's and "reset" inputs are both kept at logic level "|1>," preventing the flip-flop from altering the underlying and preserving whatever statistics are available on its output prior to the clock transition. In other words, either logic "|0>" or logic "|1>" latches the output. Table 3.1 shows the truth table of multi-valued Quantum-DNA D Flip Flop.

This proposed Multi-Valued Quantum-DNA D flip-flop operational circuit's first half of the portion is constructed using the Multi-Valued Quantum computing

TABLE 3.1
Truth Table of Multi-Valued Quantum-DNA D Flip Flop

|Clk>	|x>	Q	$\overline{Q}$	Description
↓ >> |0>	X	Q	Q	Memory no change
↑ >> |1>	|0>	TGGATC	ACCTAG	Reset Q >> 0
↑ >> |1>	|1>	ACCTAG	TGGATC	Set Q >>

principle and the rest of the portion is constructed using the Multi-Valued DNA computing principle. First of all two quantum NAND operations performed and produced two outputs which are stored in Multi-Valued Quantum cache memory. Multi-Valued Quantum cache stores the qubit into an array. Multi-Valued Quantum cache memory stores the qubit data and when required it serves the data into the NMR relaxation process. In this Multi-Valued Quantum-DNA D flip-flop, two-qubit is stored in Multi-Valued Quantum cache memory. This qubit performs the NMR Relaxation process because Multi-Valued DNA circuits need molecular sequence. NMR relaxation process removes the superposition state and makes the qubit into a molecular sequence. This molecular sequence performs Multi-Valued DNA SR latch operation. Multi-Valued DNA SR latch operation has two basic components: ternary DNA NAND operation. Ternary DNA NAND operation performs in a parallel way in a Multi-Valued DNA SR latch operation and produced two required outputs. Multi-Valued Quantum Cache memory is mainly used because Multi-Valued Quantum operation performs so fast Multi-Valued DNA operation is performed very slowly. Multi-Valued Quantum cache memory works as an intermediate process where qubit is just stored and when needed cache memory serves the qubit.

Multi-Valued Quantum Cache memory works here as the intermediate process. Multi-Valued Quantum circuit produces much heat and here two ternary Quantum NAND operations perform so it produces much heat. But Multi-Valued Quantum circuit needs to be close to zero Kelvin temperature to perform the operation. So from the Multi-Valued Quantum circuit, this proposed circuit needs to reduce the temperature to maintain the superposition state of the qubit. Hence in Multi-Valued DNA circuit operation, it needs much heat in several steps. Mainly in melting and annealing require much heat. This topic is briefly described in the first chapter named Multi-Valued Quantum-DNA circuit operation. For that reason, this proposed circuit transfers the excessive heat from the Multi-Valued Quantum circuit portion to the Multi-Valued DNA circuit portion using a heat transfer circuit. This heat transfer circuit using the junction captures the heat from the Multi-Valued Quantum circuit and using photon bath heat flows through the circuit and gives this into the Multi-Valued DNA circuit. This circuit can transfer heat maximum in the one-meter distance and in Multi-Valued Quantum-DNA flip-flop distance is less than one meter between ternary Quantum AND Multi-Valued DNA circuit. This heat transfer circuit cannot transfer full excessive heat produced from the Multi-Valued Quantum circuit and this heat is not enough to perform the Multi-Valued DNA circuit operation. But this heat transfer can optimize the cost of heat based on the needs. After all of this operation and architecture fully then Multi-Valued Quantum-DNA D flip-flop can produce two output molecular sequences.

3.2.4 Example

Suppose the qubit input is |1> is one for Multi-Valued Quantum-DNA D flip-flop operational circuit where |clk> input is high. First of all clock qubit input and |1>

will perform ternary Quantum NAND operations. So, $|X> = |1>$ and $|clk> = |1>$, then $|S> = |0>$

Then input $|1>$ performs ternary Quantum NOT gate operation and the output and clock input perform ternary Quantum NAND operation. These two ternary Quantum NAND operations parallely. So, $|X> = |0>$ and $|clk> = |1>$, then $|R> = |1>$

Hence produced these two output $|S>$ and $|R>$ will be stored in Multi-Valued Quantum cache memory. For this research used an array to store the qubit data. Then this qubit in accurate time performs the NMR Relaxation process operation first. NMR relaxation process operation makes the qubit convert into molecule sequence. These molecule sequences will perform the Multi-Valued DNA SR latch operation. So, S = TGGATC and R= ACCTAG, then the final outputs are ACCTAG and TGGATC.

This Multi-Valued Quantum output given the confirmation that produced output molecule sequence is accurate for Multi-Valued Quantum-DNA flip-flop operational circuit.

3.3　Multi-Valued Quantum-DNA SR Latch

Based on the triggering that is suited to operate it, there are two types of memory elements: one of them is a Multi-Valued Quantum-DNA SR latch, and the other one is a Multi-Valued Quantum-DNA flip-flop. Multi-Valued Quantum-DNA latches operate with enable signal, level-sensitive, whereas Multi-Valued Quantum-DNA flip-flops are edge sensitive. A Multi-Valued Quantum-DNASR latch is an asynchronous tool. It operates without the use of control signals, relying solely on the state of the $|S>$ and $|R>$ inputs. Two Multi-Valued Quantum NAND operations can make a Multi-Valued Quantum-DNA SR latch. Nevertheless, two Multi-Valued Quantum NOR operations can also make a Multi-Valued Quantum SR Latch.

In Multi-Valued Quantum-DNA SR latch, operational circuits half of the portion is constructed using the Multi-Valued Quantum computing principle and the rest of the portion is constructed by Multi-Valued DNA computing principle. This proposed Multi-Valued Quantum-DNASR latch operational circuit is made by one Multi-Valued Quantum NAND operational circuit and one Multi-Valued DNA NAND gate. Multi-Valued Quantum computing computation speed is so high compared to Multi-Valued DNA computing for that case this proposed circuit has intermediate cache memory to store the Multi-Valued Quantum qubit data. Multi-Valued Quantum computing produces qubit which performs the NMR relaxation process where EMR emit is prohibited to make qubit molecules.

In the Multi-Valued Quantum-DNA SR latch, two-qubit inputs are swapped and negated. Multi-Valued Quantum-DNASR latch can be said as SET RESET latch. In Multi-Valued Quantum-DNASR latch from two-qubit input, this circuit produced two outputs. This output is reversed to one another. This research proposed a

Multi-Valued Quantum-DNASR Latch, and two Multi-Valued Quantum NAND operation circuits designed this SR latch. This proposed Multi-Valued Quantum-DNASR latch has the input line swapped between two operational circuits where one is Multi-Valued Quantum NAND operational circuit and the other is Multi-Valued DNA NAND gate, but it is not negated. Multi-Valued Quantum-DNASR latch works as memory stuff in Multi-Valued Quantum-DNA computers, and it has several applications in a Multi-Valued Quantum-DNA processor. If this proposed circuit has designed some embedded system using the Multi-Valued Quantum-DNA operational device, then Multi-Valued Quantum-DNA SR latch will be used on this device as a memory unit. Multi-Valued Quantum-DNA SR latch is level sensitive and has few disadvantages, and this will be recovered by the Multi-Valued Quantum-DNA flip flop.

3.3.1   General Organization Multi-Valued Quantum-DNA SR Latch

The Multi-Valued Quantum-DNA SR latch is one of the most common memory devices, and it has an effect on the output as long as it is active. The essential properties of a Multi-Valued Quantum-DNA SR latch are that one qubit input behaves like a SET and another qubit input behaves like a RESET. Figure 3.3 depicts the block diagram of multi-valued Quantum-DNA SR Latch.

The Multi-Valued Quantum SR latch is made up of two fundamental processes, which are depicted in this block diagram of the Multi-Valued Quantum SR latch. There are two input lines in the Multi-Valued Quantum SR latch, one for |s> and

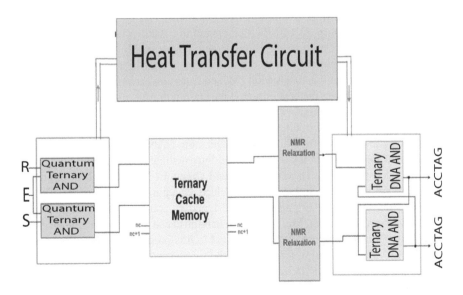

FIGURE 3.3
Block Diagram of Multi-Valued Quantum-DNA SR Latch

the other for |R>. Two outputs are obtained from this two-qubit input: |Q> and |Q>. The output of the first Multi-Valued Quantum NAND operation is used as an input in the second Multi-Valued Quantum NAND operation, and the output of the second Multi-Valued Quantum NAND operation is used as an input in the first Multi-Valued Quantum NAND operation. If the input of |S> is |1>, the SR latch is activated; however, if the input of |R> is |1>, the SR latch has no influence on the output. In a Multi-Valued Quantum SR latch, a value of |1> cannot be used to activate two inputs.

In Multi-Valued DNA computing molecular sequence needs to go through many processes during the operation. In Multi-Valued DNA computing, melting and annealing are very important and these steps require a vast amount of heat. That's why in Multi-Valued Quantum-DNASR Latch Multi-Valued Quantum portion transfers a small amount of heat produced from the Multi-Valued Quantum circuits into Multi-Valued DNA circuits. Though this amount of heat is not enough, this heat or temperature will help to perform Multi-Valued DNA computation. In Multi-Valued DNA computing Multi-Valued DNA NOR SR latch performs where two Multi-Valued DNA NOR gates perform in parallel ways and input comes from Multi-Valued Quantum circuit portion. After that operation, this proposed circuit finally found the output molecular sequence from the Multi-Valued Quantum-DNA SR latch.

3.3.2   Circuit Architecture of Multi-Valued Quantum-DNA SR Latch

Multi-Valued Quantum SR latches are level sensitive and are built using only one fundamental operation, the Multi-Valued Quantum NAND function. Figure 3.4 displays the circuit architecture of multi-valued Quantum-DNA SR Latch

Multi-Valued Quantum-DNA SR Latch operation circuit is constructed using one principle that is the first portion will be constructed by using Multi-Valued Quantum circuit and the second portion will be constructed by using Multi-Valued DNA computing circuit. This circuit also has Multi-Valued Quantum cache memory and a Heat transfer circuit to achieve the best output at the end. In this circuit, input is qubit and output is the molecular sequence. There are two inputs in the Multi-Valued Quantum NAND gate. In this circuit those inputs are |S>, |R>, and |E> as clock qubit input which is shared into the Multi-Valued Quantum NAND operational circuit and Multi-Valued DNA NAND gates through an intermediate process. Two operations are constructed parallelly. These Multi-Valued Quantum NAND operations circuit output line are connected into the Multi-Valued Quantum cache memory. This is the first portion of the circuit and is fully designed by using the principle of Multi-Valued Quantum computing. Multi-Valued Quantum cache memory is made by using some Multi-Valued Quantum shift register and in this proposed circuit it saves data in the Multi-Valued Quantum array. When the time arrives Multi-Valued Quantum cache memory supplies the data into the "NMR Relaxation" process. Multi-Valued Quantum cache memory supply line fully connected to NMR Relaxation process as well as here EMR emit is fully prohibited. Then the output line of the NMR relaxation process is connected to the Multi-Valued DNA NAND gate. Two outputs of the process are connected to two different Multi-Valued DNA NAND gates. Multi-Valued DNA NAND gates one output comes from produced output of Multi-Valued

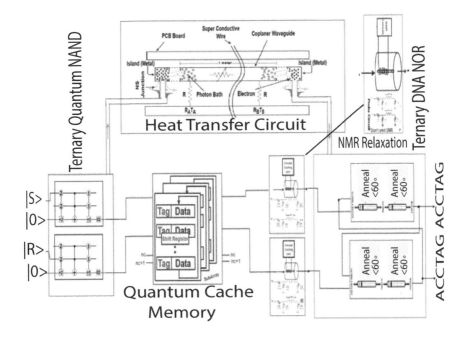

FIGURE 3.4

Circuit Architecture of Multi-Valued Quantum-DNA SR Latch

Quantum NAND gate through the NMR relaxation process and one input comes from direct qubit |R> and by the intermediate process it becomes a molecule sequence and performs Multi-Valued DNA NAND gate operation.

Multi-Valued Quantum Circuit produces more heat and the Multi-Valued DNA circuit needs to process the input. For that reason, it would be best to transfer the heat into Multi-Valued DNA circuits. Multi-Valued Quantum portion Heat transfers using the heat transfer circuit into Multi-Valued DNA circuit's portion. In the heat transfer circuit, two junctions are connected to Multi-Valued Quantum Circuits and Multi-Valued DNA circuits. Maximum one meter can transfer heat to this circuit. This heat transfer circuit has superconductive ware, photon batch, PCB board to make the architecture. This circuit mainly transfers the heat using the photon bath. With this circuit proposed Multi-Valued Quantum-DNASR Latch circuit is constructed completely.

3.3.3 Working Principle of Multi-Valued Quantum-DNA SR Latch

Multi-Valued Quantum-DNA SR latch is working using two principles: Multi-Valued Quantum computing principle and the Multi-Valued DNA computing principle. There are two inputs in the Multi-Valued Quantum NAND operations in this circuit.

In this circuit those inputs are |S>, |R>, and |E> which is shared between the Multi-Valued Quantum NAND operational circuit as well as Multi-Valued DNA NAND gate. Multi-Valued Quantum circuit operation will be happening in close to zero temperature because qubits need a coherence state. The superposition state will be stable if this circuit is full of prohibited particles from other environment particles. The Multi-Valued Quantum NAND operations and Multi-Valued DNA NAND gate operation will be performed parallelly. This Multi-Valued Quantum NAND operation produces some garbage value but in this proposed circuit this topic is avoidable. The Multi-Valued Quantum NAND Operation has one output. These qubit outputs go to the Multi-Valued Quantum Cache Memory which is basically made out of Shift Register. This Multi-Valued Quantum Cache Memory is built using the rules of Multi-Valued Quantum computing. This Cache Memory will store the qubit output from the NAND gate and will serve it when needed.

Multi-Valued Quantum cache stores the qubit into an array. Multi-Valued Quantum cache memory stores the qubit data and when required it serves the data into the "NMR relaxation" process. In this Multi-Valued Quantum-DNASR Latch, two qubits are stored in Multi-Valued Quantum cache memory. This qubit performs the "NMR relaxation" process because Multi-Valued DNA circuits need molecular sequence. This process removes the superposition state and makes the qubit into a molecular sequence. This molecular sequence performs the Multi-Valued DNA NAND operation. Multi-Valued Quantum Cache memory is mainly used because Multi-Valued Quantum operation performs so fast Multi-Valued DNA operation is performed very slowly. Multi-Valued Quantum cache memory works as an intermediate process where qubit is just stored and when needed cache memory serves the qubit.

Multi-Valued Quantum Cache memory works here as the intermediate process. Multi-Valued Quantum circuit produces much heat and here two Multi-Valued Quantum NAND operations perform, so, it produces much heat. But Multi-Valued Quantum circuit needs to be close to zero Kelvin temperature to perform the operation. So from the Multi-Valued Quantum circuit, this proposed circuit needs to reduce the temperature to maintain the superposition state of the qubit. Hence in Multi-Valued DNA circuit operation, it needs much heat in several steps. Mainly in melting and annealing require much heat. This topic is briefly described in the first chapter named Multi-Valued Quantum-DNA circuit operation. For that reason, this proposed circuit transfers the excessive heat from the Multi-Valued Quantum circuit portion to the Multi-Valued DNA circuit portion using a heat transfer circuit. This heat transfer circuit using the junction captures the heat from the Multi-Valued Quantum circuit and using photon bath heat flows through the circuit and gives this into the Multi-Valued DNA circuit. This circuit can transfer heat maximum in the one-meter distance and in Multi-Valued Quantum-DNA flip-flop distance is less than one meter between Multi-Valued Quantum and Multi-Valued DNA circuit. This heat transfer circuit cannot transfer full excessive heat produced from the Multi-Valued Quantum circuit and this heat is not enough to perform the Multi-Valued DNA circuit operation. But this heat transfer can optimize the cost of heat based on the needs. After all of this operation and architecture fully then Multi-Valued Quantum-DNASR Latch

TABLE 3.2
Truth Table of Multi-Valued Quantum-
DNA SR Latch

|S>	|R>	|Q>	|$\overline{Q}$ >
|0>	|0>	Latched	
|0>	|1>	ACCTAG	TGGATC
|1>	|0>	TGGATC	ACCTAG
|1>	|1>	Metastable	

can produce two output molecular sequences. Table 3.2 depicts the truth table of multi-valued Quantum-DNA SR Latch.

The Multi-Valued Quantum-DNA SR latch, on the other hand, is still a vital component of a CPU or Multi-Valued Quantum-DNA-based embedded device.

The Multi-Valued Quantum circuit generates a lot of heat, making it difficult to isolate the qubit into a superposition state. As a result, it needs to cool the circuit to isolate the qubit into a superposition for an able Multi-Valued Quantum circuit. Any type of external particle can disrupt the qubit's coherence and cause it to become decoherent. If all of this is preserved, the Multi-Valued Quantum-DNA SR latch can truly function.

3.3.4 Example

Presume that the qubits |0> and |1> are both present in the Multi-Valued Quantum SR Latch. One qubit input will be used for SET instructions, while the other will be used for RESET instructions. Here, |0> will be used as a SET instruction, and it will conduct the Multi-Valued Quantum NAND operation according to the suggested circuit idea. Here the initial state assumes the molecule sequence is ACCTAG for RESET. This molecule sequence performs NMR process operation and by EMR it becomes qubit. Suppose here ACCTAG = "TRUE" so the qubit will be |1>

As a result, the final output is |0>. The output principle of a Multi-Valued Quantum NAND operation is that if one of the inputs is |0>, the output will also be |1>. As a corollary, whether |Q> is |0> or |1>, the output will be |1>.

Qubit input |1> now functions as a reset instruction and is inserted into the circuit, as well as performing the NAND operation. But Another NAND operation is made by the Multi-Valued DNA computing principle. Multi-Valued DNA NAND gate needs Multi-Valued DNA molecule sequence to perform the Multi-Valued DNA NAND gate operation. For that cause qubit |1> and input |R> stores in Multi-Valued Quantum cache memory to make up the speed with Multi-Valued DNA NAND gate. Then these two qubits perform the NMR relaxation process and make them as molecule sequence. Then these molecule sequence performs the Multi-Valued DNA NAND gate operation.

Multi-Valued Quantum-DNA SR Latch operates in the same mechanism with each qubit input. However, because all of the computations took place in the Multi-Valued Quantum superposition state, a lot of heat was generated throughout the process. All computations take place in the coherence state, and the result will be decoherent, that's why heat transfers into the Multi-Valued DNA portion but this heat is not enough for the Multi-Valued DNA operational circuit according to demand.

3.4   Multi-Valued Quantum-DNA SR Flip Flop

Multi-Valued Quantum-DNA Sequential Logic circuits, unlike Multi-Valued Quantum-DNA Combinational Logic circuits, include some type of built-in "Memory" that changes state depending on the real signals supplied to its inputs at the moment. Multi-Valued Quantum-DNA SR flip-flops, for example, have a |1> qubit memory bistable. The SET and RESET inputs of the Multi-Valued Quantum-DNA SR flip flop are the same. The output of the SET input is a |1>, whereas the output of the RESET input is a |0>.

The Multi-Valued Quantum-DNA SR flip flop is often referred to as the SET RESET flip flop. The reset input is used to restore the flip flop to its starting state from the current state with an output. The ternary Quantum NAND operational circuit SR flip flop is a basic flip flop with both outputs providing feedback to its opposite input. This circuit is used to store a single data bit in a memory circuit. The three inputs are: SET, RESET, and a found output. A two qubit model will be used since Multi-Valued Quantum-DNA SR flip flops have two inputs that are mostly from the outside. Because using two-qubits, it generates more heat at first than Multi-Valued Quantum-DNA D flip flops. The computation time of this Multi-Valued Quantum-DNA SR flip-flop is determined by the fundamental gate in its middle. Multi-Valued Quantum SR flip flops may be found in a wide range of processors and embedded systems. Although the suggested flip-flop can generate some trash, an error correcting auxiliary qubit provides the desired output. The real-world implementation of the suggested Multi-Valued Quantum circuit will address a wide range of issues more quickly and effectively.

3.4.1   General Organization of Multi-Valued Quantum-DNA SR Flip-Flop

Multi-Valued Quantum-DNA SR flip-flop toggles the input independently. Multi-Valued Quantum SR flip-flop takes two qubit inputs |S> and |R>. In Multi-Valued Quantum SR flip-flop, the output of one operation is used by the other operation. Multi-Valued Quantum SR flip-flop depends on it. Figure 3.5 shows the block diagram of multi-valued quantum-DNA SR flip-flop circuits.

Multi-Valued Quantum-DNA SR flip-flop needs two-qubit input and one shared input which is also a qubit. In one ternary Quantum NAND operation |R> and in another ternary Quantum NAND operation |S> is taken as inputs. A qubit input |E> is shared by both NAND operations.

These ternary Quantum NAND operations perform in parallel way, according to the Multi-Valued Quantum computing principle. These outputs will be temporarily

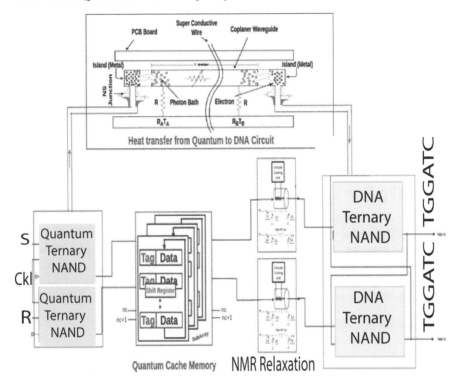

FIGURE 3.5
Block Diagram of Multi-Valued Quantum-DNA SR Flip-Flop Circuits

stored in a Multi-Valued Quantum Cache Memory. As the ternary Quantum NAND operations and the ternary DNA NAND operations are held in different times, it is necessary to store the outputs of the ternary Quantum NAND operations in a Multi-Valued Quantum Cache Memory. As this block diagram depicts, the final output will be the Multi-Valued DNA molecule sequence. So the outputs from the Multi-Valued Quantum Cache Memory will be gone through the "NMR Relaxation" process and after that one qubit will go to Multi-Valued Quantum NOR operation as input. Like the same other output, another qubit from the Multi-Valued Quantum Cache Memory will come as input in another Multi-Valued Quantum NOR operation. Then finally two Multi-Valued Quantum NOR operations will parallelly be performed and it is the last operation of Multi-Valued Quantum computing according to the Multi-Valued Quantum-DNA SR flip-flop circuit block diagram. In this portion, the conversion of the qubit happens after getting out of the ternary DNA NAND gate, the output becomes Multi-Valued DNA molecular sequence.

In Multi-Valued DNA computing molecular sequence needs to go through many processes during the operation. In Multi-Valued DNA computing, melting and annealing are very important and these steps require a vast amount of heat. That's why in Multi-Valued Quantum-DNA SR flip-flop Multi-Valued Quantum portion transfers a small amount of heat produced from the Multi-Valued Quantum circuits into Multi-Valued DNA circuits. Though this amount of heat is not enough, this heat or temperature will help to perform Multi-Valued DNA computation. In Multi-Valued DNA computing ternary DNA NAND SR flip-flop performs where two ternary DNA NAND gates perform parallelly and input comes from Multi-Valued Quantum circuit portion. After that operation, the output molecular sequence are found from the Multi-Valued Quantum-DNA SR flip-flop.

3.4.2 Circuit Architecture of Multi-Valued Quantum-DNA SR Flip-Flop

Multi-Valued Quantum-DNA SR flip-flop operation circuit is constructed using one principle, that is, this circuit's first portion will be constructed by using Multi-Valued Quantum circuit and the second portion will be constructed by using Multi-Valued DNA computing circuit. This circuit also has Multi-Valued Quantum cache memory and a Heat transfer circuit to achieve the best output at the end. Figure 3.6 depicts the circuit architecture of multi-valued Quantum-DNA SR Flip-Flop.

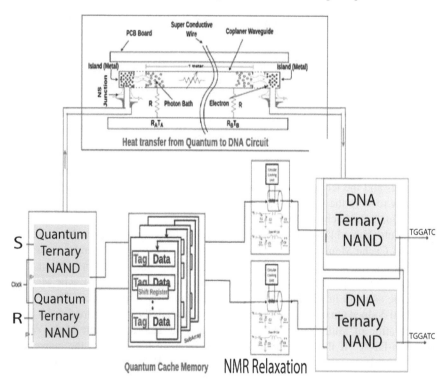

FIGURE 3.6

Circuit Architecture of Multi-Valued Quantum-DNA SR Flip-Flop

In this circuit, at first the inputs are qubits and the outputs are the molecular sequence. There are two inputs in the ternary Quantum NAND gate. In this circuit those inputs are |S>, |R>, and |E> which is shared between the two ternary Quantum NAND gates. Two Multi-Valued Quantum operations are constructed parallelly. These two ternary Quantum NAND operations circuit output lines are connected into the Multi-Valued Quantum cache memory. This is the first portion of the circuit and is fully designed by using the principle of Multi-Valued Quantum computing. Multi-Valued Quantum cache memory is made by using some Multi-Valued Quantum shift register and in this proposed circuit it saves data in the Multi-Valued Quantum array. When the time arrives Multi-Valued Quantum cache memory supplies the data into the "NMR Relaxation" process. Multi-Valued Quantum cache memory supply line fully connected to "NMR Relaxation" process as well as here EMR emission is fully prohibited. Then the output line of the "NMR Relaxation" process is connected to the ternary DNA NAND gate. Two outputs of the process are connected to two different ternary DNA NAND gates. Ternary DNA NAND gates another input come from the output which is ACCTAG and TGGATC molecular sequences. Hence this ternary DNA NAND gates' final output line is the final output of Multi-Valued Quantum-DNA SR flip-flop.

Multi-Valued Quantum Circuit produces more heat and the Multi-Valued DNA circuit needs to process the input. For that reason, it would be best to transfer the heat into Multi-Valued DNA circuits. Multi-Valued Quantum portion Heat transfers using the heat transfer circuit into Multi-Valued DNA circuit's portion. In the heat transfer circuit, two junctions are connected to Multi-Valued Quantum Circuits and Multi-Valued DNA circuits. Maximum one meter can transfer heat to this circuit. This heat transfer circuit has superconductive ware, photon batch, PCB board to make the architecture. This circuit mainly transfers the heat using the photon bath. With this circuit proposed Multi-Valued Quantum-DNA SR flip-flop circuit is constructed fully.

3.4.3   Working Principle of Multi-Valued Quantum-DNA SR Flip-Flop

Multi-Valued Quantum-DNA SR flip-flop is working using two principles: the Multi-Valued Quantum computing principle and the Multi-Valued DNA computing principle. There are two inputs in the ternary Quantum NAND gate in this circuit. In this circuit, those inputs are |S>, |R>, and |E>, which are shared between the two AND gates. Multi-Valued Quantum circuit operation will be happening in close to zero temperature because qubits need to coherence state. The superposition state will be stable if this circuit is full of prohibited from other environment particles. The two ternary Quantum NAND operations will be performed parallelly. This ternary Quantum NAND operation produce some garbage value but in this proposed circuit this topic is avoidable. Each of the ternary Quantum NAND Operation has one output. These qubit outputs go to the Multi-Valued Quantum Cache Memory which is basically made out of Shift Register. This Multi-Valued Quantum Cache Memory is built using the rules of Multi-Valued Quantum computing. This Cache Memory will store the qubit output from the AND gate and will serve it when needed.

Multi-Valued Quantum cache stores the qubit into an array. Multi-Valued Quantum cache memory stores the qubit data and when requires it serves the data into the "NMR Relaxation" process. In this Multi-Valued Quantum-DNA SR flip-flop, two qubits are stored in Multi-Valued Quantum cache memory. This qubit performs the"NMR Relaxation" process because Multi-Valued DNA circuits need molecular sequence. This process removes the superposition state and makes the qubit into a molecular sequence. This molecular sequence performs the ternary DNA NAND operation. The final output sequence and output sequence from other ternary DNA NAND operations perform ternary DNA NAND operation for each of the Multi-Valued DNA NOT gates. These two NOR operations perform parallelly. This structure is basically a ternary DNA NAND flip-flop operation. Multi-Valued Quantum Cache memory is mainly used because Multi-Valued Quantum operation performs so fast Multi-Valued DNA operation is performed very slowly. Multi-Valued Quantum cache memory works as an intermediate process where qubit is just stored and when needed cache memory serves the qubit.

Multi-Valued Quantum Cache memory works here as the intermediate process. Multi-Valued Quantum circuit produces much heat and here two ternary Quantum NAND operations perform so it produces much heat. But Multi-Valued Quantum circuit needs to be close to zero Kelvin temperature to perform the operation. So from the Multi-Valued Quantum circuit, it is needed to reduce the temperature to maintain the superposition state of the qubit. Hence in Multi-Valued DNA circuit operation, it needs much heat in several steps. Mainly melting and annealing require much heat. This topic is briefly described in the first chapter named Multi-Valued Quantum-DNA circuit operation. For that reason, it is needed to transfer the excessive heat from the Multi-Valued Quantum circuit portion to the Multi-Valued DNA circuit portion using a heat transfer circuit. This heat transfer circuit using the junction capture the heat from the Multi-Valued Quantum circuit and using photon bath heat flows through the circuit and gives this into the Multi-Valued DNA circuit. This circuit can transfer heat maximum in the one-meter distance and in Multi-Valued Quantum-DNA flip-flop distance is less than one meter between ternary Quantum NAND Multi-Valued DNA circuits. This heat transfer circuit cannot transfer full excessive heat produced from the Multi-Valued Quantum circuit and this heat is not enough to perform the Multi-Valued DNA circuit operation. But this heat transfer can optimize the cost of heat based on the needs. After all of this operation and architecture fully the Multi-Valued Quantum-DNA SR flip-flop can produce two output molecular sequences. Table 3.3 shows the truth table of multi-valued Quantum-DNA SR Flip-Flop.

TABLE 3.3

Truth Table of Multi-Valued Quantum-DNA SR Flip-Flop

$|S>$	$|R>$	$|Q>$	$|\overline{Q}>$
$|0>$	$|0>$	No Change	
$|0>$	$|1>$	TGGATC	ACCTAG
$|1>$	$|0>$	ACCTAG	TGGATC
$|1>$	$|1>$	Invalid	

3.4.4 Example

Assume that this study effort receives the inputs |0> and |1> in order to ensure that the Multi-Valued Quantum SR flip-flop operational circuit produces the right output. The |0> input end is connected to the |S> input end, while the |1> input end is connected to the |R> input end. The fact that the |R> input is |1> indicates that it is for a reset operation. Assume that the clock is activated and that the clock's input is |1>.

First, do the ternary Quantum NAND operation using the |clk> input. If just one of the inputs is |0> in a ternary Quantum NAND operation, the result is |1>.

The ternary Quantum NAND operation is then performed using another |clk> and |R> input. In this case, the |clk> input is |1>, and the |R> input is also |1>. The output is |0>.

Then, as the input to the Multi-Valued Quantum-DNA SR latch, these two input qubits, |1> and |0>, will be inputted. The Multi-Valued Quantum-DNA SR latch operation will be performed by them. A collection of ternary Quantum-DNA NAND operation circuits makes up the SR latch operation circuit. |1> and |0> will be converted to ACCTAG and TGGATC and the output will be TGGATC and ACCTAG.

3.5 Multi-Valued Quantum-DNA JK Flip-Flop

In flip-flop designs, the Multi-Valued Quantum JK flip-flop will be the most extensively utilized flip-flop. J and K are not abbreviated letters of other words, such as "S" for Set and "R" for Reset, but are independent letters chosen by the inventor Jack Kilby to identify the flip-flop design from others. Despite the fact that the digital electronics JK flip flop was created by Jack Kilby. The functioning concept of the proposed Multi-Valued Quantum JK flip flop differs from that of the digital JK flip flop.

The Multi-Valued Quantum JK flip flop's sequential operation is identical to that of the prior Multi-Valued Quantum SR flip flop, with the same "Set" and "Reset" inputs. The distinction this time is that even though S and R are both at logic "1," the "Multi-Valued Quantum JK flip flop" has no incorrect or prohibited Multi-Valued Quantum SR Latch input states. It is evident that the Multi-Valued Quantum JK flip flop does not solve the disadvantages of the Multi-Valued Quantum SR flip flop.

The Multi-Valued Quantum JK flip flop is essentially a gated Multi-Valued Quantum SR flip flop with the addition of clock qubit input circuitry to avoid the unlawful or invalid output state that can arise when both inputs |S> and |R> are equal to logic level "1." A Multi-Valued Quantum JK flip-flop has four potential input combinations due to the extra timed input: "|1>," "logic |0>," "no change," and "toggle." A Multi-Valued Quantum JK flip flop has the same symbol as a Multi-Valued Quantum SR Bistable Latch, as seen in the preceding chapter. The Multi-Valued Quantum JK flip flop, like other flip flops, generates a lot of heat, which must be dissipated in order for the operation to run properly. As compared to other Multi-Valued Quantum circuits, the Multi-Valued Quantum JK flip flop will not require as much power. The qubit may simply conduct the operation once all of the molecules are in superposition state and coherence mode. A lot of junk values are received in the Multi-Valued Quantum JK flip flop, and it is needed to do more investigation to figure out what they are. The trash value is not taken into account in this procedure.

3.5.1   General Organization of Multi-Valued Quantum-DNA JK Flip-Flop

Multi-Valued Quantum-DNA JK flip-flop toggles the input. Multi-Valued Quantum-DNA JK flip-flops have two-qubit inputs and clock qubit input. Multi-Valued Quantum JK flip-flop relies on Clock qubit input. Figure 3.7 shows the multi-valued Quantum-DNA JK Flip-Flop circuits block diagram.

Multi-Valued Quantum-DNA JK flip-flop needs one qubit input and one clock input which is also a qubit. There is a qubit input |J> and another qubit input |K> which will enter into two different AND Gates. One |clk> input is shared by both of those

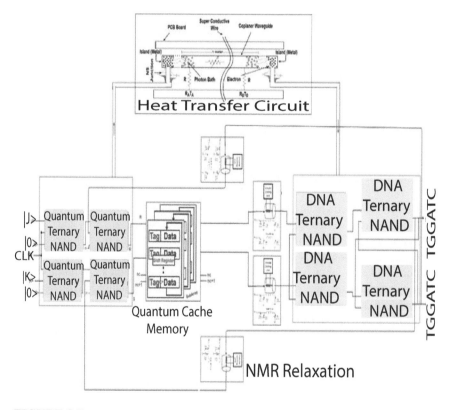

FIGURE 3.7
Block Diagram of Multi-Valued Quantum-DNA JK Flip-Flop Circuits

Multi-Valued Quantum NAND gates. These Multi-Valued Quantum NAND operations perform parallelly according to the Multi-Valued Quantum computing principle. These Multi-Valued Quantum NAND operations output lines are entered as input lines into another couple of Multi-Valued Quantum NAND operations. These Multi-Valued Quantum NAND operations' one input is the previous output qubit and the other is the final output of the Multi-Valued Quantum-DNA JK flip-flop. As this block diagram depicts, the final output will be the Multi-Valued DNA molecule sequence. So this molecular sequence will be gone through the NMR process and NMR will make this sequence a qubit, as well as this qubit, will go to Multi-Valued Quantum NAND operation as input. Like the same other output, the molecular sequence will come as input in another Multi-Valued Quantum NAND operation. Then finally two Multi-Valued Quantum NAND operations will parallelly be performed and it is the last operation of Multi-Valued Quantum computing according to the Multi-Valued Quantum-DNA jk flip-flop circuit block diagram. These two Multi-Valued Quantum NAND operations produce two-qubit which will be stored in Multi-Valued Quantum Cache memory. Then this Multi-Valued Quantum cache memory serves the qubit into the Multi-Valued DNA computing circuit's portion. These qubits need to be molecular sequence first of all so qubits are performed here first of all NMR relaxation process. In the NMR relaxation process, emitting EMR is strictly prohibited. After the NMR relaxation process qubits are relaxing their state and are converted into the molecular sequence.

In Multi-Valued DNA computing molecular sequence needs to go through many processes during the operation. In Multi-Valued DNA computing, melting and annealing is very important and these steps require a vast amount of heat. That's why in Multi-Valued Quantum-DNA JK flip-flop Multi-Valued Quantum portion transfers a small amount of heat produced from the Multi-Valued Quantum circuits into Multi-Valued DNA circuits. Though this amount of heat is not enough, this heat or temperature will help to perform Multi-Valued DNA computation. In Multi-Valued DNA computing Multi-Valued DNA NAND SR latch performs where two Multi-Valued DNA NAND gates perform parallelly and input comes from Multi-Valued Quantum circuit portion. After that operation and finally found the output molecular sequence from Multi-Valued Quantum-DNA JK flip-flop. Here, the outputs of the Multi-Valued DNA NAND gates again goes through "NMR" process which are taken as inputs for the Multi-Valued Quantum NAND operations.

3.5.2 Circuit Architecture of Multi-Valued Quantum-DNA JK Flip-Flop

Multi-Valued Quantum-DNA JK Flip-Flop operation circuit is constructed using one principle that is this circuit's first portion will be constructed by using Multi-Valued Quantum circuit and the second portion will be constructed by using Multi-Valued DNA computing circuit. This circuit also has Multi-Valued Quantum cache memory and a Heat transfer circuit to achieve the best output at the end. Figure 3.8 shows the multi-valued Quantum-DNA JK Flip-Flop operation circuit.

In this circuit, input is qubit and output is the molecular sequence. Clock input |clk> is shared into two Multi-Valued Quantum NAND operations. Two

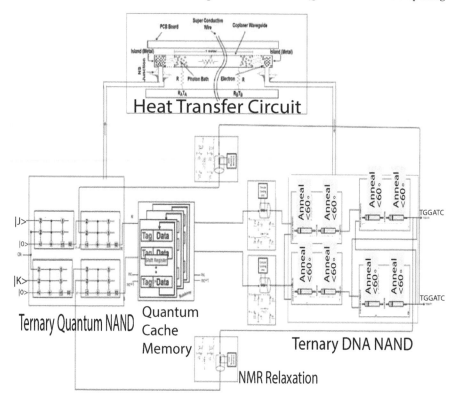

FIGURE 3.8
Multi-Valued Quantum-DNA JK Flip-Flop Operation Circuit

Multi-Valued Quantum operations are constructed parallelly. This Multi-Valued Quantum NAND operations output line and final output of Multi-Valued Quantum-DNA JK Flip-Flop are constructed into another two Multi-Valued Quantum operations parallelly. But the final two outputs are molecular sequences so these two output lines are connected to the NMR process and then after the NMR process these two-line came into use as an input line Multi-Valued Quantum NAND operation circuit. These two Multi-Valued Quantum NAND operations circuit output lines are connected into the Multi-Valued Quantum cache memory. This is the first portion of the circuit and is fully designed by using the principle of Multi-Valued Quantum computing. Multi-Valued Quantum cache memory is made by using some Multi-Valued Quantum shift register and in this proposed circuit it saves data in the Multi-Valued Quantum array. When the time arrives Multi-Valued Quantum cache memory supplies the data into the NMR Relaxation process. Multi-Valued Quantum cache memory supply line fully connected to NMR Relaxation process as well as here EMR emit is fully prohibited. Then the NMR Relaxation process output line is connected to the Multi-Valued DNA NAND gate. Two outputs of the NMR relaxation process are connected to two different Multi-Valued DNA NAND gates. Multi-Valued DNA

NAND gates another input come from the output which is Q and $\overline{Q}$ molecular sequences. Hence this Multi-Valued DNA NAND gates' final output line is the final output of Multi-Valued Quantum-DNA JK Flip-Flop. Here, the outputs of the Multi-Valued DNA NAND gates again go through the "NMR" process which are taken as inputs for the Multi-Valued Quantum NAND operations. In these "NMR" processes they emit "EMR".

Multi-Valued Quantum Circuit produces more heat and the Multi-Valued DNA circuit needs to process the input. For that reason, it would be best to transfer the heat into Multi-Valued DNA circuits. Multi-Valued Quantum portion Heat transfers using the heat transfer circuit into Multi-Valued DNA circuit's portion. In the heat transfer circuit, two junctions are connected to Multi-Valued Quantum Circuits and Multi-Valued DNA circuits. Maximum one meter can transfer heat to this circuit. This heat transfer circuit has superconductive ware, photon batch, PCB board to make the architecture. This circuit mainly transfers the heat using the photon bath. With this circuit proposed Multi-Valued Quantum-DNA JK Flip-Flop circuit is constructed fully.

3.5.3 Working Principle of Multi-Valued Quantum-DNA JK Flip-Flop

Multi-Valued Quantum-DNA JK Flip-Flop is working using two principles: the Multi-Valued Quantum computing principle and the Multi-Valued DNA computing principle. Multi-Valued Quantum -Multi-Valued DNA JK Flip-Flop has two inputs which are |J> and |K> and clock input is |clk>. Multi-Valued Quantum-DNA JK Flip-Flop won't enable if the clock input is not |0>. Hence the clock input enables this circuit to start to work and first of all, it will perform the Multi-Valued Quantum circuit operation. Multi-Valued Quantum circuit operation will be happening in close to zero temperature because qubits need a coherence state. The superposition state will be stable if this circuit is full of prohibited particles from other environment particles. The |clk> input will be shared and perform two Multi-Valued Quantum NAND operations parallelly. This Multi-Valued Quantum NAND operation produces some garbage value but in this proposed circuit this topic is avoidable. Multi-Valued Quantum NAND operations produce two output qubits in the same amount of time. This qubit one of each will perform another two Multi-Valued Quantum NAND operations parallelly. The first qubit and final output of Multi-Valued Quantum-DNA JK Flip-Flop will perform Multi-Valued Quantum NAND operation. Another qubit and another output sequence of the Multi-Valued Quantum-DNA JK Flip-Flop will perform the Multi-Valued Quantum NAND operation. These two operations will perform parallelly. But the final output of Multi-Valued Quantum-DNA JK Flip-Flop is a molecular sequence. The molecular sequence cannot perform the Multi-Valued Quantum NAND operation. For that reason, Multi-Valued Quantum NAND operations need a qubit to perform the operation. So, that's why Multi-Valued Quantum molecular sequence performs first of all NMR process and then it converts them molecular sequence into a qubit. Then these qubits and the previous output qubits perform two Multi-Valued Quantum NAND operations parallelly and produce the output of two Multi-Valued Quantum NAND operations parallelly. These two outputs of two qubits enter into the Multi-Valued Quantum cache memory.

Multi-Valued Quantum cache stores the qubit into an array. Multi-Valued Quantum cache memory stores the qubit data and when required, it serves the data into the NMR relaxation process. In this Multi-Valued Quantum-DNA JK Flip-Flop, two-qubit is stored in Multi-Valued Quantum cache memory. This qubit performs the NMR Relaxation process because Multi-Valued DNA circuits need molecular sequence. NMR relaxation process removes the superposition state and makes the qubit into a molecular sequence. This molecular sequence performs the Multi-Valued DNA NAND operation. The final output sequence and molecular sequence from cache perform Multi-Valued DNA NAND operation. Another final output sequence and previous output from cache also perform another Multi-Valued DNA NAND operation. These two NAND operations perform parallelly. This structure is basically Multi-Valued DNA NAND latch operation. Multi-Valued Quantum Cache memory is mainly used because Multi-Valued Quantum operation performs so fast and on the other hand Multi-Valued DNA operation performs very slowly. Multi-Valued Quantum cache memory works as an intermediate process where qubit is just stored and when needed cache memory serves the qubit.

Multi-Valued Quantum Cache memory works here as the intermediate process. Multi-Valued Quantum circuit produces much heat and here two Multi-Valued Quantum NAND operation performs so it produces much heat. But Multi-Valued Quantum circuit needs to be in close to zero Kelvin temperature to perform the operation. So from the Multi-Valued Quantum circuit, it is needed to reduce the temperature to maintain the superposition state of the qubit. Hence in Multi-Valued DNA circuit operation, it needs much heat in several steps. Mainly in melting and annealing require much heat. This topic is briefly described in the first chapter named Multi-Valued Quantum-DNA circuit operation. For that reason, it is needed to transfer the excessive heat from the Multi-Valued Quantum circuit portion to the Multi-Valued DNA circuit portion using a heat transfer circuit. This heat transfer circuit using the junction capture the heat from the Multi-Valued Quantum circuit and using photon bath heat flows through the circuit and gives this into the Multi-Valued DNA circuit. This circuit can transfer heat maximum in the one-meter distance and in Multi-Valued Quantum-DNA flip-flop distance is less than one meter between Multi-Valued Quantum NAND Multi-Valued DNA circuits. This heat transfer circuit cannot transfer full excessive heat produced from the Multi-Valued Quantum circuit and this heat is not enough to perform the Multi-Valued DNA circuit operation. But this heat transfer can optimize the cost of heat based on the needs. Table 3.4 shows the truth table of multi-valued quantum-DNA JK flip-flop.

3.5.4 Example

For the purpose of testing the proposed Multi-Valued Quantum JK flip-flop circuit, assume that the qubit input is $|0>$ and $|1>$. If and only if the clock input is high or $|1>$, the Multi-Valued Quantum JK flip flop will function. If the clock input is high, the Multi-Valued Quantum NAND operation will be performed by inputs $|J>= |0>$ and $|clk>=|1>$. Then Intermediate Output = $|1>$

In addition to performing the Multi-Valued Quantum NAND operation and producing the appropriate output, the inputs are $|K>= |1>$ and $|clk>= |1>$, then the Intermediate Output $=|0>$.

TABLE 3.4

Truth Table of Multi-Valued Quantum-DNA JK Flip-Flop

J	k	Q	$|\bar{Q}>$
$|0>$	$|0>$	No Change	
$|0>$	$|1>$	TGGATC	ACCTAG
$|1>$	$|0>$	ACCTAG	TGGATC
$|2>$	$|0>$	TGGATC	CGGTAC
$|1>$	$|1>$	TGGATC	ACCTAG
$|1>$	$|1>$	ACCTAG	TGGATC

These two intermediate qubit outputs are now used to conduct two independent Multi-Valued Quantum NAND operations. The Multi-Valued Quantum NAND operation is performed by the qubit $|0>$ and the final output of the Multi-Valued Quantum-DNA JK flip-flop $|Q>$. Assume that the most recent state $|Q>$ is $|1>$ and the first intermediate output $|1>$ will produce the output $|S>=|0>$.

The Multi-Valued Quantum NAND operation is then performed on the intermediate output $|0>$ and the final output of the Multi-Valued Quantum JK flip flop $|Q>=|0>$ and produce the output $|R>=|1>$.

According to the suggested circuit of Multi-Valued Quantum JK flip-flop, these qubits labeled $|S>$ and $|R>$ will now conduct the Multi-Valued DNA SR flip-flop operation and the final output will be TGGATC and ACCTAG.

3.6   Multi-Valued Quantum-DNA T Flip-Flop

Multi-Valued Quantum-DNA T flip-flop is also known as "Multi-Valued Quantum-DNA Toggle Flip-flop". To avoid the occurrence of the intermediate state in Multi-Valued Quantum-DNA SR flip-flop, only one input should be provided to the flip-flop called the Trigger qubit input or Toggle input. Toggling means 'Changing the next state output to complement the present state output'. The Multi-Valued Quantum-DNA T flip–flop can be designed by making simple modifications to the Multi-Valued Quantum-DNA JK flip-flop. The Multi-Valued Quantum-DNA T flip–flop is a single qubit input device and hence by connecting $|J>$ and $|K>$ inputs together and giving them with single input called $|T>$, a Multi-Valued Quantum-DNA JK flip-flop can be converted into a Multi-Valued Quantum-DNA T flip-flop.

Hence in Multi-Valued Quantum-DNA flip-flops have two portions based on the principle of the Multi-Valued Quantum-DNA circuit. In Multi-Valued Quantum-DNA T flip-flop circuit, some portions are made by Multi-Valued Quantum computing principle and the rest of the portions are made by Multi-Valued DNA computing principle. Multi-Valued Quantum computing and multi-valued DNA computing are connected via the NMR relaxation process. For Time consistency, it is needed to use the Multi-Valued Quantum cache memory in this circuit. The Multi-Valued Quantum circuit portion produces so much heat according to the Multi-Valued Quantum computing principle that's why this circuit can transfer a small amount of heat into the Multi-Valued DNA computing circuit portion. According to Multi-Valued DNA computing, it needs a vast amount of heat to perform the calculation. In Multi-Valued Quantum-DNA T flip-flop circuit have one input which is a qubit input named |T>. This input performs first of all Multi-Valued Quantum operations and is then stored in Multi-Valued Quantum cache memory. When the perfect time arrived it will perform the NMR relaxation and convert into Multi-Valued DNA molecular sequence as well as perform the Multi-Valued DNA computing operation. Finally, after all of this process, Multi-Valued Quantum-DNA T flip-flop produces the output as Multi-Valued DNA molecular sequence. A most vital part in this Multi-Valued Quantum T flip-flop that is from Multi-Valued DNA molecule sequence, it is needed to make a qubit first and this qubit will be entered here in the circuit as an input. Multi-Valued Quantum -Multi-Valued DNA T flip-flop toggles and it required much time compared to Multi-Valued Quantum t flip-flop but required less compared to Multi-Valued DNA T flip-flop. Multi-Valued Quantum-DNA T flip-flop circuits Multi-Valued Quantum circuit portion performs so fast but Multi-Valued DNA circuit portion much slower. In Multi-Valued Quantum-DNA T flip-flop bunch of Multi-Valued Quantum AND operations performs according to Multi-Valued Quantum computing principle and a bunch of Multi-Valued DNA NOR gate performs in Multi-Valued DNA computing portion in the circuit. Multi-Valued Quantum-DNA T flip-flop removes the problem of Multi-Valued Quantum-DNA SR flip-flop.

3.6.1 General Organization of Multi-Valued Quantum-DNA T Flip-Flop

Multi-Valued Quantum-DNA T flip-flop toggles the input and Multi-Valued Quantum-DNA Jk flip-flop's extended version is Multi-Valued Quantum-DNA T flip-flop architecturally. Multi-Valued Quantum-DNA T flip-flops have one qubit input and clock qubit input. Multi-Valued Quantum T flip-flop relies on Clock qubit input. Figure 3.9 depicts the block diagram of multi-valued Quantum-DNA T Flip-Flop.

Multi-Valued Quantum-DNA T flip-flop needs one qubit input and one clock input which is also a qubit. One qubit input |T> and |clk> are shared into two Multi-Valued Quantum AND operations. These Multi-Valued Quantum AND operations perform parallelly according to the Multi-Valued Quantum computing principle. Then, multi-valued Quantum AND operations output lines are entered as input lines into another couple of Multi-Valued Quantum AND operations. finally, multi-valued Quantum AND operations' one input is the previous output qubit and the other is the final output of the Multi-Valued Quantum-DNA T flip-flop. As this block diagram depicts, the final output will be the Multi-Valued DNA molecule sequence. So this molecular sequence will be gone through the NMR process and NMR will make this sequence a qubit, as well as this qubit, will go to Multi-Valued Quantum AND operation as input. Like the same other output, the molecular sequence

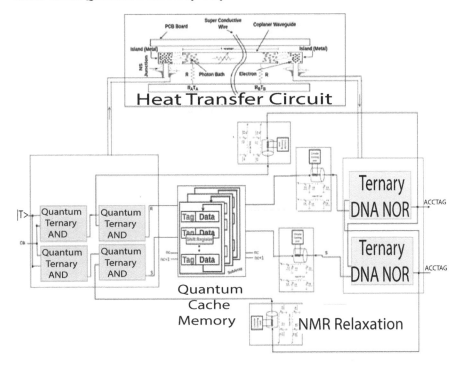

FIGURE 3.9
Block Diagram of Multi-Valued Quantum-DNA T Flip-Flop

will come as input in another Multi-Valued Quantum AND operation. Then finally two Multi-Valued Quantum AND operations will parallelly perform and it is the last operation of Multi-Valued Quantum computing according to the Multi-Valued Quantum-DNA T flip-flop circuit block diagram. These two Multi-Valued Quantum AND operations produce two-qubit which will be stored in Multi-Valued Quantum Cache memory. Then this Multi-Valued Quantum cache memory serves the qubit into the Multi-Valued DNA computing circuit's portion. These qubits need to be molecular sequence first of all so qubits are performed here first of all NMR relaxation process. In the NMR relaxation process, emitting EMR is strictly prohibited. After the NMR relaxation process qubits are relaxing their state and are converted into the molecular sequence.

In Multi-Valued DNA computing molecular sequence needs to go through many processes during the operation. In Multi-Valued DNA computing, melting and annealing are very important and these steps require a vast amount of heat. That's why in Multi-Valued Quantum-DNA T flip-flop Multi-Valued Quantum portion transfers a small amount of heat produced from the Multi-Valued Quantum circuits into Multi-Valued DNA circuits. Though this amount of heat is not enough, this heat or temperature will help to perform Multi-Valued DNA computation. In Multi-Valued DNA computing Multi-Valued DNA NOR SR latch performs where two Multi-Valued

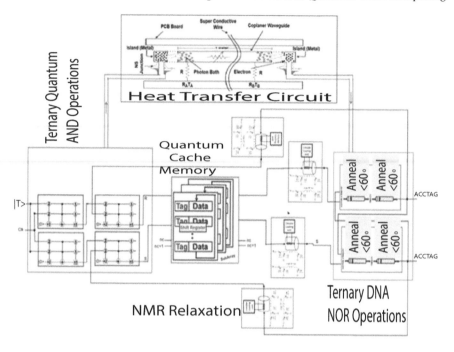

FIGURE 3.10

Multi-Valued Quantum-DNA T Flip-Flop Operation Circuit

DNA NOR gates perform parallelly and input comes from Multi-Valued Quantum circuit portion. After that operation, finally found the output molecular sequence from Multi-Valued Quantum-DNA T flip-flop.

3.6.2   Circuit Architecture of Multi-Valued Quantum-DNA T Flip-Flop

Multi-Valued Quantum-DNA T flip-flop operation circuit is constructed using one principle, this circuit's first portion will be constructed by using Multi-Valued Quantum circuit and the second portion will be constructed by using Multi-Valued DNA computing circuit. This circuit also has Multi-Valued Quantum cache memory and a Heat transfer circuit to achieve the best output at the end. Figure 3.10 depicts the multi-valued Quantum-DNA T Flip-Flop operation circuit.

In this circuit, input is qubit and output is the molecular sequence. Input |T> and clock input |clk> are shared into two Multi-Valued Quantum AND operations. Two Multi-Valued Quantum operations are constructed in parallel way. This Multi-Valued Quantum AND operations output line and final output of Multi-Valued Quantum-DNA T flip-flop are constructed into another two Multi-Valued Quantum operations in parallel way. But the final two outputs are molecular sequences so these two output lines are connected to the NMR Relaxation process and then after that, these two lines came into use as an input line Multi-Valued Quantum AND operation circuit. These two Multi-Valued Quantum AND operations circuit output lines are connected into the Multi-Valued Quantum cache memory. This is the first portion of the circuit and is fully designed by using the principle of Multi-Valued Quantum computing. Multi-Valued Quantum cache memory is made by using some Multi-Valued Quantum shift register and in this proposed circuit it saves data in the Multi-Valued Quantum array. When the time arrives Multi-Valued Quantum cache memory supplies the data into the NMR Relaxation process. Multi-Valued Quantum cache memory supply line fully connected to NMR Relaxation process as well as here EMR emit is fully prohibited. Then the NMR Relaxation process output line is connected to the Multi-Valued DNA NOR gate. Two outputs of the NMR relaxation process are connected to two different Multi-Valued DNA NOR gates. Multi-Valued DNA NOR gates another input come from the output which is Q and $\overline{Q}$ molecular sequences. Hence this Multi-Valued DNA NOR gates' final output line is the final output of Multi-Valued Quantum-DNA T flip-flop.

Multi-Valued Quantum Circuit produces more heat and the Multi-Valued DNA circuit needs to process the input. For that reason, it would be best to transfer the heat into Multi-Valued DNA circuits. Multi-Valued Quantum portion Heat transfers using the heat transfer circuit into Multi-Valued DNA circuit's portion. In the heat transfer circuit, two junctions are connected to Multi-Valued Quantum Circuits and Multi-Valued DNA circuits. Maximum one meter can transfer heat to this circuit. This heat transfer circuit has superconductive ware, photon batch, PCB board to make the architecture. This circuit mainly transfers the heat using the photon bath. With this circuit proposed Multi-Valued Quantum-DNA T flip-flop circuit is constructed fully.

3.6.3   Working Principle of Multi-Valued Quantum-DNA T Flip-Flop

Multi-Valued Quantum-DNA T flip-flop is working using two principles: Multi-Valued Quantum computing principle and the Multi-Valued DNA computing principle. Multi-Valued Quantum -Multi-Valued DNA T flip-flop has one input is named |T> and clock input is |clk>. Multi-Valued Quantum-DNA T flip-flop won't enable if the clock input is not |0>. Hence the clock input enables this circuit to start to work and first of all, it will perform the Multi-Valued Quantum circuit operation. Multi-Valued Quantum circuit operation will be happening in close to zero temperature because qubits need a coherence state. The superposition state will be stable if this circuit is full of prohibited from other environment particles. |T> and |clk> input will be shared and perform two Multi-Valued Quantum AND operations in parallel way. This Multi-Valued Quantum AND operation produces some garbage value but in this proposed circuit this topic is avoidable. Multi-Valued Quantum AND operation produces two output qubits in the same amount of time. This qubit one of each will perform another two Multi-Valued Quantum AND operations in parallel way. The first qubit and final output of Multi-Valued Quantum-DNA T flip-flop will perform Multi-Valued Quantum AND operation. Another qubit and another output sequence of the Multi-Valued Quantum-DNA T flip-flop will perform the Multi-Valued Quantum AND operation.

These two operations will perform in parallel way. But the final output of Multi-Valued Quantum-DNA T flip-flop is a molecular sequence. The molecular sequence cannot perform the Multi-Valued Quantum AND operation. For that reason, Multi-Valued Quantum AND operation need a qubit to perform the operation. So, that's why Multi-Valued Quantum molecular sequence performs first of all NMR process and then it converts them molecular sequence into a qubit. Then these qubits and the previous output qubits perform two Multi-Valued Quantum AND operations in parallel way. and produce the output of two Multi-Valued Quantum AND operations in parallel way. These two outputs of two qubits enter into the Multi-Valued Quantum cache memory.

Multi-Valued Quantum cache stores the qubit into an array. Multi-Valued Quantum cache memory stores the qubit data and when requires it serves the data into the NMR relaxation process. In this Multi-Valued Quantum-DNA T flip-flop, two-qubit is stored in Multi-Valued Quantum cache memory. This qubit performs the NMR Relaxation process because Multi-Valued DNA circuits need molecular sequence. NMR relaxation process removes the superposition state and makes the qubit into a molecular sequence. This molecular sequence performs the Multi-Valued DNA NOR operation. The final output sequence and molecular sequence from cache perform Multi-Valued DNA NOR operation. Another final output sequence and previous output from cache also perform another Multi-Valued DNA NOR operation. These two NOR operation performs in parallel way. This structure basically Multi-Valued DNA NOR latches operation. Multi-Valued Quantum Cache memory is mainly used because Multi-Valued Quantum operation performs so fast Multi-Valued DNA operation performed very slowly. Multi-Valued Quantum cache memory works as an intermediate process where qubit is just stored and when needed cache memory serves the qubit.

Multi-Valued Quantum Cache memory works here as the intermediate process. Multi-Valued Quantum circuit produces much heat and here two Multi-Valued Quantum AND operations perform so it produces much heat. But Multi-Valued Quantum circuit needs to be close to zero Kelvin temperature to perform the operation. So from the Multi-Valued Quantum circuit, it is needed to reduce the temperature to maintain the superposition state of the qubit. Hence in Multi-Valued DNA circuit operation, it needs much heat in several steps. Mainly in melting and annealing require much heat. This topic is briefly described in the first chapter named Multi-Valued Quantum-DNA circuit operation. For that reason, it is needed to transfer the excessive heat from the Multi-Valued Quantum circuit portion to the Multi-Valued DNA circuit portion using a heat transfer circuit. This heat transfer circuit using the junction capture the heat from the Multi-Valued Quantum circuit and using photon bath heat flows through the circuit and gives this into the Multi-Valued DNA circuit. This circuit can transfer heat maximum in the one-meter distance and in Multi-Valued Quantum-DNA flip-flop distance is less than one meter between Multi-Valued Quantum and Multi-Valued DNA circuit. This heat transfer circuit cannot transfer full excessive heat produced from the Multi-Valued Quantum circuit and this heat is not enough to perform the Multi-Valued DNA circuit operation. But this heat transfer can optimize the cost of heat based on the needs. After all of this operation and

TABLE 3.5
Truth Table of Multi-Valued Quantum-DNA T Flip-Flop

|T>	Q (Multi-Valued DNA Sequence)	$\overline{Q}$ (Multi-Valued DNA Sequence)
|0>	TGGATC	TGGATC
|1>	TGGATC	ACCTAG
|0>	ACCTAG	TGGATC
|1>	ACCTAG	TGGATC

architecture fully then Multi-Valued Quantum-DNA T flip-flop can produce two output molecular sequences. Table 3.5 shows truth table of multi-valued Quantum-DNA T Flip-Flop.

3.6.4 Example

Multi-Valued Quantum-DNA T flip-flop needs one qubit input and one clock input. Clock inputs value is |1> then this Multi-Valued Quantum-DNA flip-flop will be enabled. Assume clock input value is |1> here and the qubit input is |1>. These qubit input and clock input will, first of all, perform two Multi-Valued Quantum AND operations parallelly. Output will be |1>.

So, from two Multi-Valued Quantum AND operations, it is found that two-qubit which |1> respectively. Assume one output sequence is Q and the other is $\overline{Q}$. Hence here if the Q sequence is ACCTAG then $\overline{Q}$ will be the TGGATC sequence as supposed.

Qubit |1> and output sequence Q as well as another output qubit |1> and $\overline{Q}$ will perform Multi-Valued Quantum AND operation. But molecular sequence cannot perform Multi-Valued Quantum AND operation directly. For that reason, these two sequences, first of all, perform the NMR process and make them qubit, and then perform the Multi-Valued Quantum AND operation with respective other input in parallel way. ACCTAG will convert as a qubit and assume it as |1> and opposite TGGATC as |0> suppose. So |1> and ACCTAG or |1> will perform Multi-Valued Quantum AND operation and |1> and TGGATC or |0> will perform Multi-Valued Quantum AND operation in parallel way.

These outputs |1> and |0> will be stored in Multi-Valued Quantum cache memory. Multi-Valued Quantum cache memory will serve this output as input into the Multi-Valued DNA circuit when the right time arrives. These qubits will perform first of all NMR relaxation process where these qubits will become in decoherence state whereas the qubit will be relaxed and converted into molecule sequence. In the NMR relaxation process, EMR emits is fully prohibited. After performing the NMR relaxation process these output |1> and |0> will be converted as molecule sequences named ACCTAG and TGGATC respectively. These two molecule sequence performs the Multi-Valued DNA SR latch operation where two Multi-Valued DNA NOR gate performs operation in parallel way.

So, ACCTAG and TGGATC will produce the final output ACCTAG and TGGATC.

3.7   Multi-Valued Quantum-DNA Shift Register

A single qubit of two-valued qubit data ($|1>$or $|0>$) can be stored in a Multi-Valued Quantum-DNA flip flop. However, many Multi-Valued Quantum-DNA flip-flops are required to store multiple qubits of data. To store n qubits of data, N Multi-Valued Quantum-D flip flops must be coupled in a certain order. A Multi-Valued Quantum-DNA register is a gadget that stores this type of data. It consists of a sequence of Multi-Valued Quantum-DNA flip flops used to store multiple qubits of data.

Multi-Valued Quantum-DNA shift registers enable the information stored in these Multi-Valued Quantum registers to be transmitted. A Multi-Valued Quantum-DNA shift register is a collection of flip flops that stores several qubits of information. By applying clock pulses to the qubits contained in such Multi-Valued Quantum-DNA registers, they may be made to move inside them and in and out of them. By linking n Multi-Valued Quantum flip-flops and n number of Multi-Valued DNA flip-flops, each of which stores a single qubit of data or molecule sequence of data, an n Multi-Valued Quantum-DNA shift register may be built. "Multi-Valued Quantum-DNA Shift left registers" are Multi-Valued Quantum-DNA registers that will shift the qubits or molecules to the left. "Multi-Valued Quantum-DNA Shift right registers" are Multi-Valued Quantum-DNA registers that will shift the qubits or molecule sequence to the right.

Multi-Valued Quantum-DNA Shift registers are basically of 4 types. These are:

1.   Multi-Valued Quantum-DNA Serial In Serial Out shift register

2.   Multi-Valued Quantum-DNA Serial In parallel Out shift register

3.   Multi-Valued Quantum-DNA Parallel In Serial Out shift register

4.   Multi-Valued Quantum-DNA Parallel In parallel Out shift register

In this study, a shift register is built utilizing a Multi-Valued Quantum D flip-flop operational circuit and a Multi-Valued DNA D flip-flop operational circuit to convert serial data into Multi-Valued Quantum data and Multi-Valued DNA molecule sequence data. The Multi-Valued Quantum-DNA Serial-In Serial-Out shift register is a type of Multi-Valued Quantum-DNA shift register that permits serial input one qubit or one molecule sequence at a time over a single data line and outputs a serial output. The data exits the Multi-Valued Quantum-DNA shift register one qubit at a time in a serial pattern since there is only one qubit output, thus the term Multi-Valued Quantum-DNA Serial-In Serial-Out Shift Register. Four Multi-Valued Quantum-DNA D flip-flops are linked in a serial fashion in this circuit. Because the same clock signal is supplied to each Multi-Valued Quantum-DNA flip flop, they are all synchronized with one another. The circuit below in the architecture section is an example of a Multi-Valued Quantum-DNA shift right register, which accepts serial data from the Multi-Valued Quantum flip flop's left side and Multi-Valued DNA flip-flops left side.

In Multi-Valued Quantum-DNA flip-flop qubit is entered into as an input in the circuit and performs the required operation. Multi-Valued Quantum computing computation speed is much higher than Multi-Valued DNA computing computational speed. For that case qubits are stored in Multi-Valued Quantum cache memory. When actually the time arrives then qubit first of all performs the NMR relaxation process and makes the qubit into molecule sequence. Then this molecule sequence performs the Multi-Valued DNA flip-flop operation. Multi-Valued Quantum circuit produces much heat and heat needs to be removed to maintain the coherence state and Multi-Valued DNA computing circuit needs huge heat to enable the circuit to perform. In that case, this circuit used a heat transfer circuit to transfer the heat into the Multi-Valued DNA circuits.

3.7.1 General Organization of Multi-Valued Quantum-DNA Shift Register

Three Multi-Valued Quantum D flip-flop operational circuits and One Multi-Valued DNA D flip-flop circuits are used to make a Multi-Valued Quantum-DNA shift register. As a fundamental component, a Multi-Valued Quantum-DNA Shift register is utilized for data shift, and a Multi-Valued Quantum D flip-flop and Multi-Valued DNA D flip-flop is used to make it happen. Figure 3.11 depicts the block diagram of Multi-Valued Quantum-DNA Shift Register.

Two inputs are used in Multi-Valued Quantum D flip flops: one for data and the other for the clock. The outputs of Multi-Valued Quantum D flip flops are logically opposite one another. The circuit's synchronization with an external signal is aided by the clock input. A Multi-Valued Quantum D flip flop's output can have two possible values. Data input is directed to a Multi-Valued Quantum NAND operation circuit in this block diagram, while data input reverse is routed to another Multi-Valued Quantum NAND operation circuit. Both Multi-Valued Quantum NAND operations procedures use the clock pulse input. Multi-Valued Quantum SR Latch receives the result of two Multi-Valued Quantum NAND operations. The D flip flop is constructed using the SR latch. This property is used to induce a delay in the circuit's data flow. Multi-Valued Quantum SR Latch is made up of two Multi-Valued Quantum NAND Operations. The final two outputs of the Multi-Valued Quantum SR Latch function were uncovered. A Multi-Valued Quantum D flip flop may provide two types of output, one of which is logically inverse to the other. The Multi-Valued Quantum D flip flop will continue to function if the clock is enabled; otherwise, the Multi-Valued Quantum D flip flop will stop working.

Multi-Valued Quantum D flip-flop operational circuits are also coupled through serial connection in the Multi-Valued Quantum shift register block diagram.

The Multi-Valued Quantum D flip-flop operational circuit is the fundamental component of the Multi-Valued Quantum shift register. After processing the qubit input in the Multi-Valued Quantum Shift Register output of the shift register is stored in Multi-Valued Quantum Cache Memory.

In Multi-Valued DNA computing molecular sequence needs to go through many processes during the operation. In Multi-Valued DNA computing, melting and annealing are very important and these steps require a vast amount of heat. That's why in Multi-Valued Quantum-DNA Shift Register Multi-Valued Quantum portion transfers a small amount of heat produced from the Multi-Valued Quantum circuits into

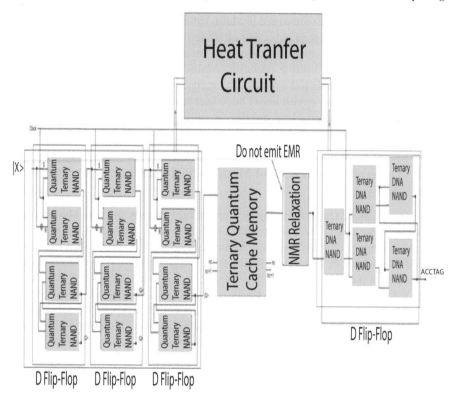

FIGURE 3.11
Block Diagram of Multi-Valued Quantum-DNA Shift Register

Multi-Valued DNA circuits. Though this amount of heat is not enough, this heat or temperature will help to perform Multi-Valued DNA computation. In Multi-Valued DNA computing Multi-Valued DNA NOR SR latch performs where two Multi-Valued DNA NOR gates perform in parallel way. and input comes from Multi-Valued Quantum circuit portion. After that operation and finally found the output molecular sequence from Multi-Valued Quantum-DNA Shift Register.

3.7.2   Circuit Architecture of Multi-Valued Quantum-DNA Shift Register

Four D flip-flops and a Multi-Valued Quantum AND Gate are used in the Multi-Valued Quantum Shift Register. Shift Register generates four qubit outputs using these. The D flip-flop has a single qubit input and is developed using Multi-Valued Quantum NAND and Multi-Valued Quantum SR latch operations. Figure 3.12 depicts the circuit architecture of multi-valued Quantum-DNA Shift Register.

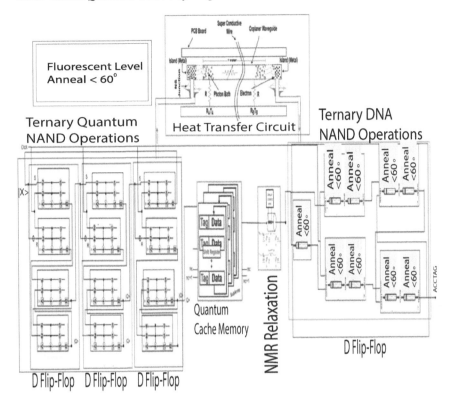

FIGURE 3.12
Circuit Architecture of Multi-Valued Quantum-DNA Shift Register

The clock qubit input is required for the Multi-Valued Quantum D flip-flop. It can be seen from the diagram that the circuit has one qubit input. One line of this qubit input will be directed into a Multi-Valued Quantum NAND operation termed S input in Circuit. Multi-Valued Quantum NAND is performed here using S qubit input and Clock qubit input.

In this circuit, input is qubit and output is a molecular sequence. Input |X> and clock input |clk> are shared into two Multi-Valued Quantum NAND operations. Two Multi-Valued Quantum operations are constructed parallelly. These Multi-Valued Quantum NAND operations output line and final output of Multi-Valued Quantum-DNA Shift Register are constructed into another two Multi-Valued Quantum operations parallelly. But the final two outputs are molecular sequences so these two output lines are connected to the NMR Relaxation process and then after that process, these two lines came into use as an input line Multi-Valued Quantum NAND operation circuit. These two Multi-Valued Quantum NAND operations circuit output lines are connected into the Multi-Valued Quantum cache memory. All of these Multi-Valued Quantum NAND Operations are the part of Multi-Valued Quantum D flip flop which is used in this system. This is the first portion of the circuit and is fully designed by using the principle of Multi-Valued Quantum computing. Multi-Valued Quantum cache memory is made by using some Multi-Valued Quantum shift register and in this proposed circuit it saves data in the Multi-Valued Quantum array. When the time arrives Multi-Valued Quantum cache memory supplies the data into the NMR Relaxation process. Multi-Valued Quantum cache memory supply line fully connected to NMR Relaxation process as well as here EMR emit is fully prohibited. Then the NMR Relaxation process output line is connected to the Multi-Valued DNA NOT gate. After going through the Multi-Valued DNA NOT operation the output is taken as input in the Multi-Valued DNA NAND operation. In these Multi-Valued DNA NAND operations 1st clock input is taken as |clk> input. These inputs are shared into two Multi-Valued DNA NAND gates. Then the outputs of these operations are connected to another set of Multi-Valued DNA NAND gates. Multi-Valued DNA NAND gates another input come from the output which is ACCTAG molecular sequence. Hence this Multi-Valued DNA NOR gates' final output line is the final output of Multi-Valued Quantum-DNA Shift Register.

Multi-Valued Quantum Circuit produces more heat and the Multi-Valued DNA circuit needs to process the input. For that reason, it would be best to transfer the heat into Multi-Valued DNA circuits. Multi-Valued Quantum portion Heat transfers using the heat transfer circuit into Multi-Valued DNA circuit's portion. In the heat transfer circuit, two junctions are connected to Multi-Valued Quantum Circuits and Multi-Valued DNA circuits. Maximum one meter can transfer heat to this circuit. This heat transfer circuit has superconductive ware, photon batch, PCB board to make the architecture. This circuit mainly transfers the heat using the photon bath. With this circuit proposed Multi-Valued Quantum-DNA Shift Register circuit is constructed fully.

3.7.3   Working Principle of Multi-Valued Quantum-DNA Shift Register

Multi-Valued Quantum-DNA shift registers are a kind of registers where both qubit data loading, as well as data retrieval to/from the Multi-Valued Quantum-DNA shift register, occurs in serial mode sometimes. This research synchronous Multi-Valued Quantum SISO shift register sensitive to the positive edge of the clock pulse. Here the data word which is to be stored is fed bit-by-bit at the qubit input of the first Multi-Valued Quantum flip-flop. Further, it is seen that the qubit inputs of all other flip-flops are driven by the outputs of the preceding ones, for example, the input of Multi-Valued Quantum D Flip-flop number- 2 is driven by the output of Multi-Valued Quantum D flip-flop number-1. At last, the data stored within the Multi-Valued Quantum shift register is obtained at the output pin of the nth Multi-Valued Quantum D flip-flop in serial fashion.

Initially, all the Multi-Valued Quantum flip-flops in the Multi-Valued Quantum register are cleared by applying high on their clear pins. Next, the input data word is fed serially to Multi-Valued Quantum D Flip-flop number-1.

This causes the qubit appearing at the first pin to be stored into Multi-Valued Quantum D flip-flop number-1 as soon as the first leading edge of the clock appears. Further at the second clock tick, B1 gets stored into Multi-Valued Quantum D flip-flop number-2 while a new bit enters into Multi-Valued Quantum flip-flop number-2.

TABLE 3.6
Truth Table of Multi-Valued Quantum-DNA Shift Register

|Clk>	|x>	Q	$\overline{Q}$	Description
↓ >> |0>	X	Q	Q	Memoryno change
↑ >> |1>	|0>	TGGATC	ACCTAG	Reset Q >> 0
↑ >> |1>	|1>	ACCTAG	TGGATC	Set Q >>

This kind of shift in data qubits continues for every rising edge of the clock pulse. This indicates that for every single clock pulse, the data within the Multi-Valued Quantum register moves towards the right by a single bit. Following the qubit data transmission, as explained, one can note that the first qubit of an input word appears at the output of nth flip-flop for the nth clock tick. On applying further clock cycles, one gets the next successive qubits of the qubit input data word as the serial output. Table 3.6 shows the truth table of multi-valued Quantum-DNA Shift Register.

After processing the inputs in a Multi-Valued Quantum D flip-flop, it creates one output. The output is used as a clock input for the next Multi-Valued Quantum D flip-flop. Thus, a Multi-Valued Quantum Shift Register creates one final qubit output.

3.7.4 Example

In the Multi-Valued Quantum-DNA shift register, there is only one qubit input needed to perform the required operation. Quantum shift register will enable if and only if when the clock input is enabled. This research proposed quantum shift register shift one qubit data into the right position.

Assume |1>|1>|1>|0> data is going to perform the Quantum-DNA shift register operation. First of all 0 will perform the quantum D flip-flop number-1 operation and produce the output is |0>. Then this |0> will shift into the right side as well as every qubit will shift right side once.

Now |1> will be entered into the quantum shift register and perform the midst quantum D flip-flop operation and then again like the previous one the data will be shifted once on the right side.

Then like these two step by step |1>, |1> also entered into the quantum shift register in the relevant period and perform the midst quantum D flip-flop operation. Then again for each data will be shifted once.

Every time qubit data shift once on the right side. In quantum shift register operational circuit will perform n+1 operation then final data will be like the table below (Table 3.7).

Here in n+1 operation time, no data is available that's why |0> is appeared in the truth table and according to the quantum Shift register principle one data qubit shifted once on the right side. After n+1 operation, final data is |0>|1>|1>|1>. These data purely proved that the multi-valued quantum-DNA shift register gives the correct required output by maintaining the quantum and DNA computing principle.

TABLE 3.7

Data Shifting Process in Multi-Valued Quantum-DNA Shift Register

|clk>	|x>	Q	Q1	Q2	Q3
|0>	|0>	TGGTAC	TGGTAC	TGGTAC	TGGTAC
|1>	|0>	TGGTAC	TGGTAC	TGGTAC	TGGTAC
|1>	|1>	ACCTAG	TGGTAC	TGGTAC	TGGTAC
|1>	|1>	ACCTAG	ACCTAG	TGGTAC	TGGTAC
|1>	|1>	ACCTAG	ACCTAG	ACCTAG	TGGTAC
|1>		ACCTAG	ACCTAG	ACCTAG	

Multi-Valued Quantum-DNA shift register can produce some garbage value but in this research, this is not the part of the study.

3.8   Multi-Valued Quantum-DNA Ripple Counter

A Multi-Valued Quantum counter is basically used to count the number of clock pulses applied to a Multi-Valued Quantum flip-flop. It can also be used for Multi-Valued Quantum Frequency divider, Multi-Valued Quantum time measurement, Multi-Valued Quantum frequency measurement, Multi-Valued Quantum distance measurement and also for generating square waveforms. In this, the Multi-Valued Quantum flip-flops are Multi-Valued Quantum asynchronous counters and are supplied with different clock signals, there may be a delay in producing output. Also, a few numbers of Multi-Valued Quantum logic gates are needed to design asynchronous counters. So they are elementary in design and also are less expensive.

An n-bit ripple counter can count up to 2n states. It is also known as MOD n counter. It is known as a ripple counter because of the way the clock pulse ripples its way through the flip-flops. It is an asynchronous counter. Different Ripple Counters are used with a different clock pulse. All the flip-flops are used in toggle mode. Only one flip-flop is applied with an external clock pulse and another flip-flop clock is obtained from the output of the previous flip-flop. The flip-flop applied with an external clock pulse acts as LSB (Least Significant Bit) in the counting sequence. A counter may be an up counter that counts upwards or can be a down counter that counts downwards or can do both i.e. count up as well as count downwards depending on the input control. The sequence of counting usually gets repeated after a limit

Multi-Valued Quantum-DNA Ripple Counter is made out of four JK flip flops. Using these Multi-Valued Quantum JK Flip flops, the Multi-Valued Quantum-DNA Ripple Counter creates four qubit outputs. Here in JK flip flop, J and K are not shortened abbreviated letters of other words, such as "S" for Set and "R" for Reset, but are autonomous letters chosen by its inventor Jack Kilby to distinguish the flip-flop design from other types. Though Jack Kilby invented the digital electronics JK flip

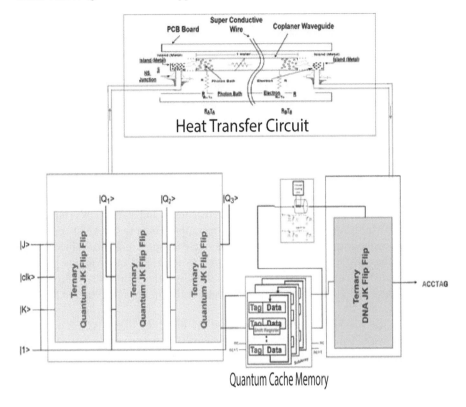

FIGURE 3.13

Block Diagram of Multi-Valued Quantum-DNA Ripple Counter

flop. Multi-Valued Quantum-DNA Ripple Counter is an asynchronous counter. It is created using Multi-Valued Quantum JK flip flops and these flip flops are only controlled by clock pulse input.

Multi-Valued Quantum-DNA Ripple Counter produces much heat to produce the molecule's superposition state and also produces some garbage value.

3.8.1 General Organization of Multi-Valued Quantum-DNA Ripple Counter

Multi-Valued Quantum-DNA Ripple Counter uses three Multi-Valued Quantum JK flip flop to operate two qubit inputs. Multi-Valued Quantum JK flip flop has the two-qubit input named as |J> and |K>. Figure 3.13 shows the block diagram of multi-valued Quantum-DNA Ripple Counter.

Multi-Valued Quantum JK flip-flop has the qubit input |J> and |k>. This Multi-Valued Quantum JK flip flop consists of many Multi-Valued Quantum NAND operations. At first basic Multi-Valued Quantum NAND operation performs a couple of

then this operation output is entered into the SR flip flop as well as got the output |Q>
and $|\overline{Q}>$. First of all |J> and |clk> input performs the Multi-Valued Quantum NAND
operation. The output of the Multi-Valued Quantum NAND operation and output of
the Multi-Valued Quantum JK flip flop $|\overline{Q}>$ performs another Multi-Valued Quan-
tum NAND operation and produces the output named as |S>. |clk> input is shared, so
|K> and |clk> input also performs the Multi-Valued Quantum NAND operation. This
Multi-Valued Quantum NAND operations output and |Q> output of the Multi-Valued
Quantum JK flip flop performs another Multi-Valued Quantum NAND operation as
well as produces the output named as |R>.

These |S> and |R> entered into the Multi-Valued Quantum SR flip flop and pro-
duce two outputs |Q> and $|\overline{Q}>$. In Multi-Valued Quantum SR flip flop, it has four
Multi-Valued Quantum NAND operations. |S> input and |clk> input performs Multi-
Valued Quantum NAND operation as well as |R> input and shared |clk> input also
performs the Multi-Valued Quantum NAND operation. These two NAND operations
output entered the Multi-Valued Quantum SR latch as input. In Multi-Valued Quan-
tum SR latches two |Q> and one input as well as $|\overline{Q}>$ and other inputs perform
the Multi-Valued Quantum NAND operation. Finally, after all of this Multi-Valued
Quantum operation, got the Multi-Valued Quantum JK flip flops final output |Q> and
$|\overline{Q}>$.

After processing the qubit inputs in three JK flip flops like this, the output is
stored in a Multi-Valued Quantum Cache Memory, which is made out of Multi-
Valued Quantum Shift Register. Then this Multi-Valued Quantum cache memory
serves the qubit into the Multi-Valued DNA computing circuit's portion. These qubits
need to be molecular sequence first of all so qubits are performed here NMR relaxation process. In the NMR relaxation process, emitting EMR is strictly prohibited.
After the NMR relaxation process qubits are relaxing their state and are converted
into the molecular sequence.

In Multi-Valued DNA computing molecular sequence needs to go through many
processes during the operation. In Multi-Valued DNA computing, melting and annealing are very important and these steps require a vast amount of heat. That's
why in Multi-Valued Quantum-DNA Ripple Counter Multi-Valued Quantum portion transfers a small amount of heat produced from the Multi-Valued Quantum circuits into Multi-Valued DNA circuits. Though this amount of heat is not enough,
this heat or temperature will help to perform Multi-Valued DNA computation. In
Multi-Valued DNA computing Multi-Valued DNA NOR SR latch performs where
two Multi-Valued DNA NOR gates perform in parallel way and input comes from
Multi-Valued Quantum circuit portion. After that operation, finally found the output
molecular sequence from Multi-Valued Quantum-DNA Ripple Counter.

3.8.2   Circuit Architecture of Multi-Valued Quantum-DNA Ripple Counter

Multi-Valued Quantum-DNA Ripple Counter uses three Multi-Valued Quantum JK
flip flops and one Multi-Valued DNA JK flip flop to create three qubit outputs. Multi-
Valued Quantum JK flip flop has a two-qubit input and one clock shared input. Multi-
Valued Quantum JK flip flops also depend on the clock. If the clock is enabled then

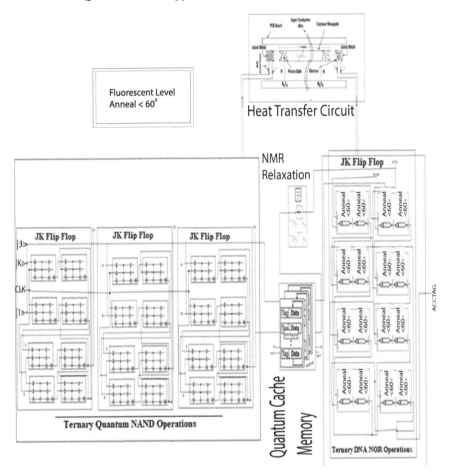

FIGURE 3.14
Circuit Diagram of Multi-Valued Quantum-DNA Ripple Counter

the circuit will enable, otherwise not. Figure 3.14 depicts the circuit diagram of multi-valued Quantum-DNA Ripple Counter.

In Multi-Valued Quantum-DNA Ripple Counter there's a one clock input and one logic input which is shared in both |J> and |K> input port. Input |J> and input |K> both the value will perform the Multi-Valued Quantum NAND operation with the shared |clk> input differently. |J> and |clk> input performs the Multi-Valued Quantum NAND operation. NAND operation made by using Multi-Valued Quantum basic gates. Basic gates in mainly Multi-Valued Quantum computing are V, V+, and CNOT gates. For error correction here used an ancillary bit. In Quant NAND operation the value of an ancillary bit in |1>. After |J> and |clk> perform the Multi-Valued Quantum NAND operation produced an output qubit and this output qubit and the output of the flip flop |Q̄> perform the Multi-Valued Quantum NAND operation and produce the output |S>. Like in the same procedure |K>, |clk> and |Q> input produces the |R> output. These operation circuits are mainly Multi-Valued Quantum NAND operations. This |S> and |R> input is performed the Multi-Valued Quantum SR flip-flop operation. In Multi-Valued Quantum SR flip flop operational circuit architecture also made by the proposed basic component of Multi-Valued Quantum computing is Multi-Valued Quantum NAND operation.

After processing the inputs in a Multi-Valued Quantum JK flip flop, the output of the flip flop is going to be stored as an output of the Multi-Valued Quantum-DNA Ripple Counter. Multi-Valued Quantum-DNA Ripple Counter intermediate architectures four Multi-Valued Quantum JK flip-flops connected in serial connection using the logical qubit input. Every clock input as |clk> enters in every qubit and from 2nd Multi-Valued Quantum JK flip-flop |clk> input is previous Multi-Valued Quantum Jk flip-flops first output |Q> . With the same architecture Multi-Valued Quantum-DNA Ripple Counter can be performed as a up counter or as a down counter but clock pulse as |clk> need to sometimes have a positive edge triggered and sometimes a negative edge triggered.

The JK flip flop's circuit output lines are connected into the Multi-Valued Quantum cache memory. This is the first portion of the circuit and is fully designed by using the principle of Multi-Valued Quantum computing. Multi-Valued Quantum cache memory is made by using some Multi-Valued Quantum shift register and in this proposed circuit it saves data in the Multi-Valued Quantum array. When the time arrives Multi-Valued Quantum cache memory supplies the data into the NMR Relaxation process. Multi-Valued Quantum cache memory supply line fully connected to NMR Relaxation process as well as here EMR emit is fully prohibited. Then the NMR Relaxation process output line is connected to the Multi-Valued DNA NAND gate. Two outputs of the NMR relaxation process are connected to two different Multi-Valued DNA NOR gates. Multi-Valued DNA NOR gates another input come from the output which is Q and $\overline{Q}$ molecular sequences. Hence this Multi-Valued DNA NAND gates' final output line is the final output of Multi-Valued Quantum-DNA Ripple Counter.

Multi-Valued Quantum Circuit produces more heat and the Multi-Valued DNA circuit needs to process the input. For that reason, it would be best to transfer the heat into Multi-Valued DNA circuits. Multi-Valued Quantum portion Heat transfers using the heat transfer circuit into Multi-Valued DNA circuit's portion. In the heat transfer circuit, two junctions are connected to Multi-Valued Quantum Circuits and Multi-Valued DNA circuits. Maximum one meter can transfer heat to this circuit. This heat transfer circuit has superconductive ware, photon batch, PCB board to make the architecture. This circuit mainly transfers the heat using the photon bath. With this circuit proposed Multi-Valued Quantum-DNA Ripple Counter circuit is constructed fully.

3.8.3   Working Principle of Multi-Valued Quantum-DNA Ripple Counter

In the Multi-Valued Quantum-DNA Ripple Counter there are four Multi-Valued Quantum JK flip-flop operational circuits. These Multi-Valued Quantum JK flip-flop operational circuits are connected in serial connection. In this Multi-Valued

Quantum-DNA Ripple Counter there is one clock input and a logic input which shared into the port |J> and |K> of Multi-Valued Quantum JK flip-flop operational circuit. The inputs |J> and |K> of a Multi-Valued Quantum Jk flip-flop conducts two Multi-Valued Quantum processes in parallel. The Multi-Valued Quantum NAND operation is performed using |J> and shared input |clk>. The Multi-Valued Quantum NAND operations are then completed, and one of the Multi-Valued Quantum JK flip flop's outputs executes the Multi-Valued Quantum NAND operation, producing |S>. The |K> and |clk > inputs are performed first in the Multi-Valued Quantum NAND operation. The output of the Multi-Valued Quantum JK flip-flop |Q>, as well as the result of the Multi-Valued Quantum first NAND operation, are then used to perform another Multi-Valued Quantum NAND operation, yielding |R>. The steps for creating |S> and |R> are carried out simultaneously. It is known that one of the distinctive properties of Multi-Valued Quantum operations is that they may do several operations at the same time, and this is exactly what is happening. The Multi-Valued Quantum SR flip flop operation is then conducted on these |S> and |R> inputs. The Multi-Valued Quantum NAND operation, which is employed here, is also used to make Multi-Valued Quantum SR flip flops. Two outputs are discovered after conducting the Multi-Valued Quantum SR flip-flop operation. The opposite of one output is the opposite of the other.

Here if the |clk> is |0>, then the Multi-Valued Quantum JK flip-flop will not be triggered, but if |clk> is |1> the Multi-Valued Quantum JK flip-flop will be triggered and it will toggle the output. First of all Multi-Valued Quantum JK flip-flop operational circuit getting |clk> value is |1> and toggles the output value from the previous state value. Then the output of the initial Multi-Valued Quantum JK flip-flop will be |clk> input of next Multi-Valued Quantum JK flip-flop. If the |clk> value is |1>, then the output value will be toggled, otherwise the output will be the previous state output. Maintaining the same procedure every Multi-Valued Quantum JK flip-flop operated in the Multi-Valued Quantum-DNA Ripple Counter. Multi-Valued Quantum JK flip-flop is toggled very much, that's why in the Multi-Valued Quantum-DNA Ripple Counter Multi-Valued Quantum JK flip-flop operational circuit here is used as a basic component. Table 3.8 depicts the truth table of multi-valued Quantum-DNA Ripple Counter

The JK flip flop's circuit output lines are connected into the Multi-Valued Quantum cache memory. This is the first portion of the circuit and is fully designed by using the principle of Multi-Valued Quantum computing. Multi-Valued Quantum cache memory is made by using some Multi-Valued Quantum shift register and in this proposed circuit it saves data in the Multi-Valued Quantum array. When the time arrives

TABLE 3.8

Truth Table of Multi-Valued Quantum-DNA Ripple Counter

clk	$|Q_0 >$	$|Q_1 >$	$|Q_2 >$	$|Q_3 >$
|0>	|0>	|0>	|0>	|0>
|1>	|1>	|1>	|1>	|1>
|1>	|0>	|0>	|0>	|0>

Multi-Valued Quantum cache memory supplies the data into the NMR Relaxation process. Multi-Valued Quantum cache memory supply line fully connected to NMR Relaxation process as well as here EMR emit is fully prohibited. Then the NMR Relaxation process output line is connected to the Multi-Valued DNA NAND gate. Two outputs of the NMR relaxation process are connected to two different Multi-Valued DNA NOR gates. Multi-Valued DNA NOR gates another input come from the output which is Q and $\overline{Q}$ molecular sequences. Hence this Multi-Valued DNA NAND gates' final output line is the final output of Multi-Valued Quantum-DNA Ripple Counter.

Multi-Valued Quantum Cache memory works here as the intermediate process. Multi-Valued Quantum circuit produces much heat and here two Multi-Valued Quantum AND operations perform so it produces much heat. But Multi-Valued Quantum circuit needs to be close to zero Kelvin temperature to perform the operation. So from the Multi-Valued Quantum circuit, it is needed to reduce the temperature to maintain the superposition state of the qubit. Hence in Multi-Valued DNA circuit operation, it needs much heat in several steps. Mainly in melting and annealing require much heat. This topic is briefly described in the first chapter named Multi-Valued Quantum-DNA circuit operation. For that reason, it is needed to transfer the excessive heat from the Multi-Valued Quantum circuit portion to the Multi-Valued DNA circuit portion using a heat transfer circuit. This heat transfer circuit using the junction captures the heat from the Multi-Valued Quantum circuit and using photon bath heat flows through the circuit and gives this into the Multi-Valued DNA circuit. This circuit can transfer heat maximum in the one-meter distance and in Multi-Valued Quantum-DNA flip-flop distance is less than one meter between Multi-Valued Quantum and Multi-Valued DNA circuit. This heat transfer circuit cannot transfer full excessive heat produced from the Multi-Valued Quantum circuit and this heat is not enough to perform the Multi-Valued DNA circuit operation. But this heat transfer can optimize the cost of heat based on the needs. After all of this operation and architecture fully then Multi-Valued Quantum-DNA Ripple Counter can produce two output molecular sequences.

3.8.4   Example

To check the Multi-Valued Quantum-DNA Ripple Counter proposed circuit assumed a clock signal is |1> and the logical qubit is high. So, if initially clock signal |1> is delivered then for the first Multi-Valued Quantum JK flip-flop operational circuit, the output will be |1>.

According to the work principle of Multi-Valued Quantum-DNA Ripple Counter if clock input value is |1> then the previous state value will be toggled. Then now from the first Multi-Valued Quantum JK flip-flop got one output which is |1>. This |1> will be the clock input for the second Multi-Valued Quantum JK flip-flop circuit according to the architecture of the Multi-Valued Quantum-DNA Ripple Counter.

As like the previous Multi-Valued Quantum JK flip-flop for third and fourth Multi-Valued Quantum flip-flops the principle is the same. For both of the flip-flops the clock value will be the previous Multi-Valued Quantum JK flip-flops output. Now, third and fourth will be DNA sequences.

Hence from the truth table, it is seen that the proposed Multi-Valued Quantum-DNA Ripple Counter is correct theoretically. Multi-Valued Quantum JK flip-flop used here for toggling. This Multi-Valued Quantum-DNA Ripple Counter produces much heat when the molecule's normal state becomes a superposition state. Multi-Valued Quantum-DNA Ripple Counters full operation needs to be an environment where other particles are totally prohibited to maintain the coherence state.

3.9 Multi-Valued Quantum-DNA Synchronous Counter

A Multi-Valued Quantum counter is a Multi-Valued Quantum device which can count any particular event on the basis of how many times the particular event(s) has occurred. In a Multi-Valued Quantum logic system or computers, this Multi-Valued Quantum counter can count and store the number of times any particular event or process has occurred, depending on a Multi-Valued Quantum clock signal. Most common type of Multi-Valued Quantum counter is a sequential Multi-Valued Quantum logic circuit with a single clock input and multiple qubit outputs. The qubit outputs represent two valued decimal numbers. Each clock pulse either increases the number or decreases the number.

Multi-Valued Quantum-DNA Synchronous Circuit generally refers to something which is coordinated with others based on time. Multi-Valued Quantum-DNA Synchronous Signals occur at same clock rate and all the clocks follow the same reference clock. Multi-Valued Quantum asynchronous Counter have shown that the qubit output of that Multi-Valued Quantum counter is directly connected to the input of next subsequent counter and making a chain system, and due to this chain system propagation delay appears during counting stage and create counting delays. In a Multi-Valued Quantum-DNA Synchronous Counter, the clock qubit input across all the Multi-Valued Quantum flip-flops use the same source and create the same clock signal at the same time. So, a Multi-Valued Quantum counter which is using the same clock signal from the same source at the same time is called a Multi-Valued Quantum-DNA Synchronous Counter.

Multi-Valued Quantum-DNA Synchronous Counter is made out of four JK flip flops and two Multi-Valued Quantum AND Gates. Using these Multi-Valued Quantum JK Flip flops and Multi-Valued Quantum AND Gates, the Multi-Valued Quantum-DNA Synchronous Counter creates four qubit outputs. A Multi-Valued Quantum-DNA Synchronous Counter produces much heat and this circuit operation needs to happen in the required environment of Multi-Valued Quantum computing.

3.9.1 General Organization of Multi-Valued Quantum-DNA Synchronous Counter

Multi-Valued Quantum-DNA Synchronous Counter uses three Multi-Valued Quantum JK flip flops and one Multi-Valued DNA JK flip flop to create molecular sequence outputs from qubit outputs. Multi-Valued Quantum JK flip flop has the two-

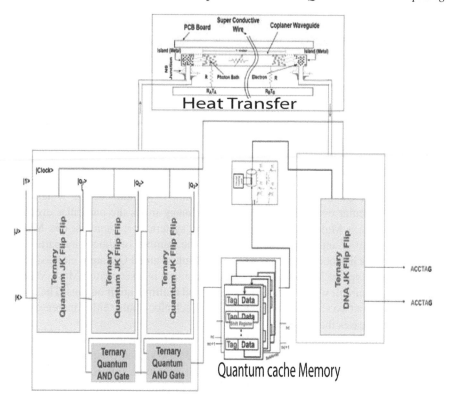

FIGURE 3.15
Block Diagram of Multi-Valued Quantum-DNA Synchronous Counter

qubit input named as |J> and |K>. Figure 3.15 depicts the block diagram of multi-valued Quantum-DNA Synchronous Counter.

Multi-Valued Quantum JK flip-flop has the qubit input |J> and |k>. This Multi-Valued Quantum JK flip flop consists of many Multi-Valued Quantum NAND operations. At first basic Multi-Valued Quantum NAND operation performs a couple of then this operation output is entered into the SR flip flop as well as got the output |Q> and $|\overline{Q}>$. First of all |J> and |clk> input performs the Multi-Valued Quantum NAND operation. The output of the Multi-Valued Quantum NAND operation and output of the Multi-Valued Quantum JK flip flop $|\overline{Q}>$ performs another Multi-Valued Quantum NAND operation and produces the output named as |S>. |clk> input is shared so |K> and |clk> input also performs the Multi-Valued Quantum NAND operation. This Multi-Valued Quantum NAND operations output and |Q> output of the Multi-Valued Quantum JK flip flop performs another Multi-Valued Quantum NAND operation as well as produces the output named as |R>.

These |S> and |R> entered into the Multi-Valued Quantum SR flip flop and produce two outputs |Q> and $|\overline{Q}>$. In Multi-Valued Quantum SR flip flop, it has four Multi-Valued Quantum NAND operations. |S> input and |clk> input performs

Multi-Valued Quantum NAND operation as well as |R> input and shared |clk> input also performs the Multi-Valued Quantum NAND operation. These two NAND operations output entered the Multi-Valued Quantum SR latch as input. In Multi-Valued Quantum SR latches two |Q> and one input as well as $|\overline{Q}>$ and other inputs perform the Multi-Valued Quantum NAND operation. Finally, after all of these Multi-Valued Quantum operations, the Multi-Valued Quantum JK flip flop's final outputs |Q> and $|\overline{Q}>$ are obtained.

After processing the inputs in a Multi-Valued Quantum JK flip flop, the output of the flip flop is going to be stored as an output of the Multi-Valued Quantum-DNA Synchronous Counter. Thus, Multi-Valued Quantum-DNA Synchronous Counter creates four qubit outputs using four Multi-Valued Quantum JK flip flops. The clock inputs for all of the four Multi-Valued Quantum Jk flip flops come from the same source. For this, all of the flip flops work synchronously. One output of each of the second and third flip flops go through Multi-Valued Quantum AND Gates.

Then the output of the Multi-Valued Quantum portion is stored in the Multi-Valued Quantum Cache Memory. Whenever needed the output then goes through the "NMR Relaxation" process in which the emitting of the EMR is strictly prohibited. Thus the qubits of the superstrate become the molecular sequence and are used as inputs for the Multi-Valued DNA JK flip flop.

In Multi-Valued DNA computing molecular sequence needs to go through many processes during the operation. In Multi-Valued DNA computing, melting and annealing are very important and these steps require a vast amount of heat. That's why in Multi-Valued Quantum-DNA Synchronous Counter Multi-Valued Quantum portion transfers a small amount of heat produced from the Multi-Valued Quantum circuits into Multi-Valued DNA circuits. Though this amount of heat is not enough, this heat or temperature will help to perform multi-valued DNA computation. In Multi-Valued DNA computing Multi-Valued DNA NOR SR latch performs where two Multi-Valued DNA NOR gates perform parallelly and input comes from Multi-Valued Quantum circuit portion. After that operation, finally found the output molecular sequence from Multi-Valued Quantum-DNA Synchronous Counter.

3.9.2 Circuit Architecture of Multi-Valued Quantum-DNA Synchronous Counter

Multi-Valued Quantum-DNA Synchronous Counter uses three Multi-Valued Quantum JK flip flops and one Multi-Valued DNA JK flip flops to create molecular sequence outputs. Multi-Valued Quantum JK flip flop has a two-qubit input and one clock shared input. Multi-Valued Quantum JK flip flops also depend on the clock. If the clock is enabled then the circuit will enable, otherwise not. Figure 3.16 shows the diagram of multi-valued Quantum-DNA Synchronous Counter.

Input |J> and input |K> both the value will perform the Multi-Valued Quantum NAND operation with the shared |clk> input differently. |J> and |clk> input performs the Multi-Valued Quantum NAND operation. NAND operation made by using Multi-Valued Quantum basic gates. Basic gates in mainly Multi-Valued Quantum computing are v, V+, and CNOT gates. For error correction here used an ancillary bit. In

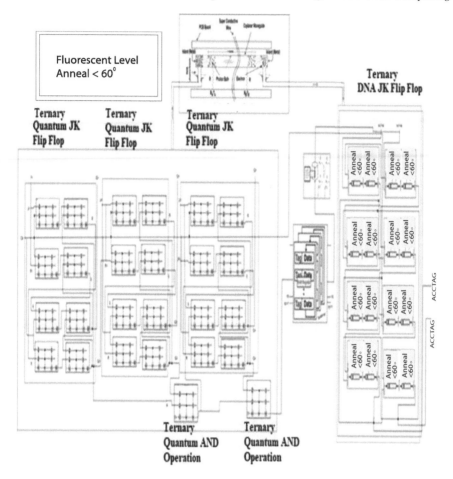

FIGURE 3.16
Diagram of Multi-Valued Quantum-DNA Synchronous Counter

Quant NAND operation the value of an ancillary bit in |1>. After |J> and |clk> per-
form the Multi-Valued Quantum NAND operation, an output qubit is obtained and
this output qubit and the output of the flip flop $|\overline{Q}>$ perform the Multi-Valued Quan-
tum NAND operation and produce the output |S>. Like in the same procedure |K>,
|clk> and |Q> input produces the |R> output. These operational circuits are mainly
Multi-Valued Quantum NAND operations. This |S> and |R> input is performed the
Multi-Valued Quantum SR flip-flop operation. In Multi-Valued Quantum SR flip flop
operation circuit architecture also made by the proposed basic component of Multi-
Valued Quantum computing is Multi-Valued Quantum NAND operation.

After processing the inputs in a Multi-Valued Quantum JK flip flop, the output
of the flip flop is going to be stored as an output of the Multi-Valued Quantum-
DNA Synchronous Counter. Multi-Valued Quantum-DNA Synchronous Counter

intermediate architectures three Multi-Valued Quantum JK flip-flops connected in serial connection using the logical qubit input. Every clock input as |clk> enters in every qubit and from 2nd Multi-Valued Quantum JK flip-flop |clk> input is previous Multi-Valued Quantum JK flip-flops first output |Q>. With the same architecture Multi-Valued Quantum-DNA Synchronous Counter can be performed as a up counter or as a down counter but clock pulse as |clk> need to sometimes have a positive edge triggered and sometimes a negative edge triggered.

The JK flip flop's circuit output lines are connected into the Multi-Valued Quantum cache memory. This is the first portion of the circuit and is fully designed by using the principle of Multi-Valued Quantum computing. Multi-Valued Quantum cache memory is made by using some Multi-Valued Quantum shift register and in this proposed circuit it saves data in the Multi-Valued Quantum array. When the time arrives Multi-Valued Quantum cache memory supplies the data into the NMR Relaxation process. Multi-Valued Quantum cache memory supply line fully connected to NMR Relaxation process as well as here EMR emit is fully prohibited. Then the NMR Relaxation process output line is connected to the Multi-Valued DNA NAND gate. Two outputs of the NMR relaxation process are connected to two different Multi-Valued DNA NAND gates. Multi-Valued DNA NAND gates another input come from the output which is Q and $\overline{Q}$ molecular sequences. Hence this Multi-Valued DNA NAND gates' final output line is the final output of Multi-Valued Quantum-DNA Synchronous Counter.

Multi-Valued Quantum Circuit produces more heat and the Multi-Valued DNA circuit needs to process the input. For that reason, it would be best to transfer the heat into Multi-Valued DNA circuits. Multi-Valued Quantum portion Heat transfers using the heat transfer circuit into Multi-Valued DNA circuit's portion. In the heat transfer circuit, two junctions are connected to Multi-Valued Quantum Circuits and Multi-Valued DNA circuits. Maximum one meter can transfer heat to this circuit. This heat transfer circuit has superconductive ware, photon batch, PCB board to make the architecture. This circuit mainly transfers the heat using the photon bath. With this circuit proposed Multi-Valued Quantum-DNA Synchronous Counter circuit is constructed fully.

3.9.3 Working Principle of Multi-Valued Quantum-DNA Synchronous Counter

Multi-Valued Quantum-DNA Synchronous counter has one logical qubit input which is shared and one clock input. Multi-Valued Quantum-DNA Synchronous counter is constructed by the basic component as a Multi-Valued Quantum JK flip-flop operational circuit. Multi-Valued Quantum JK flip flop is the two-qubit circuit. Multi-Valued Quantum JK flip flop is mostly used flip-flop in Multi-Valued Quantum computing. Multi-Valued Quantum Jk flip-flop's input |J> and |K> perform two Multi-Valued Quantum operations in parallel way. |J> and shared input |clk> performs the Multi-Valued Quantum NAND operation. Then this Multi-Valued Quantum NAND operations result and one output of the Multi-Valued Quantum JK flip flop performs the Multi-Valued Quantum NAND operation and produce |S>. Like Multi-Valued Quantum NAND operation performs |K> and |clk > input first. Then the output of

Multi-Valued Quantum JK flip-flop |Q> and the result of Multi-Valued Quantum first NAND operation executes again another Multi-Valued Quantum NAND operation and produce |R>. The procedure of producing |S> and |R> are executed in parallel. It is known that the Multi-Valued Quantum operations have one of the iconic characteristics that perform multiple operations parallelly and here it is happening. These |S> and |R> inputs are performed then Multi-Valued Quantum SR flip flop operation. Multi-Valued Quantum SR flip flop is also basically made by the Multi-Valued Quantum NAND operation that is used here. After performing the Multi-Valued Quantum SR flip-flop operation, two outputs are found. One output is the opposite of another. This Multi-Valued Quantum JK flip flop actually removes the problem of Multi-Valued Quantum SR flip flop. Multi-Valued Quantum SR flip-flop describes briefly in the 5th chapter.

Multi-Valued Quantum JK flip-flop is triggered when the value of clock is |1> . So, in Multi-Valued Quantum-DNA Synchronous the counter clock needs to be always high. According to the principle counters need to be toggled, that's why Multi-Valued Quantum JK flip-flop is perfect for Multi-Valued Quantum-DNA Synchronous counters. Multi-Valued Quantum JK flip-flop operational circuit triggered then output of Multi-Valued Quantum JK flip-flop will be shared input of second Multi-Valued Quantum JK flip-flop operational circuit. Then first Multi-Valued Quantum JK flip-flops output and second Multi-Valued Quantum JK flip-flops output enters in the Multi-Valued Quantum AND operation circuit and performs the Multi-Valued Quantum AND operation. The produced output from Multi-Valued Quantum AND operation performs the Multi-Valued Quantum JK flip-flop operation. Then again produced output from the first Multi-Valued Quantum AND operation and the third Multi-Valued Quantum JK flip-flop operation's output performs another Multi-Valued Quantum AND operation. Hence the previous Multi-Valued Quantum AND operations output is shared into the Multi-Valued Quantum JK flip-flops as two inputs as well as performs the Multi-Valued Quantum JK flip-flop operational circuit. Multi-Valued Quantum-DNA Synchronous counter circuit mainly performs as a finite counter. Table 3.9 shows the truth table of multi-valued Quantum-DNA Synchronous Counter.

Between the second and the third Multi-Valued Quantum JK flip flop, Multi-Valued Quantum AND Gate is used. The JK flip flop's circuit output lines are connected into the Multi-Valued Quantum cache memory. This is the first portion of the circuit and is fully designed by using the principle of Multi-Valued Quantum computing. Multi-Valued Quantum cache memory is made by using some Multi-Valued Quantum shift register and in this proposed circuit it saves data in the Multi-Valued Quantum array. When the time arrives Multi-Valued Quantum cache memory supplies the data into the NMR Relaxation process. Multi-Valued Quantum cache memory supply line fully connected to NMR Relaxation process as well as here EMR emit is fully prohibited. Then the NMR Relaxation process output line is connected to the Multi-Valued DNA NAND gate. Two outputs of the NMR relaxation process are connected to two different Multi-Valued DNA NOR gates. Multi-Valued DNA NOR gates another input come from the output which is Q and $\overline{Q}$ molecular sequences. Hence this Multi-Valued DNA NAND gates' final output line is the final output of Multi-Valued Quantum-DNA Ripple Counter.

TABLE 3.9

Truth Table of Multi-Valued Quantum-DNA Synchronous Counter

clk	$|Q_0>$	$|Q_1>$	$|Q_2>$	$|Q_3>$
$|0>$	TGGATC	TGGATC	TGGATC	TGGATC
$|1>$	TGGATC	TGGATC	TGGATC	ACCTAG
$|1>$	TGGATC	TGGATC	ACCTAG	TGGATC
$|1>$	TGGATC	TGGATC	ACCTAG	ACCTAG
$|1>$	TGGATC	ACCTAG	TGGATC	TGGATC
$|1>$	TGGATC	ACCTAG	TGGATC	ACCTAG
$|1>$	TGGATC	ACCTAG	ACCTAG	TGGATC
$|1>$	TGGATC	ACCTAG	ACCTAG	ACCTAG
$|1>$	ACCTAG	TGGATC	TGGATC	TGGATC
$|1>$	ACCTAG	TGGATC	TGGATC	ACCTAG
$|1>$	ACCTAG	TGGATC	ACCTAG	TGGATC
$|1>$	ACCTAG	TGGATC	ACCTAG	ACCTAG
$|1>$	ACCTAG	ACCTAG	TGGATC	TGGATC
$|1>$	ACCTAG	ACCTAG	TGGATC	ACCTAG
$|1>$	ACCTAG	ACCTAG	ACCTAG	TGGATC
$|1>$	ACCTAG	ACCTAG	ACCTAG	ACCTAG
$|1>$	TGGATC	TGGATC	TGGATC	TGGATC

Multi-Valued Quantum Cache memory works here as the intermediate process. Multi-Valued Quantum circuit produces much heat and here two Multi-Valued Quantum AND operations perform so it produces much heat. But Multi-Valued Quantum circuit needs to be close to zero Kelvin temperature to perform the operation. So from the Multi-Valued Quantum circuit, the temperature is needed to be reduced to maintain the superposition state of the qubit. Hence in Multi-Valued DNA circuit operation, it needs much heat in several steps. Mainly in melting and annealing require much heat. This topic is briefly described in the first chapter named Multi-Valued Quantum-DNA circuit operation. For that reason, it is needed to transfer the excessive heat from the Multi-Valued Quantum circuit portion to the Multi-Valued DNA circuit portion using a heat transfer circuit. This heat transfer circuit using the junction captures the heat from the Multi-Valued Quantum circuit and using photon bath heat flows through the circuit and gives this into the Multi-Valued DNA circuit. This circuit can transfer heat maximum in the one-meter distance and in Multi-Valued Quantum-DNA flip-flop distance is less than one meter between Multi-Valued Quantum and Multi-Valued DNA circuit. This heat transfer circuit cannot transfer full excessive heat produced from the Multi-Valued Quantum circuit and this heat is not enough to perform the Multi-Valued DNA circuit operation. But this heat transfer can optimize the cost of heat based on the needs. After all of this operation and architecture fully then Multi-Valued Quantum-DNA Ripple Counter can produce two output molecular sequences.

3.9.4   Example

As suppose the input qubit is |1> entered into the Multi-Valued Quantum-DNA Synchronous counter. Multi-Valued Quantum-DNA Synchronous counters basic component is Multi-Valued Quantum JK flip-flop operational circuit so for triggered the circuit clock value is also |1> . Now first Multi-Valued Quantum JK flip-flop both |J> input and |K> input is |1>. If two value is |1> the flip-flop triggered and value toggles. Now $|\overline{Q_0}>$ is |1> from previous state |0>.

But after the first Multi-Valued Quantum JK flip-flop operation $|\overline{Q_0}>$'s value became |0>. Then $|\overline{Q_0}>$ entered as an input in the second Multi-Valued Quantum JK flip-flop operation and this circuit won't be triggered. Then again $|\overline{Q_1}>$ and $|\overline{Q_0}>$ performs the Multi-Valued Quantum AND operation. As Multi-Valued Quantum AND operation depicts that if any input is |0> then the produced output will |0>. Then |0> will be input for the third Multi-Valued Quantum JK flip-flop operational circuit. As this circuit Multi-Valued Quantum JK flip-flop depicts that if two input is |0> then Multi-Valued Quantum Jk flip-flop won't be triggered. Again third Multi-Valued Quantum flip-flop's output $|\overline{Q_2}>$ and previous Multi-Valued Quantum AND operations produced output performs Multi-Valued Quantum AND operation again. As this propose circuit depicts previous Multi-Valued Quantum AND operation output is |0> so, this Multi-Valued Quantum AND operations output will be |0>. Hence |0> will be input of |J> and |k> for the fourth Multi-Valued Quantum JK flip-flop operation as well as finally it produces |0>, which means TGGATC.

3.10   Summary

As the name implies, the multi-valued quantum-DNA sequential circuits contain flip-flops which are all in sync with each other i.e. their clock qubit inputs are connected together and are triggered by the same external clock signal. This implies that all the quantum-DNA flip-flops update their values at the same time. As the name suggests, Quantum Synchronous counters perform "counting" such as time and electronic pulses (external source like infrared light). They are widely used in lots of other designs as well such as quantum-DNA computing processors, quantum-DNA calculators, real time clocks, etc. Alarm Clock, Set AC Timer, Set time in camera to take the picture, flashing light indicator in automobiles, car parking control, etc. can be constructed using a quantum-DNA synchronous counter.

Bibliography

[1] Isailovic, N., Patel, Y., Whitney, M., & Kubiatowicz, J. (2006, June). Interconnection networks for scalable quantum computers. In 33rd International Symposium on Computer Architecture (ISCA'06) (pp. 366-377). IEEE.

[2] Thaker, D. D., Metodi, T. S., Cross, A. W., Chuang, I. L., & Chong, F. T. (2006, June). Quantum memory hierarchies: Efficient designs to match available parallelism in quantum computing. In 33rd International Symposium on Computer Architecture (ISCA'06) (pp. 378-390). IEEE.

[3] Metodi, T. S., Thaker, D. D., Cross, A. W., Chong, F. T., & Chuang, I. L. (2005, November). A quantum logic array microarchitecture: Scalable quantum data movement and computation. In 38th Annual IEEE/ACM International Symposium on Microarchitecture (MICRO'05) (pp. 12-pp). IEEE.

[4] Adleman, L. M. (1994). Molecular computation of solutions to combinatorial problems. Science, 266(5187), 1021-1024.

[5] Smith, L. M., Sanders, J. Z., Kaiser, R. J., Hughes, P., Dodd, C., & Connell, C. R. (1986). C. Heiner, S. B. Kent. and L. E. Hood. Nature, 321(674), 7679.

[6] Zheng, X., Yang, J., Zhou, C., Zhang, C., Zhang, Q., & Wei, X. (2019). Allosteric DNAzyme-based DNA logic circuit: Operations and Dynamic Analysis. Nucleic Acids Research, 47(3), 1097-1109.

4

Multi-Valued Sequential Circuits in DNA-Quantum Computing

4.1 Introduction

Some people believe that molecular computers would solve problems that would cause existing machines to struggle because of biology's massive parallelism. For example, a small drop of water can contain trillions of DNA strands, and they perform biological operations because of parallelism. Adleman gave the idea of DNA computing when he noticed that DNA replication was notably similar to an early theoretical computer developed in the 1930s by Alan Turing. During reproduction, DNA polymerase slides along a single DNA strand and reads each base. After that, write its complement on the new strand. In a Turing machine, a mechanism moves along a pair of tapes where read instructions come from an "input tape" and write out the result from the output tape.

Multi-Valued DNA-Quantum computing is one of the most intriguing new scientific topics to have emerged in recent years: it is a combination of multi-valued DNA and multi-valued quantum computing, and it is one of the most intriguing new topic. The construction of Multi-Valued DNA-Quantum computers at present time would be incredibly challenging, even if it were conceivable. This research is aimed at the development of Multi-Valued DNA-Quantum processors, memory devices, and other devices that are all based on the technology under consideration. The Multi-Valued DNA-Quantum sequential circuit is one of those chapters. It is necessary to implement the functionality of this circuit by using the fundamental gates AND, OR, and NOT of the Multi-Valued DNA-Quantum system. The numerous features of general design, construction, working principle, and examples are discussed here. Before large-scale Multi-Valued DNA-Quantum computing can be implemented, a number of technical difficulties must be overcome. For computer architecture to function properly, it is important to conform to precise geometrical limitations in order to ensure that it runs smoothly. This book is an attempt to totally remove a stumbling block to multi-valued DNA-quantum computing, to put it another way.

DOI: 10.1201/9781003381921-4

4.2   Multi-Valued DNA-Quantum D Flip Flop

A Multi-Valued DNA-Quantum D flip flop is essentially a two-state timed flip flop. In one clock cycle, the molecular sequence inputs of a Multi-Valued DNA D-type flip flop are actuated with a delay. A delay flip flop is another term for the Multi-Valued DNA-Quantum D flip flop.

The indeterminate input condition of SET = "ACCTAG" and RESET = "ACC-TAG" is banned in the basic Multi-Valued DNA SR NAND Gate Bistable circuit, which is one of its fundamental drawbacks. This condition forces both Multi-Valued Quantum qubit outputs to logic "$|0>$," overriding the feedback latching action, and whichever input goes to logic "$|0>$" first loses control, while the other input, which is still at logic "$|1>$," controls the latch's final state. However, an inverter may be connected between the "SET" and "RESET" molecular sequence inputs to create a Multi-Valued DNA-Quantum Data Latch, Multi-Valued DNA-Quantum Delay flip flop, Multi-Valued DNA-Quantum D-type Bistable, Multi-Valued DNA-Quantum D-type Flip Flop, or simply a Multi-Valued DNA-Quantum D Flip Flop as it is most often known.

By far the most essential of all the Multi-Valued DNA-Quantum timed flip-flops is the Multi-Valued DNA-Quantum D Flip Flop. The S and R inputs become complements of each other when a Multi-Valued DNA inverter (Multi-Valued DNA NOT gate) is added between the Set and Reset inputs, ensuring that the two inputs S and R are never equal (ACCTAG or TGGATC) to each other at the same time, allowing us to control the toggle action of the flip-flop with just one $|D>$ (Data) input.

The Data input, labeled "$|D>$," is then utilized in place of the "Set" signal, and the inverter is used to create the complementary "Reset" input, resulting in a level-sensitive Multi-Valued DNA D-type flip-flop from a level-sensitive SR-latch, with S = $|D>$ and R = $|D>$.

The proposed Multi-Valued DNA-Quantum D flip-flop circuit has just one molecular sequence input, and the molecular sequence input must be in a coherence state in order to conduct the Multi-Valued DNA computational function. As a result, the circuit must exist in an environment that does not exist. If any particle emerges, the coherence state will be disrupted. The Multi-Valued DNA D flip-flop will generate heat, which must be removed quickly in order to cool down the circuit and stabilize the coherence state.

Hence in Multi-Valued DNA-Quantum flip-flops have two portions based on the principle of the Multi-Valued DNA-Quantum circuit. Multi-Valued DNA-Quantum D flip-flop operational circuit's first portion constructed using Multi-Valued DNA principle and second portion constructed using Multi-Valued Quantum computing principle. The Multi-Valued DNA circuit's portion produced molecular sequence and stored into Multi-Valued DNA cache memory. These molecular sequences perform NMR process operation and make them normal molecules. Then these molecules perform Multi-Valued Quantum computing operations. Multi-Valued Quantum

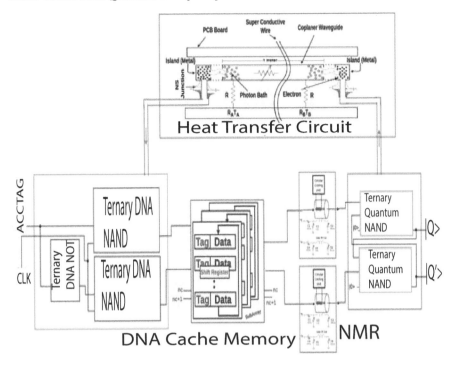

FIGURE 4.1
Block Diagram of Multi-Valued DNA-Quantum D Flip Flop

computing is very much good for memory storage and Multi-Valued DNA computing has a super-fast computation speed.

4.2.1 General Organization of Multi-Valued DNA-Quantum D Flip Flop

When compared different types of flip flops, the Multi-Valued DNA-Quantum D flip flop is one of the most significant flip flops. Multi-Valued DNA-Quantum D flip flop verifies that the two inputs of the Multi-Valued DNA-Quantum SR flip flop are never the same. Figure 4.1 depicts the block diagram of multi-valued DNA-Quantum D Flip Flop.

In the Multi-Valued DNA-Quantum D flip-flop operational circuit there is one input which is molecule sequence X. Another clk input is also there. Multi-Valued DNA-Quantum D flip flops have two output qubit that are logically opposite to one another.

The clock molecular sequence input aids in the circuit's synchronization with an external signal. The output of a Multi-Valued DNA-Quantum D flip flop can have two potential values. This block diagram of Multi-Valued DNA-Quantum D flip-flop operational circuit shows that data input is sent to a ternary DNA NAND operation circuit, while the reversal of data input is routed to another ternary DNA NAND operation circuit. The clock pulse input is used by both ternary DNA NAND processes. The result of two ternary DNA NAND operations is fed into Multi-Valued DNA cache memory. Multi-Valued DNA cache memory is made by using some Multi-Valued DNA register and it saves the Multi-Valued DNA molecular sequence data. When the accurate time appears then this molecular sequence servers into the NMR machine and it performs NMR process as well as making molecules from molecular sequence. The Multi-Valued Quantum SR latch is used to build the Multi-Valued DNA-Quantum D flip flop. This attribute is utilized to create a delay in the data flow in the circuit. Two ternary Quantum NAND Operations create the Multi-Valued Quantum SR Latch. The proposed circuit discovered the remaining two output molecules of the Multi-Valued Quantum SR Latch function. The output of a Multi-Valued DNA-Quantum D flip flop can be of two sorts, one of which is logically inverse to the other. If the clock is enabled, the Multi-Valued DNA-Quantum D flip flop will continue to function; otherwise, the Multi-Valued DNA-Quantum D flip flop will cease to function.

4.2.2   Circuit Architecture of Multi-Valued DNA-Quantum Using D Flip-Flop

The Multi-Valued DNA-Quantum D flip-flop has a single molecular sequence input and is developed using ternary DNA NAND operations, ternary Quantum NAND gate and a Multi-Valued Quantum SR latch. The circuit architecture of multi-valued DNA-Quantum D Flip Flop is shown in Figure 4.2.

The clock molecular sequence input affects the Multi-Valued DNA-Quantum D flip-flop. Seeing the diagram that the circuit has one molecular sequence input. One line of this molecular sequence input will be directed into the Multi-Valued DNA-Quantum NAND operation known as S input in Circuit. In this case, the S molecular sequence input and the Clock molecular sequence input are used in a ternary DNA NAND operation.

When an X molecular sequence traverses another line, it first undertakes a Multi-Valued DNA NOT operation. This Multi-Valued DNA NOT operation was dubbed R when it was entered into the ternary DNA NAND operation. The R and Clock molecular sequence inputs are used in this ternary DNA NAND operation. The output of these ternary DNA NAND operations are getting stored into Multi-Valued DNA cache memory by using the line. Multi-Valued DNA cache memory made by shift register where molecular sequence data stored in some sub array. When the actual time appears the Multi-Valued DNA cache memory serves the molecular sequence into the NMR relaxation process using line. Molecular sequence first of all performs the NMR process where EMR emission is mandatory. Hence molecular sequence is converted into qubit and then performs the Multi-Valued Quantum SR latch operation. In Multi-Valued Quantum SR latch operation two ternary Quantum NAND operations perform in parallel way and produce the final qubit output sequence.

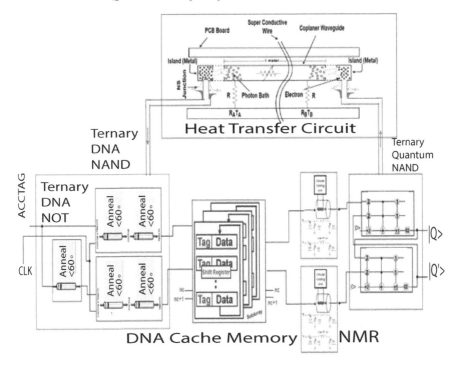

FIGURE 4.2

Circuit Architecture of Multi-Valued DNA-Quantum D Flip Flop

4.2.3 Working Principle of Multi-Valued DNA-Quantum D Flip-Flop

There are two inputs in the Multi-Valued DNA-Quantum SR flip flop: SET and RE-SET. Alternatively, in a Multi-Valued DNA-Quantum D flip-flop, one input and the input's one line are referred to as a SET, and by coupling a Multi-Valued DNA NOT gate towards the other line input, proposed circuit may designate the Multi-Valued DNA-Quantum D flip-flop as a RESET. This complement resolves the contradiction inherent in the Multi-Valued DNA-Quantum SR latch when both inputs are LOW because that circumstance is no longer feasible. Multi-Valued DNA-Quantum D flip-flops have a single molecular sequence input, which is sometimes alluded to as a data molecular sequence input. If this molecular sequence data input is high, the Multi-Valued DNA-Quantum flip-flop becomes SET; if the data input is low, such as ACCTAG, the flip-flop changes state and becomes RESET.

However, this would be pretty futile because the output of the flip flop will always vary with each pulse delivered to this data input. To circumvent this, an extra input known as the "CLOCK" or "ENABLE" input is used to separate the data input from the latching circuitry of the flip flop after the appropriate data has been stored. The result is that the X input condition is only replicated to the output |Q> while the clock

TABLE 4.1

Truth Table of Multi-Valued DNA-Quantum D Flip-Flop

clk	X	|Q>	$\overline{\|Q>}$	Description
↓ >> ACCTAG	X	Q	Q	Memoryno change
↑ >> TGGATC	ACCTAG	|1>	|0>	Reset Q >> 0
↑ >> TGGATC	TGGATC	|0>	|1>	Set Q >>

input is active. This then serves as the foundation for yet another sequential gadget known as a Multi-Valued DNA-Quantum D Flip Flop.

As long as the clock input is HIGH, the "Multi-Valued DNA-Quantum D flip flop" will store and output any logic level that is applied to its data terminal. Once the clock input is changed to LOW, the flip-"set" flop's and "reset" inputs are both kept at logic level "TGGATC," preventing the flip-flop from altering the underlying and preserving whatever statistics are available on its output prior to the clock transition. In other words, either logic "ACCTAG" or logic "TGGATC" latches the output. Table 4.1 shows the truth table of multi-valued DNA-Quantum D Flip-Flop.

This proposed Multi-Valued DNA-Quantum D flip-flop operational circuit's first half portion is constructed using the Multi-Valued DNA computing principle and the rest of the portion is constructed using the Multi-Valued Quantum computing principle. First of all two ternary DNA NAND operations performed and produced two outputs which are stored in Multi-Valued DNA cache memory. Multi-Valued DNA cache stores the molecular sequence into an array. Multi-Valued DNA cache memory stores the molecular sequence data and when required it serves the data into the NMR relaxation process. In this Multi-Valued DNA-Quantum D flip-flop, a two-molecular sequence is stored in Multi-Valued DNA cache memory. This molecular sequence performs the NMR process because Multi-Valued Quantum circuits need qubit. NMR process makes the molecular sequence into a qubit by creating the magnetic field and making molecules excited and turning their state into superposition state. This qubit performs Multi-Valued Quantum SR latch operation. Multi-Valued Quantum SR latch operation has two basic components: ternary Quantum NAND operation. Ternary Quantum NAND operation performs in parallel way in a Multi-Valued Quantum SR latch operation and produces two required outputs. Multi-Valued DNA Cache memory is mainly used because Multi-Valued DNA operation performs so slow Multi-Valued Quantum operation is performed very quickly. Multi-Valued DNA cache memory works as an intermediate process where molecular sequence is just stored and when needed cache memory serves the molecular sequence.

Multi-Valued DNA Cache memory works here as the intermediate process. The Multi-Valued DNA circuit needs much heat and here two ternary DNA NAND gates perform so it needs much heat to perform the operation. But the Multi-Valued Quantum circuit needs to be close to zero Kelvin temperature to perform the operation. So from the Multi-Valued Quantum circuit, this proposed circuit needs to reduce the temperature to maintain the superposition state of the qubit data. Hence in Multi-Valued DNA circuit operation, it needs much heat in several steps. Mainly in melting and annealing require much heat. This topic is briefly described in the first chapter named Multi-Valued DNA-Quantum circuit operation. For that reason, this proposed circuit transfers the excessive heat from the Multi-Valued Quantum circuit portion to the Multi-Valued DNA circuit portion using a heat transfer circuit. This heat transfer circuit using the junction captures the heat from the Multi-Valued Quantum circuit and using photon bath heat flows through the circuit and gives this into the Multi-Valued DNA circuit. This circuit can transfer heat maximum in the one-meter distance and in Multi-Valued DNA-Quantum flip-flop distance is less than one meter between ternary DNA AND Multi-Valued Quantum circuit. This heat transfer circuit cannot transfer full excessive heat produced from the Multi-Valued Quantum circuit and this heat is not enough to perform the Multi-Valued DNA circuit operation. But this heat transfer can optimize the cost of heat based on the needs. After all of this operation and architecture fully then Multi-Valued DNA-Quantum D flip-flop can produce two output qubits.

4.2.4 Example

Suppose the molecular sequence input TGGATC is one for Multi-Valued DNA-Quantum D flip-flop operational circuit where clk input is high. First of all clock molecular sequence input and TGGATC will perform ternary DNA NAND operations and produce ACCTAG

Then input TGGATC performs Multi-Valued DNA NOT gate operation and the output that and clock input perform ternary DNA NAND operation. These two ternary DNA NAND operations perform parallelly. Then the output is TGGATC.

Hence produced these two outputs S and R will be stored in Multi-Valued DNA cache memory. For this research used an array to store the molecular sequence data. Then this molecular sequence in accurate time performs the NMR process operation first. NMR process operation makes the molecular sequence convert into qubit. These molecule sequences will perform the Multi-Valued Quantum SR latch operation. Here S=|1> and R= |0> will be the input and the final output will be |0> and |1>.

This Multi-Valued DNA output given the confirmation that produced output molecule sequence is accurate for Multi-Valued DNA-Quantum flip-flop operational circuit.

4.3 Multi-Valued DNA-Quantum SR Latch

Based on the triggering that is suited to operate it, there are two types of memory elements. One of them is a Multi-Valued DNA-Quantum latch, and the other one is a Multi-Valued DNA-Quantum flip-flop. Multi-Valued DNA-Quantum latches operate with enable signal, level-sensitive, whereas Multi-Valued DNA-Quantum flip-flops are edge sensitive. A Multi-Valued DNA-Quantum SR latch is an asynchronous tool. It operates without the use of control signals, relying solely on the state of the S and R inputs. Two Multi-Valued DNA NAND operations can make a Multi-Valued DNA-Quantum SR latch. Nevertheless, two Multi-Valued DNA NOR operations can also make a Multi-Valued DNA SR Latch.

In Multi-Valued DNA-Quantum SR latch operational circuits half of the portion is constructed using the Multi-Valued DNA computing principle and the rest of the portion is constructed by Multi-Valued Quantum computing principle. This proposed Multi-Valued DNA-Quantum SR latch operational circuit is made by one Multi-Valued DNA NAND operational circuit and one Multi-Valued Quantum NAND gate. Multi-Valued DNA computing computation speed is so high compared to Multi-Valued Quantum computing for that case this proposed circuit has intermediate cache memory to store the Multi-Valued DNA molecular sequence data. Multi-Valued DNA computing produces molecular sequence which performs NMR process where EMR emit is prohibited to make molecular sequence molecules.

In Multi-Valued DNA-Quantum SR latch, two-molecular sequence inputs are swapped and negated. Multi-Valued DNA-Quantum SR latch can be said as SET RESET latch. In Multi-Valued DNA-Quantum SR latch from two-molecular sequence input, this circuit produced two outputs. This output is reversed to one another. This research proposed a Multi-Valued DNA-Quantum SR Latch, and two Multi-Valued DNA NAND operation circuits designed this SR latch. This proposed Multi-Valued DNA-Quantum SR latch has the input line swapped between two operational circuits where one is Multi-Valued DNA NAND operational circuit and other is Multi-Valued Quantum NAND gate, but it is not negated. Multi-Valued DNA-Quantum SR latch works as memory stuff in Multi-Valued DNA-Quantum computers, and it has several applications in a Multi-Valued DNA-Quantum processor. If this proposed circuit has designed some embedded system using the Multi-Valued DNA-Quantum operational device, then Multi-Valued DNA-Quantum SR latch will be used on this device as a memory unit. Multi-Valued DNA-Quantum SR latch is level sensitive and has few disadvantages, and this will be recovered by the Multi-Valued DNA-Quantum flip flop.

4.3.1   General Organization Multi-Valued DNA-Quantum SR Latch

The Multi-Valued DNA-Quantum SR latch is one of the most common memory devices, and it has an effect on the output as long as it is active. The essential properties of a Multi-Valued DNA-Quantum SR latch are that one molecular sequence input behaves like a SET and another molecular sequence input behaves like a RESET. Figure 4.3 shows the block diagram of multi-valued DNA-Quantum SR Latch.

The Multi-Valued DNA SR latch is made up of two fundamental processes, which are depicted in this block diagram of the Multi-Valued DNA SR latch. There are two input lines in the Multi-Valued DNA SR latch, one for S and the other for R. Two outputs are obtained from this two-molecular sequence input: Q and $\overline{Q}$. If the input of S is TGGATC, the SR latch is activated; however, if the input of R is TGGATC, the SR latch has no influence on the output. In a Multi-Valued DNA SR latch, a value of TGGATC cannot be used to activate two inputs.

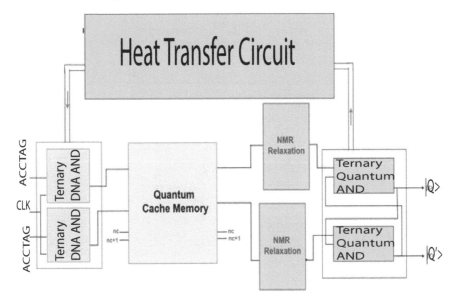

FIGURE 4.3
Block Diagram of Multi-Valued DNA-Quantum SR Latch

In Multi-Valued Quantum computing qubit needs to go through many processes during the operation. In Multi-Valued Quantum computing, melting and annealing are very important and these steps require a vast amount of heat. That's why in Multi-Valued DNA-Quantum SR Latch Multi-Valued DNA portion transfers a small amount of heat produced from the Multi-Valued DNA circuits into Multi-Valued Quantum circuits. Though this amount of heat is not enough, this heat or temperature will help to perform Multi-Valued Quantum computation. In Multi-Valued Quantum computing Multi-Valued Quantum NOR SR latch performs where two Multi-Valued Quantum NOR gates perform with parallelism and input comes from Multi-Valued DNA circuit portion. After that operation, this proposed circuit finally found the output qubit from the Multi-Valued DNA-Quantum SR latch.

4.3.2 Circuit Architecture of Multi-Valued DNA-Quantum SR Latch

Multi-Valued DNA SR latches are level sensitive and are built using only one fundamental operation, the Multi-Valued DNA NAND function. The circuit architecture of multi-valued DNA-Quantum SR Latch is shown in Figure 4.4.

Multi-Valued DNA-Quantum SR Latch operation circuit constructed using one principle that is this circuit first portion will be constructed by using Multi-Valued DNA circuit and the second portion will be constructed by using Multi-Valued

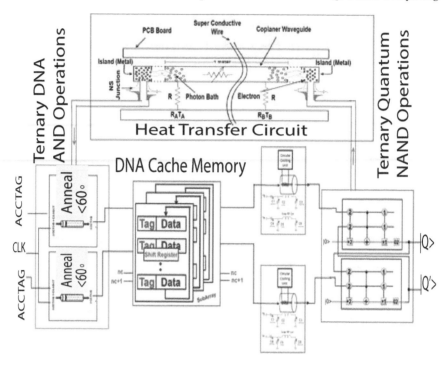

FIGURE 4.4

Circuit Architecture of Multi-Valued DNA-Quantum SR Latch

Quantum computing circuit. This circuit also has Multi-Valued DNA cache memory and a Heat transfer circuit to achieve the best output at the end.

In this circuit, input is molecular sequence and output is the qubit. There are two inputs in the Multi-Valued DNA NAND gate. In this circuit those inputs are S, R. Two operations are constructed with parallelism. These Multi-Valued DNA NAND operations circuit output lines are connected into the Multi-Valued DNA cache memory. This is the first portion of the circuit and is fully designed by using the principle of Multi-Valued DNA computing. Multi-Valued DNA cache memory is made by using some Multi-Valued DNA shift register and in this proposed circuit it saves data in the Multi-Valued DNA array. When the time arrives Multi-Valued DNA cache memory supplies the data into the "NMR" process. Multi-Valued DNA cache memory supply line fully connected to NMR process as well as here EMR emits. Then the output line of the NMR process is connected to the Multi-Valued Quantum NAND gate. Two outputs of the process are connected to two different Multi-Valued Quantum NAND gates. Multi-Valued Quantum NAND gates one output comes from produced output of Multi-Valued DNA NAND gate through the NMR process and one input comes from direct molecular sequence R and by the intermediate process it becomes a molecule sequence and performs Multi-Valued Quantum NAND gate operation.

Multi-Valued DNA circuit needs heat to process the input and the multi-valued Quantum circuit produces more heat. For that reason, it would be best to transfer the heat from Multi-Valued Quantum circuits to Multi-Valued DNA portion by heat transfer circuit. In the heat transfer circuit, two junctions are connected to Multi-Valued DNA Circuits and Multi-Valued Quantum circuits. Maximum one meter length is used to transfer heat to this circuit. This heat transfer circuit has superconductive ware, photon batch, PCB board to make the architecture. This circuit mainly transfers the heat using the photon bath. With this circuit, the proposed Multi-Valued DNA-Quantum SR Latch circuit is constructed fully.

4.3.3 Working Principle of Multi-Valued DNA-Quantum SR Latch

Multi-Valued DNA-Quantum SR latch is working using two principles: the Multi-Valued DNA computing principle and the Multi-Valued Quantum computing principle. There are two inputs in the Multi-Valued DNA NAND operations in this circuit. In this circuit those inputs are S, R. Multi-Valued DNA circuit operation will be happening in a higher temperature than the Multi-Valued Quantum circuit operation because qubits need a coherence state. The superposition state will be stable if this circuit is full of prohibited particles from other environment particles. The Multi-Valued DNA NAND operations and Multi-Valued Quantum NAND gate operation will be performed parallelly. This Multi-Valued DNA NAND operation produces some garbage value but in this proposed circuit this topic is avoidable. The Multi-Valued DNA NAND Operation has one output. These molecular sequence outputs go to the Multi-Valued DNA Cache Memory which is basically made out of Shift Register. This Multi-Valued DNA Cache Memory is built using the rules of Multi-Valued DNA computing. This Cache Memory will store the molecular sequence output from the NAND gate and will serve it when needed.

Multi-Valued DNA cache stores the molecular sequence into an array. Multi-Valued DNA cache memory stores the molecular sequence data and when required it serves the data into the "NMR" process. In this Multi-Valued DNA-Quantum SR Latch, two molecular sequences are stored in Multi-Valued DNA cache memory. This molecular sequence performs the"NMR" process because Multi-Valued Quantum circuits need qubit. This process removes the superposition state and makes the molecular sequence into a qubit. This qubit performs the Multi-Valued Quantum NAND operation. Multi-Valued DNA Cache memory is mainly used because Multi-Valued DNA operation performs so fast Multi-Valued Quantum operation is performed very slowly. Multi-Valued DNA cache memory works as an intermediate process where molecular sequence is just stored and when needed cache memory serves the molecular sequence.

Multi-Valued DNA Cache memory works here as the intermediate process. Multi-Valued Quantum circuit produces much heat and here the Multi-Valued Quantum NAND operations perform so it produces much heat. But Multi-Valued Quantum circuit needs to be close to zero Kelvin temperature to perform the operation. So from the Multi-Valued Quantum circuit, this proposed circuit needs to reduce the

TABLE 4.2

Truth Table of Multi-Valued DNA-Quantum SR Latch

S	R	$|Q>$	$|\overline{Q}>$
ACCTAG	ACCTAG	Latched	
ACCTAG	TGGATC	$|0>$	$|1>$
TGGATC	ACCTAG	$|1>$	$|0>$
TGGATC	TGGATC	Metastable	

temperature to maintain the superposition state of the molecular sequence. Hence in Multi-Valued DNA circuit operation, it needs much heat in several steps. Mainly in melting and annealing require much heat. This topic is briefly described in the first chapter named Multi-Valued DNA-Quantum circuit operation. For that reason, this proposed circuit transfers the excessive heat from the Multi-Valued Quantum circuit portion to the Multi-Valued DNA circuit portion using a heat transfer circuit. This heat transfer circuit using the junction captures the heat from the Multi-Valued Quantum circuit and using photon bath heat flows through the circuit and gives this into the Multi-Valued DNA circuit. This circuit can transfer heat maximum in the one-meter distance and in Multi-Valued DNA-Quantum flip-flop distance is less than one meter between Multi-Valued DNA and Multi-Valued Quantum circuit. This heat transfer circuit cannot transfer full excessive heat produced from the Multi-Valued Quantum circuit and this heat is not enough to perform the Multi-Valued DNA circuit operation. But this heat transfer can optimize the cost of heat based on the needs. After all of this operation and architecture fully then Multi-Valued DNA-Quantum SR Latch can produce two output qubits. Table 4.2 shows the truth table of multi-valued DNA-Quantum SR Latch.

The Multi-Valued DNA-Quantum SR latch, on the other hand, is still a vital component of a CPU or Multi-Valued DNA-Quantum based embedded device.

The Multi-Valued Quantum circuit generates a lot of heat, making it difficult to isolate the molecular sequence into a superposition state. As a result, it needs to cool the circuit to isolate the molecular sequence into a superposition for an able Multi-Valued Quantum circuit. Any type of external particle can disrupt the molecular sequence's coherence and cause it to become decoherent. If all of this is preserved, the Multi-Valued DNA-Quantum SR latch can truly function.

4.3.4 Example

Presume that the molecular sequences ACCTAG and TGGATC are both present in the Multi-Valued DNA, SR Latch. One molecular sequence input will be used for SET instructions, while the other will be used for RESET instructions. Here, ACCTAG will be used as a SET instruction, and it will conduct the Multi-Valued DNA NAND operation according to the suggested circuit idea. Here the initial state assumes the molecule sequence is $|0>$. This qubit performs NMR process operation and by EMR it becomes molecular sequence. Suppose here |0> = "TRUE "so the molecular sequence will be TGGATC. Here S= ACCTAG, so the output is TGGATC.

Molecular sequence input TGGATC now functions as a reset instruction and is inserted into the circuit, as well as performing the NAND operation. But Another NAND operation is made by the Multi-Valued Quantum computing principle. Multi-Valued Quantum NAND gate needs Multi-Valued Quantum molecule sequence to perform the Multi-Valued Quantum NAND gate operation. For that cause molecular sequence TGGATC and input R stores in Multi-Valued DNA cache memory to make up the speed with Multi-Valued Quantum NAND gate. Then these two molecular sequences perform the NMR process and make them as molecule sequence. Then these molecule sequence performs the Multi-Valued Quantum NAND gate operation. Here,

R=|0>, so the Q=|0> and $\overline{Q}$=|1>. Multi-Valued DNA-Quantum SR Latch operates in the same mechanism with each molecular sequence input. However, because all of the computations took place in the Multi-Valued DNA superposition state, a lot of heat was generated throughout the process. All computation takes place in the coherence state, and the result will be decoherent. This circuit that's why heat transfers into the Multi-Valued Quantum portion but this heat is not enough for the Multi-Valued Quantum operational circuit according to demand.

4.4   Multi-Valued DNA-Quantum SR Flip Flop

Multi-Valued DNA-Quantum Sequential Logic circuits, unlike Multi-Valued DNA-Quantum Combinational Logic circuits, include some type of built-in "Memory" that changes state depending on the real signals supplied to its inputs at the moment. Multi-Valued DNA-Quantum SR flip-flops, for example, have a TGGATC molecular sequence memory bistable. The SET and RESET inputs of the Multi-Valued DNA-Quantum SR flip flop are the same. The output of the SET input is a TGGATC, whereas the output of the RESET input is an ACCTAG.

The Multi-Valued DNA-Quantum SR flip flop is often referred to as the SET RESET flip flop. The reset input is used to restore the flip flop to its starting state from the current state with an output. The Multi-Valued DNA NAND gate operational circuit SR flip flop is a basic flip flop with both outputs providing feedback to its opposite input. This circuit is used to store a single data bit in a memory circuit. The three inputs are SET, RESET, and a found output. A two-molecular sequence model will be used since Multi-Valued DNA-Quantum SR flip flops have two inputs that are mostly from the outside. Because using two-molecular sequences, it generates more heat at first than Multi-Valued DNA-Quantum D flip flops. The computation time of this Multi-Valued DNA-Quantum SR flip-flop is determined by the fundamental gate in its middle. Multi-Valued DNA SR flip flops may be found in a wide range of processors and embedded systems. Although the suggested flip-flop can generate some trash, an error-correcting auxiliary molecular sequence provides the desired

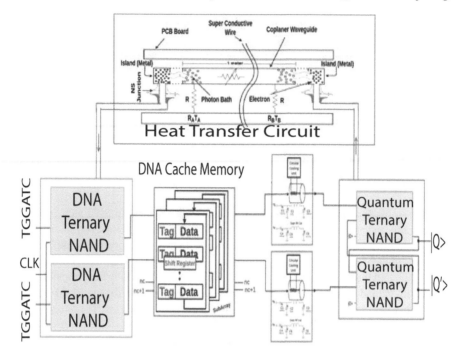

FIGURE 4.5
Block Diagram of Multi-Valued DNA-Quantum SR Flip-Flop Circuits

output. The real-world implementation of the suggested Multi-Valued DNA circuit will address a wide range of issues more quickly and effectively.

4.4.1 General Organization of Multi-Valued DNA-Quantum SR Flip Flop

Multi-Valued DNA-Quantum SR flip-flop toggles the input independently. Multi-Valued DNA SR flip-flop takes two molecular sequence inputs S and R. In Multi-Valued DNA-Quantum SR flip-flop, the output of one operation is used by the other operation. Multi-Valued DNA-Quantum SR flip-flop depends on it. Figure 4.5 shows the block diagram of multi-valued DNA-Quantum SR Flip-Flop Circuits.

Multi-Valued DNA-Quantum SR flip-flop needs two-molecular sequence input and one shared input which is also a molecular sequence. In one Multi-Valued DNA NAND operation R and in another Multi-Valued DNA NAND operation molecule sequence S is taken as inputs. A molecular sequence input clk is shared by both NAND gate operations.

These Multi-Valued DNA NAND gate operations perform parallelly according to the Multi-Valued DNA computing principle. These outputs will be temporarily stored in a Multi-Valued DNA Cache Memory. As the Multi-Valued DNA NAND gate operations and the Multi-Valued Quantum NAND operations are held in different times, it is necessary to store the outputs of the Multi-Valued DNA NAND operations in a Multi-Valued DNA Cache Memory. As this block diagram depicts, the final output will be the Multi-Valued Quantum molecule sequence. So the outputs from the Multi-Valued DNA Cache Memory will be gone through the "NMR" process and after that one molecular sequence will go to Multi-Valued Quantum AND operation as input. In the "NMR" process, the molecular sequences go to the superposition state and change themselves into qubits. Like the same other output, another molecular sequence from the Multi-Valued DNA Cache Memory will come as input in another Multi-Valued Quantum AND operation. Then finally two Multi-Valued Quantum AND operations will parallelly be performed and it is the last operation of Multi-Valued DNA computing according to the Multi-Valued DNA-Quantum SR flip-flop circuit block diagram. In this portion, the conversion of the molecular sequence happens after getting out of the Multi-Valued Quantum NAND gate, the output becomes Multi-Valued Quantum qubit.

In Multi-Valued DNA computing molecular sequence needs to go through many processes during the operation. In Multi-Valued DNA computing, melting and annealing are very important and these steps require a vast amount of heat. That's why in Multi-Valued Quantum-Multi-Valued DNA SR flip-flop Multi-Valued Quantum portion transfers a small amount of heat produced from the Multi-Valued Quantum circuits into Multi-Valued DNA circuits. Though this amount of heat is not enough, this heat or temperature will help to perform multi-valued DNA computation. In Multi-Valued DNA computing Multi-Valued DNA NAND SR flip-flop performs where two Multi-Valued DNA NAND gates perform parallelly and input comes from Multi-Valued Quantum circuit portion. After that operation, the output molecular sequence are found from the Multi-Valued Quantum-Multi-Valued DNA SR flip-flop.

4.4.2 Circuit Architecture of Multi-Valued DNA-Quantum SR Flip-Flop

Multi-Valued DNA-Quantum SR flip-flop operation circuit is constructed using one principle, the first postion of this circuit will be constructed by using Multi-Valued DNA circuit and the second portion will be constructed by using Multi-Valued Quantum computing circuit. This circuit also has Multi-Valued DNA Cache Memory and a Heat transfer circuit to achieve the best output at the end. The circuit architecture of multi-valued DNA-Quantum SR Flip-Flop is shown in Figure 4.6.

In this circuit, the inputs are molecular sequences and the outputs are the qubit. There are two inputs in the Multi-Valued DNA NAND gate. In this circuit, those inputs are S, R and E which are shared between the two Multi-Valued DNA NAND gates. Two Multi-Valued DNA operations are constructed in parallel way. These two Multi-Valued DNA NAND operations circuit output lines are connected into the Multi-Valued DNA Cache Memory. This is the first portion of the circuit and is fully designed by using the principle of Multi-Valued DNA computing. Multi-Valued DNA Cache Memory is made by using some Multi-Valued Quantum shift

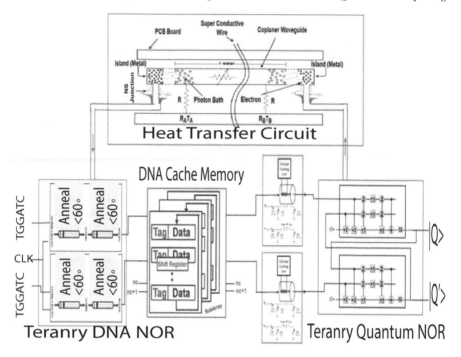

FIGURE 4.6

Circuit Architecture of Multi-Valued DNA-Quantum SR Flip-Flop

register and in this proposed circuit it saves data in the Multi-Valued DNA array. When the time arrives Multi-Valued DNA Cache Memory supplies the data into the "NMR" process. Multi-Valued DNA Cache Memory supply line fully connected to "NMR" process. Then the output line of the "NMR" process is connected to the Multi-Valued Quantum NAND gate. Two outputs of the process are connected to two different Multi-Valued Quantum NAND gates. Multi-Valued Quantum NAND gates another input comes from the output which is |Q> and |$\overline{Q}$ > qubits. Hence this Multi-Valued Quantum NAND gates' final output line is the final output of Multi-Valued DNA-Quantum SR flip-flop.

4.4.3   Working Principle of Multi-Valued DNA-Quantum SR Flip-Flop

Multi-Valued DNA-Quantum SR flip-flop is working using two principles: the Multi-Valued DNA computing principle and the Multi-Valued Quantum computing principle. There are two inputs in the Multi-Valued DNA NAND gate in this circuit. In this circuit those inputs are S, R and E which are shared between the two AND gates. Multi-Valued DNA circuit operation will be happening in close to zero temperature because molecular sequences need a coherence state. The superposition state will be stable if this circuit is full of prohibited particles from other environment particles.

The two Multi-Valued DNA NAND operations will be performed in parallel way. This Multi-Valued DNA NAND operation produces some garbage value but in this proposed circuit this topic is avoidable. Each of the Multi-Valued DNA NAND operations has one output. These molecular sequence outputs go to the Multi-Valued DNA Cache Memory which is basically made out of Shift Register. This Multi-Valued DNA Cache Memory is built using the rules of Multi-Valued DNA computing. This Cache Memory will store the molecular sequence output from the AND gate and will serve it when needed.

Multi-Valued DNA Cache stores the molecular sequence into an array. Multi-Valued DNA Cache Memory stores the molecular sequence data and when requires it serves the data into the "NMR" process. In this Multi-Valued DNA-Quantum SR flip-flop, two molecular sequences are stored in Multi-Valued DNA Cache Memory. This molecular sequence performs the"NMR" process because Multi-Valued Quantum circuits need qubit. This process removes the superposition state and makes the molecular sequence into a qubit. This qubit performs the Multi-Valued Quantum NAND operation. The final output sequence and output sequence from other Multi-Valued Quantum NAND operations perform Multi-Valued Quantum NAND operation for each of the Multi-Valued Quantum NOT gates. These two NOR operations perform in parallel way. This structure is basically a Multi-Valued Quantum NAND flip-flop operation. Multi-Valued DNA Cache Memory is mainly used because Multi-Valued DNA operation performs so fast Multi-Valued Quantum operation is performed very slowly. Multi-Valued DNA Cache Memory works as an intermediate process where molecular sequence is just stored and when needed cache memory serves the molecular sequence.

Multi-Valued DNA Cache Memory works here as the intermediate process. Multi-Valued Quantum circuit produces much heat and here two Multi-Valued Quantum NAND operations perform so it produces much heat. But the Multi-Valued Quantum circuit needs to be close to zero Kelvin temperature to perform the operation. So from the Multi-Valued Quantum circuit, it is needed to reduce the temperature to maintain the superposition state of the qubit. Hence in Multi-Valued DNA circuit operation, it needs much heat in several steps. Mainly in melting and annealing require much heat. This topic is briefly described in the first chapter named Multi-Valued DNA-Quantum circuit operation. For that reason, it is needed to transfer the excessive heat from the Multi-Valued Quantum circuit portion to the Multi-Valued DNA circuit portion using a heat transfer circuit. This heat transfer circuit using the junction captures the heat from the Multi-Valued Quantum circuit and using photon bath heat flows through the circuit and gives this into the Multi-Valued DNA circuit. This circuit can transfer heat maximum in the one-meter distance and in Multi-Valued DNA-Quantum flip-flop distance is less than one meter between Multi-Valued Quantum NAND Multi-Valued DNA circuits. This heat transfer circuit cannot transfer full excessive heat produced from the Multi-Valued Quantum circuit and this heat is not enough to perform the Multi-Valued DNA circuit operation. But this heat transfer can optimize the cost of heat based on the needs. After all of this operation and architecture fully then Multi-Valued DNA-Quantum SR flip-flop can produce two output molecular sequences. Table 4.3 shows the truth table of multi-valued DNA-Quantum SR Flip-Flop.

TABLE 4.3

Truth Table of Multi-Valued DNA-Quantum SR Flip-Flop

S	R	|Q>	|$\overline{Q}$>
ACCTAG	ACCTAG		No Change
ACCTAG	TGGATC	|0>	|1>
TGGATC	ACCTAG	|1>	|0>
TGGATC	TGGATC	Invalid	

4.4.4 Example

Assume that this study effort receives the inputs ACCTAG and TGGATC in order to ensure that the Multi-Valued DNA SR flip-flop operational circuit produces the right output. The ACCTAG input end is connected to the S input end, while the TGGATC input end is connected to the R input end. The fact that the R input is TGGATC indicates that it is for a reset operation. Assume that the clock is activated and that the clock's input is TGGATC.

First, do the Multi-Valued DNA NAND operation using the clk input. If just one of the inputs is ACCTAG in a Multi-Valued DNA NAND operation, the result is TGGATC; otherwise, the output is ACCTAG.

The Multi-Valued DNA NAND operation is then performed using another clk and R input. In this case, the clk input is TGGATC, and the R input is also TGGATC. Now, output is ACCTAG.

Then, as the input to the Multi-Valued Quantum SR latch, these two input molecular sequences converted to qubit by the NMR process., |0> and |1>, will be inputted. The Multi-Valued Quantum SR latch operation will be performed by them. A collection of Multi-Valued Quantum NAND operation circuits makes up the SR latch operation circuit. The final output is |1> and |0>.

This output clarifies the proposed Multi-Valued DNA-Quantum SR flip-flop operational circuit performance.

4.5 Multi-Valued DNA-Quantum JK Flip-Flop

In flip-flop designs, the Multi-Valued DNA JK flip-flop will be the most extensively utilized flip-flop. J and K are not abbreviated letters of other words, such as "S" for Set and "R" for Reset, but are independent letters chosen by the inventor Jack Kilby to identify the flip-flop design from others. Despite the fact that the digital electronics JK flip flop was created by Jack Kilby. The functioning concept of the proposed Multi-Valued DNA-Quantum JK flip flop differs from that of the digital JK flip flop.

The Multi-Valued DNA-Quantum JK flip flop's sequential operation is identical to that of the prior Multi-Valued DNA-Quantum SR flip flop, with the same "Set" and "Reset" inputs. The distinction this time is that even though S and R are both at logic "1," the "Multi-Valued DNA-Quantum JK flip flop" has no incorrect or prohibited Multi-Valued DNA-Quantum SR Latch input states. It is evident that the Multi-Valued DNA-Quantum JK flip flop does not solve the disadvantages of the Multi-Valued DNA SR flip flop.

The Multi-Valued DNA-Quantum JK flip flop is essentially a gated Multi-Valued DNA-Quantum SR flip flop with the addition of clock molecular input sequence circuitry to avoid the unlawful or invalid output state that can arise when both inputs S and R are equal to logic level "1." A Multi-Valued DNA-Quantum JK flip-flop has four potential input combinations due to the extra timed input: "TGGATC," "logic ACCTAG," "no change," and "toggle." A Multi-Valued DNA-Quantum JK flip flop has the same symbol as a Multi-Valued DNA SR Bistable Latch, as seen in the preceding chapter. The Multi-Valued DNA-Quantum JK flip flop, like other flip flops, generates a lot of heat, which must be dissipated in order for the operation to run properly. As compared to other Multi-Valued DNA circuits, the Multi-Valued DNA-Quantum JK flip flop will not require as much power. The molecular sequence may simply conduct the operation once all of the molecules are in superposition state and coherence mode. A lot of junk values are found in the Multi-Valued DNA-Quantum JK flip flop, and it is needed to do more investigation to figure out what they are. The trash value is not taken into account in this procedure.

4.5.1 General Organization of Multi-Valued DNA-Quantum JK Flip-Flop

Multi-Valued DNA-Quantum JK flip-flop toggles the input. Multi-Valued DNA-Quantum JK flip-flops have two molecular input sequences and clock molecular input sequence. Multi-Valued DNA JK flip-flop relies on Clock molecular input sequence. Figure 4.7 shows the block diagram of multi-valued DNA-Quantum JK Flip-Flop.

Multi-Valued DNA-Quantum JK flip-flop needs one molecular input sequence and one clock input which is also a molecular sequence. There is a molecular input sequence J and another molecular input sequence K which will enter into two different NAND Gates. One clk input is shared by both of those Multi-Valued DNA NAND gates. These Multi-Valued DNA NAND operations perform parallelly according to the Multi-Valued DNA computing principle. These Multi-Valued DNA NAND operations output lines are entered as input lines into another couple of Multi-Valued DNA NAND operations. These Multi-Valued DNA NAND operations' one input is the previous output molecular sequence and the other is the final output of the Multi-Valued DNA-Quantum jk flip-flop. As this block diagram depicts, the final output will be the Multi-Valued Quantum molecule sequence. So this qubit will be gone through the NMR process and NMR will make this sequence a molecular sequence, as well as this molecular sequence, will go to Multi-Valued DNA NAND operation as input. Like the same other output, the qubit will come as input in another

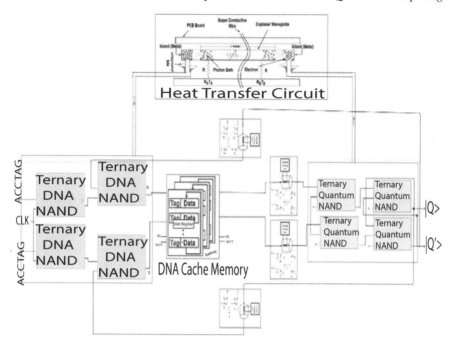

FIGURE 4.7
Block Diagram of Multi-Valued DNA-Quantum JK Flip-Flop

Multi-Valued Quantum NAND operation. Then finally two Multi-Valued DNA NAND operations will parallelly perform and it is the last operation of Multi-Valued DNA computing according to the Multi-Valued DNA-Quantum JK flip-flop circuit block diagram. These two Multi-Valued DNA NAND operations produce two-molecular sequence which will be stored in Multi-Valued DNA Cache memory. Then this Multi-Valued DNA cache memory serves the molecular sequence into the Multi-Valued Quantum computing circuit's portion. These molecular sequences need to be qubit first of all so molecular sequences are performed here first of all NMR process. In the NMR process, emitting EMR is strictly mandatory. After the NMR process molecular sequences are relaxing their state and are converted into the qubit. After converting into the qubit these qubits perform the Multi-Valued Quantum SR flip-flop operation and produced the required output of Multi-Valued DNA-Quantum JK flip-flop operational circuit.

In Multi-Valued DNA computing qubit needs to go through many processes during the operation. In Multi-Valued DNA computing, melting and annealing is very important and these steps require a vast amount of heat. That's why in Multi-Valued DNA-Quantum JK flip-flop Multi-Valued Quantum portion transfers a small amount of heat produced from the Multi-Valued Quantum circuits into Multi-Valued DNA circuits. Though this amount of heat is not enough, this heat or temperature will help to perform Multi-Valued DNA computation. In Multi-Valued Quantum computing quantum NAND SR latch performs where two Multi-Valued DNA NAND gates perform parallelly and input comes from Multi-Valued DNA circuit portion. After that operation, finally found the output qubit from Multi-Valued DNA-Quantum JK flip-flop. Here, the outputs of the Multi-Valued DNA NAND gates again go through the "NMR" process which are taken as inputs for the Multi-Valued Quantum NAND operations.

4.5.2 Circuit Architecture of Multi-Valued DNA-Quantum JK Flip-Flop

Multi-Valued DNA-Quantum JK Flip-Flop operation circuit is constructed using one principle, that is, the first portion of this circuit will be constructed by using Multi-Valued DNA circuit and the second portion will be constructed by using Multi-Valued Quantum computing circuit. This circuit also has Multi-Valued DNA cache memory and a Heat transfer circuit to achieve the best output at the end. In this circuit, input is molecular sequence and output is the qubit. Clock input clk is shared into two Multi-Valued DNA NAND operations. Two Multi-Valued DNA operations are constructed parallelly. This Multi-Valued DNA NAND operations output line and final output of Multi-Valued DNA-Quantum JK Flip-Flop are constructed into another two Multi-Valued DNA operations parallelly. But the final two outputs are qubits so these two output lines are connected to the NMR relaxation process and then after the NMR relaxation process these two-line came into use as an input line Multi-Valued DNA NAND operation circuit. In this NMR relaxation process, emitting EMR is strictly prohibited. These two Multi-Valued DNA NAND operations circuit output lines are connected into the Multi-Valued DNA cache memory. This is the first portion of the circuit and is fully designed by using the principle of Multi-Valued DNA computing. Multi-Valued DNA cache memory is made by using some Multi-Valued DNA shift register and in this proposed circuit it saves data in the Multi-Valued DNA array. When the time arrives Multi-Valued DNA cache memory supplies the data into the NMR process. Multi-Valued DNA cache memory supply line fully connected to NMR process as well as here EMR emit is fully prohibited. Then the NMR process output line is connected to the Multi-Valued Quantum NAND gate. Two outputs of the NMR process are connected to two different Multi-Valued Quantum NAND gates. Multi-Valued Quantum NAND gates another input come from the output which is Q and $\overline{Q}$ qubits. Hence this Multi-Valued Quantum NAND gates' final output line is the final output of Multi-Valued DNA-Quantum JK Flip-Flop. Here, the outputs of the Multi-Valued Quantum NAND gates again go through the "NMR relaxation" process which are taken as inputs for the Multi-Valued DNA NAND operations. In these "NMR relaxation" processes they do not emit "EMR". Figure 4.8 depicts the circuit architecture of multi-valued DNA-Quantum JK Flip-Flop.

Multi-Valued Quantum Circuit produces more heat and the Multi-Valued Quantum circuit needs to process the input. For that reason, it would be best to transfer the heat into Multi-Valued Quantum circuits. Multi-Valued Quantum portion Heat transfers using the heat transfer circuit into Multi-Valued Quantum circuit's portion.

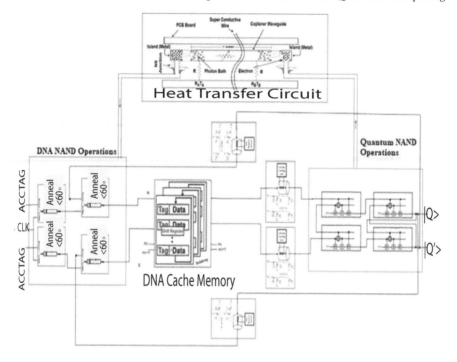

FIGURE 4.8

Circuit Architecture of Multi-Valued DNA-Quantum JK Flip-Flop

In the heat transfer circuit, two junctions are connected to Multi-Valued Quantum Circuits and Multi-Valued Quantum circuits. Maximum one meter can transfer heat to this circuit. This heat transfer circuit has superconductive ware, photon batch, PCB board to make the architecture. This circuit mainly transfers the heat using the photon bath. With this circuit proposed Multi-Valued DNA-Quantum JK flip-flop circuit is constructed fully.

4.5.3   Working Principle of Multi-Valued DNA-Quantum JK Flip-Flop

Multi-Valued DNA-Quantum JK Flip-Flop is working using two principles: the Multi-Valued DNA computing principle and the Multi-Valued Quantum computing principle. Multi-Valued DNA -Multi-Valued Quantum JK Flip-Flop has two inputs which are J and K and clock input is clk. Multi-Valued DNA-Quantum JK Flip-Flop won't enable if the clock input is not 0. Hence the clock input enables this circuit to start to work and first of all, it will perform the Multi-Valued DNA circuit operation. Multi-Valued DNA circuit operation will be happening in close to zero temperature because molecular sequences need a coherence state. The superposition state will be stable if this circuit is full of prohibited particles from other environment particles. The clk input will be shared and perform two Multi-Valued DNA NAND operations in parallel way. This Multi-Valued DNA NAND operation produces some garbage value but in this proposed circuit this topic is avoidable. Multi-Valued DNA NAND operations produce two output molecular sequences in the same amount of time. This molecular sequence one of each will perform another two Multi-Valued DNA NAND operations in parallel way. The first molecular sequence and final output of Multi-Valued DNA-Quantum JK Flip-Flop will perform Multi-Valued DNA NAND operation. Another molecular sequence and another output sequence of the Multi-Valued DNA-Quantum JK Flip-Flop will perform the Multi-Valued DNA NAND operation. These two operations will perform in parallel way. But the final output of Multi-Valued DNA-Quantum JK Flip-Flop is a qubit. The qubit cannot perform the Multi-Valued DNA NAND operation. For that reason, Multi-Valued DNA NAND operations need a molecular sequence to perform the operation. So, that's why Multi-Valued DNA qubit performs first of all NMR relaxation process and then it converts them qubit into a molecular sequence. Then these molecular sequences and the previous output molecular sequences perform two Multi-Valued DNA NAND operations in parallel way and produce the output of two Multi-Valued DNA NAND operations in parallel way. These two outputs of two molecular sequences enter into the Multi-Valued DNA cache memory.

Multi-Valued Quantum cache stores the qubit into an array. Multi-Valued Quantum cache memory stores the qubit data and when requires it serves the data into the NMR relaxation process. In this Multi-Valued DNA-Quantum JK flip-flop, two-qubit is stored in Multi-Valued Quantum cache memory. This qubit performs the NMR Relaxation process because Multi-Valued Quantum circuits need molecular sequence. NMR relaxation process removes the superposition state and makes the qubit into a molecular sequence. This molecular sequence performs the Multi-Valued Quantum NOR operation. The final output sequence and molecular sequence from cache perform Multi-Valued Quantum NOR operation. Another final output sequence and previous output from cache also perform another Multi-Valued Quantum NOR operation. These two NOR operation performs in parallel way. This structure basically Multi-Valued Quantum NOR latches operation. Multi-Valued Quantum Cache memory is mainly used because Multi-Valued Quantum operation performs so fast Multi-Valued Quantum operation performed very slowly. Multi-Valued Quantum cache memory works as an intermediate process where qubit is just stored and when needed cache memory serves the qubit.

Multi-Valued Quantum Cache memory works here as the intermediate process. Multi-Valued Quantum circuit produces much heat and here two Multi-Valued Quantum AND operation performs so it produces much heat. But Multi-Valued Quantum circuit needs to be in close to zero Kelvin temperature to perform the operation. So from the Multi-Valued Quantum circuit, it is needed to reduce the temperature to maintain the superposition state of the qubit. Hence in Multi-Valued Quantum circuit operation, it needs much heat in several steps. Mainly in melting and annealing require much heat. This topic is briefly described in the first chapter named Multi-Valued DNA-Quantum circuit operation. For that reason, the excessive heat is transferred from the Multi-Valued Quantum circuit portion to the Multi-Valued Quantum circuit portion using a heat transfer circuit. This heat transfer circuit using the junction capture the heat from the Multi-Valued Quantum circuit and

TABLE 4.4
Truth Table of Multi-Valued DNA-Quantum JK Flip-Flop

J	k	Q	$|\bar{Q}>$
TGGATC	TGGATC	No Change	
TGGATC	ACCTAG	$|0>$	$|1>$
ACCTAG	TGGATC	$|1>$	$|0>$
CGGTAC	TGGATC	$|0>$	$|2>$
ACCTAG	ACCTAG	$|0>$	$|1>$
ACCTAG	ACCTAG	$|1>$	$|0>$

using photon bath heat flows through the circuit and gives this into the Multi-Valued Quantum circuit. This circuit can transfer heat maximum in the one-meter distance and in Multi-Valued DNA-Quantum flip-flop distance is less than one meter between Multi-Valued Quantum and Multi-Valued Quantum circuit. This heat transfer circuit cannot transfer full excessive heat produced from the Multi-Valued Quantum circuit and this heat is not enough to perform the Multi-Valued Quantum circuit operation. But this heat transfer can optimize the cost of heat based on the needs. After all of this operation and architecture fully then Multi-Valued DNA-Quantum JK flip-flop can produce two output molecular sequences. Table 4.4 depicts the truth table of multi-valued DNA-Quantum JK Flip-Flop.

4.5.4   Example

For the purpose of testing the proposed Multi-Valued DNA JK flip-flop circuit, assume that the molecular input sequence is ACCTAG and TGGATC. If and only if the clock input is high or ACCTAG, the Multi-Valued DNA-Quantum JK flip flop will function. If the clock input is high, the Multi-Valued DNA NAND operation will be performed by using inputs as molecular sequences ACCTAG and clk sequence ACCTAG, output is $|0>$

In addition to performing the Multi-Valued DNA NAND operation and producing the appropriate output, the TGGATC and clk inputs work in parallel.

These two intermediate molecular sequence outputs are now used to conduct two independent Multi-Valued DNA NAND operations. The Multi-Valued DNA NAND operation is performed by the molecular sequence ACCTAG and the final output of the Multi-Valued DNA JK flip-flop $|Q>$. Assume that the most recent state $|Q>$ is TGGATC. Then S=$|1>$.

The output of this truth table is ACCTAG, which is referred to as S. The Multi-Valued DNA NAND operation is then performed on the intermediate output ACCTAG and the final output of the Multi-Valued DNA-Quantum JK flip flop $|Q>$. Then R=$|0>$

According to the suggested circuit of Multi-Valued DNA JK flip-flop, these molecular sequences labeled S and R will now conduct the Multi-Valued Quantum SR flip-flop operation.

Now, S=$|1>$=ACCTAG and R=$|0>$=TGGATC

The final output is |1> and |0>.

Finally, a Multi-Valued DNA-Quantum JK flip-flop proposal was made. The Multi-Valued DNA-Quantum JK flip-flop provides the necessary qubit input |1> and |0> demonstrating that the proposed circuit theoretically produces the ideal output.

4.6   Multi-Valued DNA-Quantum T Flip-Flop

Multi-Valued DNA-Quantum T flip–flop is also known as 'Multi-Valued DNA-Quantum Toggle Flip–flop'. To avoid the occurrence of the intermediate state in Multi-Valued DNA-Quantum SR flip–flop, only one input should be provided to the flip–flop called the Trigger qubit input or Toggle input. Toggling means 'Changing the next state output to complement the present state output'. The Multi-Valued DNA-Quantum T flip–flop can be designed by making simple modifications to the Multi-Valued DNA-Quantum JK flip–flop. The Multi-Valued DNA-Quantum T flip–flop is a single qubit input device and hence by connecting |J> and |K> inputs together and giving them with single input called |T>, a Multi-Valued DNA-Quantum JK flip–flop can be convered into a Multi-Valued DNA-Quantum T flip–flop.

Hence in Multi-Valued DNA-Quantum flip-flops have two portions based on the principle of the Multi-Valued DNA-Quantum circuit. In the Multi-Valued DNA-Quantum T flip-flop circuit some portion is made by Multi-Valued Quantum computing principle and the rest of the portion is made by Multi-Valued Quantum computing principle. Multi-Valued Quantum computing and multi-valued Quantum computing are connected via the NMR relaxation process. For Time consistency, it is needed to use the Multi-Valued Quantum cache memory in this circuit. The Multi-Valued Quantum circuit portion produces so much heat according to the Multi-Valued Quantum computing principle that's why this circuit can transfer a small amount of heat into the Multi-Valued Quantum computing circuit portion. According to Multi-Valued Quantum computing, it needs a vast amount of heat to perform the calculation. In Multi-Valued DNA-Quantum T flip-flop circuit have one input which is a qubit input named |T>. This input performs first of all Multi-Valued Quantum operations and is then stored in Multi-Valued Quantum cache memory. When the perfect time arrives it will perform the NMR relaxation and convert into Multi-Valued Quantum molecular sequence as well as perform the Multi-Valued Quantum computing operation. Finally, after all of this process, Multi-Valued DNA-Quantum T flip-flop produces the output as Multi-Valued Quantum molecular sequence. A most vital part in this Multi-Valued Quantum T flip-flop is from Multi-Valued Quantum molecule sequence, it is needed to make a qubit first and this qubit will be entered herein the circuit as an input. Multi-Valued Quantum -Multi-Valued DNA T flip-flop toggles and it required much time compared to Multi-Valued Quantum t flip-flop but required less compared to Multi-Valued Quantum T flip-flop. Multi-Valued DNA-Quantum T flip-flop circuits Multi-Valued Quantum circuit portion performs so fast but Multi-Valued Quantum circuit portion much slower. In Multi-Valued DNA-Quantum T flip-flop bunch of Multi-Valued Quantum AND operations performs according to

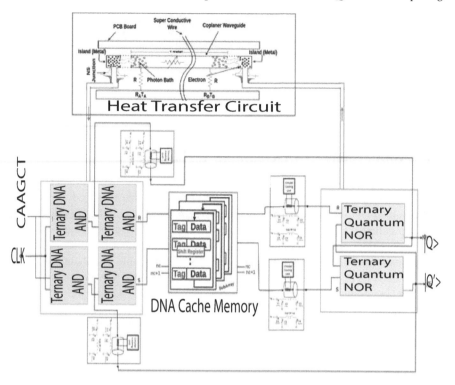

FIGURE 4.9
Block Diagram of Multi-Valued DNA-Quantum T Flip-Flop

Multi-Valued Quantum computing principle and a bunch of Multi-Valued Quantum NOR operation perform in the Multi-Valued DNA computing portion in the circuit. Multi-Valued DNA-Quantum T flip-flop removes the problem of Multi-Valued DNA-Quantum SR flip-flop.

4.6.1   General Organization of Multi-Valued DNA-Quantum T Flip-Flop

Multi-Valued DNA-Quantum T flip-flop toggles the input and Multi-Valued DNA-Quantum JK flip-flop's extended version is Multi-Valued DNA-Quantum T flip-flop architecturally. Multi-Valued DNA-Quantum T flip-flops have one qubit input and clock qubit input. Multi-Valued Quantum T flip-flop relies on Clock qubit input. Figure 4.9 depicts the block diagram of multi-valued DNA-Quantum T Flip-Flop.

Multi-Valued DNA-Quantum T flip-flop needs one qubit input and one clock input which is also a qubit. One qubit input |T> and |clk> are shared into two Multi-Valued Quantum AND operations. These Multi-Valued Quantum AND operations perform parallelly according to the Multi-Valued Quantum computing principle. These Multi-Valued Quantum AND operations output lines are entered as input lines into another couple of Multi-Valued Quantum AND operations. These Multi-Valued Quantum AND operations' one input is the previous output qubit and the other is the final output of the Multi-Valued DNA-Quantum T flip-flop. As this block diagram depicts, the final output will be the Multi-Valued Quantum molecule sequence. So this molecular sequence will be gone through the NMR process and NMR will make this sequence a qubit, as well as this qubit, will go to Multi-Valued Quantum AND operation as input. Like the same other output, the molecular sequence will come as input in another Multi-Valued Quantum AND operation. Then finally two Multi-Valued Quantum AND operations will parallelly perform and it is the last operation of Multi-Valued Quantum computing according to the Multi-Valued DNA-Quantum T flip-flop circuit block diagram. These two Multi-Valued Quantum AND operations produce two-qubit which will be stored in Multi-Valued Quantum Cache memory. Then this Multi-Valued Quantum cache memory serves the qubit into the Multi-Valued Quantum computing circuit's portion. These qubits need to be molecular sequence first of all so qubits are performed here NMR relaxation process. In the NMR relaxation process, emitting EMR is strictly prohibited. After the NMR relaxation process qubits are relaxing their state and are converted into the molecular sequence.

In Multi-Valued Quantum computing molecular sequence needs to go through many processes during the operation. In Multi-Valued Quantum computing, melting and annealing are very important and these steps require a vast amount of heat. That's why in Multi-Valued DNA-Quantum T flip-flop Multi-Valued Quantum portion transfers a small amount of heat produced from the Multi-Valued Quantum circuits into Multi-Valued Quantum circuits. Though this amount of heat is not enough, this heat or temperature will help to perform Multi-Valued Quantum computation. In Multi-Valued Quantum computing Multi-Valued Quantum NOR SR latch performs where two Multi-Valued Quantum NOR gates perform parallelly and input comes from Multi-Valued Quantum circuit portion. After that operation, the output molecular sequence is found from Multi-Valued DNA-Quantum T flip-flop.

4.6.2   Circuit Architecture of Multi-Valued DNA-Quantum T Flip-Flop

Multi-Valued DNA-Quantum T flip-flop operation circuit is constructed using one principle, that is, the first portion of this circuit will be constructed by using Multi-Valued Quantum circuit and the second portion will be constructed by using Multi-Valued Quantum computing circuit. This circuit also has Multi-Valued Quantum cache memory and a Heat transfer circuit to achieve the best output at the end. The multi-valued DNA-Quantum T Flip-Flop operation circuit is shown in Figure 4.10.

In this circuit, input is qubit and output is the molecular sequence. Input |T> and clock input |clk> are shared into two Multi-Valued Quantum AND operations. Two Multi-Valued Quantum operations are constructed parallelly. This Multi-Valued Quantum AND operations output line and final output of Multi-Valued DNA-Quantum T flip-flop are constructed into another two Multi-Valued Quantum operations parallelly. But the final two outputs are molecular sequences so these two output lines are connected to the NMR process and then after the NMR process this two-line came into use as an input line Multi-Valued Quantum AND operation

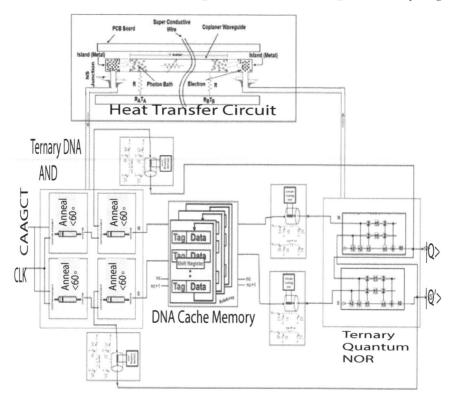

FIGURE 4.10
Multi-Valued DNA-Quantum T Flip-Flop Operation Circuit

circuit. These two Multi-Valued Quantum AND operations circuit output lines are connected into the Multi-Valued Quantum cache memory. This is the first portion of the circuit and is fully designed by using the principle of Multi-Valued Quantum computing. Multi-Valued Quantum cache memory is made by using some Multi-Valued Quantum shift register and in this proposed circuit it saves data in the Multi-Valued Quantum array. When the time arrives Multi-Valued Quantum cache memory supplies the data into the NMR Relaxation process. Multi-Valued Quantum cache memory supply line fully connected to NMR Relaxation process as well as here EMR emit is fully prohibited. Then the NMR Relaxation process output line is connected to the Multi-Valued Quantum NOR gate. Two outputs of the NMR relaxation process are connected to two different Multi-Valued Quantum NOR gates. Multi-Valued Quantum NOR gates another input come from the output which is Q and $\overline{Q}$ molecular sequences. Hence this Multi-Valued Quantum NOR gates' final output line is the final output of Multi-Valued DNA-Quantum T flip-flop.

Multi-Valued Quantum Circuit produces more heat and the Multi-Valued Quantum circuit needs to process the input. For that reason, it would be best to transfer the heat into Multi-Valued Quantum circuits. Multi-Valued Quantum portion Heat transfers using the heat transfer circuit into Multi-Valued Quantum circuit's portion. In the heat transfer circuit, two junctions are connected to Multi-Valued Quantum Circuits and Multi-Valued Quantum circuits. Maximum one meter can transfer heat to this circuit. This heat transfer circuit has superconductive ware, photon batch, PCB board to make the architecture. This circuit mainly transfers the heat using the photon bath. With this circuit proposed Multi-Valued DNA-Quantum T flip-flop circuit is constructed fully.

4.6.3 Working Principle of Multi-Valued DNA-Quantum T Flip-Flop

Multi-Valued DNA-Quantum T flip-flop is working using two principles those are Multi-Valued Quantum computing principle and the Multi-Valued Quantum computing principle. Multi-Valued Quantum -Multi-Valued DNA T flip-flop has one input is named |T> and clock input is |clk>. Multi-Valued DNA-Quantum T flip-flop won't enable if the clock input is not |0>. Hence the clock input enables this circuit to start to work and first of all, it will perform the Multi-Valued Quantum circuit operation. Multi-Valued Quantum circuit operation will be happening in close to zero temperature because qubits need to coherence state. The superposition state will be stable if this circuit is full of prohibited from other environment particles. |T> and |clk> input will be shared and perform two Multi-Valued Quantum AND operations in parallel way. This Multi-Valued Quantum AND operation produce some garbage value but in this proposed circuit this topic is avoidable. Multi-Valued Quantum AND operation produce two output qubits in the same amount of time. This qubit one of each will perform another two Multi-Valued Quantum AND operations in parallel way. The first qubit and final output of Multi-Valued DNA-Quantum T flip-flop will perform Multi-Valued Quantum AND operation. Another qubit and another output sequence of the Multi-Valued DNA-Quantum T flip-flop will perform the Multi-Valued Quantum AND operation. These two operations will perform in parallel way. But the final output of Multi-Valued DNA-Quantum T flip-flop is a molecular sequence. The molecular sequence cannot perform the Multi-Valued Quantum AND operation. For that reason, Multi-Valued Quantum AND operation need a qubit to perform the operation. So, that's why Multi-Valued Quantum molecular sequence performs first of all NMR process and then it converts them molecular sequence into a qubit. Then these qubits and the previous output qubits perform two Multi-Valued Quantum AND operations parallelly and produce the output of two Multi-Valued Quantum AND operations parallelly. These two outputs of two qubits enter into the Multi-Valued Quantum cache memory.

Multi-Valued Quantum cache stores the qubit into an array. Multi-Valued Quantum cache memory stores the qubit data and when requires it serves the data into the NMR relaxation process. In this Multi-Valued DNA-Quantum T flip-flop, two-qubit is stored in Multi-Valued Quantum cache memory. This qubit performs the NMR Relaxation process because Multi-Valued Quantum circuits need molecular sequence. NMR relaxation process removes the superposition state and makes the qubit into a molecular sequence. This molecular sequence performs the Multi-Valued Quantum NOR operation. The final output sequence and molecular sequence from cache perform Multi-Valued Quantum NOR operation. Another final output sequence and previous output from cache also perform another Multi-Valued Quantum NOR operation. These two NOR operation performs parallelly. This structure basically Multi-Valued Quantum NOR latches operation. Multi-Valued Quantum Cache memory is mainly used because Multi-Valued Quantum operation performs so fast Multi-Valued Quantum operation performed very slowly. Multi-Valued Quantum cache memory works as an intermediate process where qubit is just stored and when needed cache memory serves the qubit.

Multi-Valued Quantum Cache memory works here as the intermediate process. Multi-Valued Quantum circuit produces much heat and here two Multi-Valued Quantum AND operation performs so it produces much heat. But Multi-Valued Quantum circuit needs to be in close to zero Kelvin temperature to perform the operation. So from the Multi-Valued Quantum circuit, it is needed to reduce the temperature to maintain the superposition state of the qubit. Hence in Multi-Valued Quantum circuit operation, it needs much heat in several steps. Mainly in melting and annealing require much heat. This topic is briefly described in the first chapter named Multi-Valued DNA-Quantum circuit operation. For that reason, the excessive heat can be transferred from the Multi-Valued Quantum circuit portion to the Multi-Valued Quantum circuit portion using a heat transfer circuit. This heat transfer circuit using the junction capture the heat from the Multi-Valued Quantum circuit and using photon bath heat flows through the circuit and gives this into the Multi-Valued Quantum circuit. This circuit can transfer heat maximum in the one-meter distance and in Multi-Valued DNA-Quantum flip-flop distance is less than one meter between Multi-Valued Quantum and Multi-Valued Quantum circuit. This heat transfer circuit cannot transfer full excessive heat produced from the Multi-Valued Quantum circuit and this heat is not enough to perform the Multi-Valued Quantum circuit operation. But this heat transfer can optimize the cost of heat based on the needs. After all of this operation and architecture fully then Multi-Valued DNA-Quantum T flip-flop can produce two output molecular sequences. Table 4.5 shows the truth table of multi-valued DNA-Quantum T Flip-Flop.

TABLE 4.5

Truth Table of Multi-Valued DNA-Quantum T Flip-Flop

|T>	Q (**Multi-Valued Quantum Sequence**)	$\overline{Q}$ (**Multi-Valued Quantum Sequence**)
TGGATC	|0>	|0>
ACCTAG	|0>	|1>
TGGATC	|1>	|0>
ACCTAG	|1>	|0>

4.6.4 Example

Multi-Valued DNA-Quantum T flip-flop needs one qubit input and one clock input. Clock inputs value is ACCTAG then this Multi-Valued DNA-Quantum flip-flop will be enabled. Assume clock input value is ACCTAG here and then the output is ACCTAG.

These qubit input and clock input will, first of all, perform two Multi-Valued Quantum AND operations parallelly.

So, from two Multi-Valued Quantum AND operations, two-qubit are found which |1> respectively. Assume one output sequence is Q and the other is $\overline{Q}$. Hence here if the Q sequence is ACCTAG then $\overline{Q}$ will be the TGGATC sequence as supposed.

Qubit |1> and output sequence Q as well as another output qubit |1> and $\underline{Q}$ will perform Multi-Valued Quantum AND operation. But molecular sequence cannot perform Multi-Valued Quantum AND operation directly. For that reason, these two sequences, first of all, perform the NMR process and make them qubit, and then perform the Multi-Valued Quantum AND operation with respective other input parallelly. ACCTAG will convert as a qubit and assumed it as |1> and opposite TGGATC as |0> suppose. So |1> and ACCTAG or |1> will perform Multi-Valued Quantum AND operation and |1> and TGGATC or |0> will perform Multi-Valued Quantum AND operation parallelly. Then the outputs will be ACCTAG and TGGATC.

These outputs will be stored in Multi-Valued DNA cache memory. Multi-Valued DNA cache memory will serve this output as input into the Multi-Valued DNA circuit when the right time arrives. These DNA sequences will perform first of all NMR process where these DNA sequences will become the qubit. Now, |1> and |0> will produce the final output as |1> and |0>.

Here found the final output of the Multi-Valued DNA-Quantum T flip-flop. Output can be toggle here according to condition. These Multi-Valued DNA-Quantum T flip-flops are mainly used for toggling.

4.7 Multi-Valued DNA-Quantum Shift Register

A single molecular sequence of two-valued molecular sequence data (TGGATC or ACCTAG) can be stored in a Multi-Valued DNA-Quantum flip flop. However, many Multi-Valued DNA-Quantum flip-flops are required to store multiple molecular sequences of data. To store n molecular sequences of data, N Multi-Valued DNA-D flip flops must be coupled in a certain order. A Multi-Valued DNA-Quantum register is a gadget that stores this type of data. It consists of a sequence of Multi-Valued DNA-Quantum flip flops used to store multiple molecular sequences of data.

Multi-Valued DNA-Quantum shift registers enable the information stored in these Multi-Valued DNA registers to be transmitted. A Multi-Valued DNA-Quantum shift register is a collection of flip flops that stores several molecular sequences of information. By applying clock pulses to the molecular sequences contained in such Multi-Valued DNA-Quantum registers, they may be made to move inside them and in and out of them. By linking n Multi-Valued DNA flip-flops and n number of Multi-Valued Quantum flip-flops, each of which stores a single molecular sequence of data or molecule sequence of data, an n Multi-Valued DNA-Quantum shift register may be built. "Multi-Valued DNA-Quantum Shift left registers" are Multi-Valued DNA-Quantum registers that will shift the molecular sequences or molecules to the left. "Multi-Valued DNA-Quantum Shift right registers" are Multi-Valued DNA-Quantum registers that will shift the molecular sequences or molecule sequence to the right.

Multi-Valued DNA-Quantum Shift registers are basically of 4 types. These are:

1. Multi-Valued DNA-Quantum Serial In Serial Out shift register

2. Multi-Valued DNA-Quantum Serial In parallel Out shift register

3. Multi-Valued DNA-Quantum Parallel In Serial Out shift register

4. Multi-Valued DNA-Quantum Parallel In parallel Out shift register

In this study, a shift register is built utilizing a Multi-Valued DNA D flip-flop operational circuit and a Multi-Valued Quantum D flip-flop operational circuit to convert serial data into Multi-Valued DNA data and Multi-Valued Quantum molecule sequence data. The Multi-Valued DNA-Quantum Serial-In Serial-Out shift register is a type of Multi-Valued DNA-Quantum shift register that permits serial input one molecular sequence or one molecule sequence at a time over a single data line and outputs a serial output. The data exits the Multi-Valued DNA-Quantum shift register one molecular sequence at a time in a serial pattern since there is only one molecular sequence output, thus the term Multi-Valued DNA-Quantum Serial-In Serial-Out Shift Register. Four Multi-Valued DNA-Quantum D flip-flops are linked in a serial fashion in this circuit. Because the same clock signal is supplied to each Multi-Valued DNA-Quantum flip flop, they are all synchronized with one another. The circuit below in the architecture section is an example of a Multi-Valued DNA-Quantum shift right register, which accepts serial data from the Multi-Valued DNA flip flop's left side and Multi-Valued Quantum flip-flops left side.

In Multi-Valued DNA-Quantum flip-flop molecular sequence is entered into as an input in the circuit and performs the required operation. Multi-Valued DNA computing computation speed is much higher than Multi-Valued Quantum computing computational speed. For that case molecular sequences are stored in Multi-Valued DNA cache memory. When actually the time arrives then molecular sequence first of all performs the NMR process and makes the molecular sequence into molecule sequence. Then this molecule sequence performs the Multi-Valued Quantum flip-flop operation. Multi-Valued DNA circuit produces much heat and heat needs to be removed to maintain the coherence state and Multi-Valued Quantum computing circuit needs huge heat to enable the circuit to perform. In that case, this circuit uses a heat transfer circuit to transfer the heat into the Multi-Valued Quantum circuits.

4.7.1 General Organization of Multi-Valued DNA-Quantum Shift Register

Three Multi-Valued DNA D flip-flop operational circuits and One Multi-Valued Quantum D flip-flop circuits are used to make a Multi-Valued DNA-Quantum shift register. As a fundamental component, a Multi-Valued DNA-Quantum Shift register is utilized for data shift, and a Multi-Valued DNA D flip-flop and Multi-Valued Quantum D flip-flop is used to make it happen.

Two inputs are used in Multi-Valued DNA D flip flops: one for data and one for the clock. The outputs of Multi-Valued DNA D flip flops are logically opposite one another. The circuit's synchronization with an external signal is aided by the clock input. A Multi-Valued DNA D flip flop's output can have two possible values. Data input is directed to a Multi-Valued DNA NAND operation circuit in this block diagram, while data input reverse is routed to another Multi-Valued DNA NAND operation circuit. Both Multi-Valued DNA NAND operations procedures use the clock pulse input. Multi-Valued DNA SR Latch receives the result of two Multi-Valued DNA NAND operations. The D flip flop is constructed using the SR latch. This property is used to induce a delay in the circuit's data flow. Multi-Valued DNA SR Latch is made up of two Multi-Valued DNA NAND Operations. The final two outputs of the Multi-Valued DNA SR Latch function were uncovered. A Multi-Valued DNA D flip flop may provide two types of output, one of which is logically inverse to the other. The Multi-Valued DNA D flip flop will continue to function if the clock is enabled; otherwise, the Multi-Valued DNA D flip flop will stop working.

Multi-Valued DNA D flip-flop operational circuits are also coupled through serial connection in the Multi-Valued DNA shift register block diagram.

The Multi-Valued DNA D flip-flop operational circuit is the fundamental component of the Multi-Valued DNA shift register. After processing the molecular sequence input in the Multi-Valued DNA Shift Register output of the shift register is stored in Multi-Valued DNA Cache Memory.

In Multi-Valued Quantum computing molecular sequence needs to go through many processes during the operation. In Multi-Valued Quantum computing, melting and annealing are very important and these steps require a vast amount of heat. That's why in Multi-Valued DNA-Quantum T flip-flop Multi-Valued Quantum portion transfers a small amount of heat produced from the Multi-Valued Quantum circuits into Multi-Valued Quantum circuits. Though this amount of heat is not enough, this heat or temperature will help to perform Multi-Valued Quantum computation. In Multi-Valued Quantum computing Multi-Valued Quantum NAND SR latch performs where two Multi-Valued Quantum NAND gates perform parallelly and input comes from Multi-Valued Quantum circuit portion. After that operation, the output molecular sequence are found from Multi-Valued DNA-Quantum T flip-flop.

4.7.2 Circuit Architecture of Multi-Valued DNA-Quantum Shift Register

Four D flip-flops and a Multi-Valued DNA AND Gate are used in the Multi-Valued DNA portion and a Multi-Valued Quantum D flip-flop in the Multi-Valued DNA-Quantum Shift Register. Shift Register generates four molecular sequence outputs using these. The D flip-flop has a single molecular sequence input and is developed

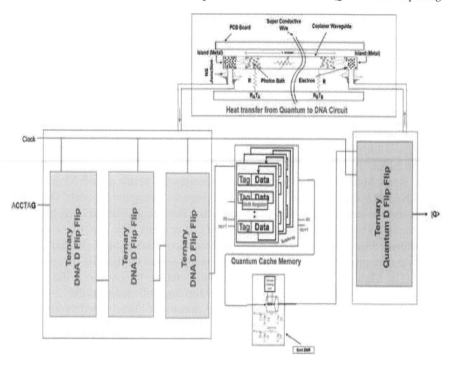

FIGURE 4.11

Block Diagram of Multi-Valued DNA-Quantum Shift Register

using Multi-Valued DNA NAND and Multi-Valued DNA SR latch operations. Figure 4.11 shows the block diagram of multi-valued DNA-Quantum Shift Register.

The clock molecular sequence input is required for the Multi-Valued DNA D flip-flop. It can be seen from the diagram that the circuit has one molecular sequence input. One line of this molecular sequence input will be directed into a Multi-Valued DNA NAND operation termed S input in Circuit. Multi-Valued DNA NAND is performed here using S molecular sequence input and Clock molecular sequence input.

In this circuit, input is molecular sequence and output is a qubit. Clock input clk is shared into a Multi-Valued DNA NAND operation and a Multi-Valued DNA NOT operation. Two Multi-Valued DNA operations are constructed parallelly. These Multi-Valued DNA NAND operations output line and final output of Multi-Valued DNA-Quantum Shift Register are constructed into another two Multi-Valued DNA operations parallelly. But the final two outputs are qubits so these two output lines are connected to the NMR process and then after that process, these two lines came into use as an input line Multi-Valued DNA NAND operation circuit. These two Multi-Valued DNA NAND operations circuit output lines are connected into the Multi-Valued DNA cache memory. All of these Multi-Valued DNA NAND Operations are the part of Multi-Valued DNA D flip flop which is used in this system. This is the first portion of the circuit and is fully designed by using the principle of

Multi-Valued DNA computing. Multi-Valued DNA cache memory is made by using some Multi-Valued DNA shift register and in this proposed circuit it saves data in the Multi-Valued DNA array. When the time arrives Multi-Valued DNA cache memory supplies the data into the NMR process. Multi-Valued DNA cache memory supply line fully connected to NMR process as well as here EMR is emitted. Then the NMR process output line is connected to the Multi-Valued Quantum NOT gate. After going through the Multi-Valued Quantum NOT operation the output is taken as input in the Multi-Valued Quantum NAND operation. In these Multi-Valued Quantum NAND operations 1st clock input is taken as |clk> input. These inputs are shared into two Multi-Valued Quantum NAND gates. Then the outputs of these operations are connected to another set of Multi-Valued Quantum NAND gates. Multi-Valued Quantum NAND gates another input come from the output which is |0> qubit. Hence this Multi-Valued Quantum NAND gates' final output line is the final output of Multi-Valued DNA-Quantum Shift Register.

Multi-Valued Quantum Circuit produces more heat and the Multi-Valued Quantum circuit needs to process the input. For that reason, it would be best to transfer the heat into Multi-Valued Quantum circuits. Multi-Valued Quantum portion Heat transfers using the heat transfer circuit into Multi-Valued Quantum circuit's portion. In the heat transfer circuit, two junctions are connected to Multi-Valued Quantum Circuits and Multi-Valued Quantum circuits. Maximum one meter can transfer heat to this circuit. This heat transfer circuit has superconductive ware, photon batch, PCB board to make the architecture. This circuit mainly transfers the heat using the photon bath. With this circuit proposed Multi-Valued DNA-Quantum T flip-flop circuit is constructed fully.

4.7.3 Working Principle of Multi-Valued DNA-Quantum Shift Register

Multi-Valued DNA-Quantum shift registers are a kind of registers where both molecular sequence data loading, as well as data retrieval to/from the Multi-Valued DNA-Quantum shift register, occurs in serial mode sometimes. This research synchronous Multi-Valued DNA SISO shift register sensitive to the positive edge of the clock pulse. Here the data word which is to be stored is fed bit-by-bit at the molecular sequence input of the first Multi-Valued DNA flip-flop. Further, it is seen that the molecular sequence inputs of all other flip-flops are driven by the outputs of the preceding ones, for example, the input of Multi-Valued DNA D Flip-flop number- 2 is driven by the output of Multi-Valued DNA D flip-flop number-1. At last, the data stored within the Multi-Valued DNA shift register is obtained at the output pin of the nth Multi-Valued DNA D flip-flop in serial fashion.

Initially, all the Multi-Valued DNA flip-flops in the Multi-Valued DNA register are cleared by applying high on their clear pins. Next, the input data word is fed serially to Multi-Valued DNA D Flip-flop number-1.

This causes the molecular sequence appearing at the first pin to be stored into Multi-Valued DNA D flip-flop number-1 as soon as the first leading edge of the clock appears. Further at the second clock tick, B1 gets stored into Multi-Valued DNA D flip-flop number-2 while a new bit enters into Multi-Valued DNA flip-flop number-2.

TABLE 4.6

Truth Table of Multi-Valued DNA-Quantum Shift Register

Clk	|x>	|Q>	$\overline{\|Q>}$	Description
↓ >> ACCTAG	X	Q	Q	Memoryno change
↑>> TGGATC	|0>	|0>	|1>	Reset Q >>0
↑>> TGGATC	|1>	|1>	|0>	Set Q >>

This kind of shift in data molecular sequences continues for every rising edge of the clock pulse. This indicates that for every single clock pulse the data within the Multi-Valued DNA register moves towards the right by a single bit. Following the molecular sequence data transmission, as explained, one can note that the first molecular sequence of an input word appears at the output of nth flip-flop for the nth clock tick. On applying further clock cycles, one gets the next successive molecular sequences of the molecular sequence input data word as the serial output. Table 4.6 shows the truth table of multi-valued DNA-Quantum Shift Register.

Multi-Valued Quantum cache stores the qubit into an array. Multi-Valued Quantum cache memory stores the qubit data and when requires it serves the data into the NMR relaxation process. In this Multi-Valued DNA-Quantum T flip-flop, two-qubit is stored in Multi-Valued Quantum cache memory. This qubit performs the NMR Relaxation process because Multi-Valued Quantum circuits need molecular sequence. NMR relaxation process removes the superposition state and makes the qubit into a molecular sequence. This molecular sequence performs the Multi-Valued Quantum NAND operation. The final output sequence and molecular sequence from cache perform Multi-Valued Quantum NAND operation. Another final output sequence and previous output from cache also perform another Multi-Valued Quantum NAND operation. These two NAND operation performs parallelly. This structure basically Multi-Valued Quantum NAND latches operation. Multi-Valued Quantum Cache memory is mainly used because Multi-Valued Quantum operation performs so fast Multi-Valued Quantum operation performed very slowly. Multi-Valued Quantum cache memory works as an intermediate process where qubit is just stored and when needed cache memory serves the qubit.

Multi-Valued Quantum Cache memory works here as the intermediate process. Multi-Valued Quantum circuit produces much heat and here two Multi-Valued Quantum AND operation performs so it produces much heat. But Multi-Valued Quantum circuit needs to be in close to zero Kelvin temperature to perform the operation. So from the Multi-Valued Quantum circuit, it is needed to reduce the temperature to maintain the superposition state of the qubit. Hence in Multi-Valued Quantum circuit operation, it needs much heat in several steps. Mainly in melting and annealing require much heat. This topic is briefly described in the first chapter named Multi-Valued DNA-Quantum circuit operation. For that reason, the excessive heat can be transferred from the Multi-Valued Quantum circuit portion to the Multi-Valued Quantum circuit portion using a heat transfer circuit. This heat transfer circuit using the junction capture the heat from the Multi-Valued Quantum circuit and using photon bath heat flows through the circuit and gives this into the Multi-Valued Quantum circuit. This circuit can transfer heat maximum in the one-meter distance and in Multi-Valued DNA-Quantum flip-flop distance is less than one meter between Multi-Valued Quantum and Multi-Valued Quantum circuit. This heat transfer circuit cannot transfer full excessive heat produced from the Multi-Valued Quantum circuit and this heat is not enough to perform the Multi-Valued Quantum circuit operation. But this heat transfer can optimize the cost of heat based on the needs. After all of this operation and architecture fully then Multi-Valued DNA-Quantum T flip-flop can produce two output molecular sequences.

4.7.4   Example

In the Multi-Valued DNA shift register, only one molecular sequence input is needed to perform the required operation. Multi-Valued DNA shift register will enable if and only if when the clock input is enabled. This research proposed Multi-Valued DNA shift register shift one molecular sequence data into the right position.

Assume TGGATC data is going to perform the Multi-Valued DNA shift register operation. First of all 0 will perform the Multi-Valued DNA D flip-flop number-1 operation and produce the output is ACCTAG. Then this ACCTAG will shift into the right side as well as every molecular sequence will shift right side once.

Now TGGATC will be entered into the Multi-Valued DNA shift register and perform the midst Multi-Valued DNA D flip-flop operation and then again like the previous one the data will be shifted once on the right side.

Then like these two step by step TGGATC,TGGATC also entered into the Multi-Valued DNA shift register in the relevant period and perform the midst Multi-Valued DNA D flip-flop operation. Then again for each data will be shifted once.

This truth table, it illustrates that every time molecular sequence data shift once on the right side. In Multi-Valued DNA shift register operational circuit will perform n+1 operation then truth final data will be like the table below (Table 4.7).

Here in n+1 operation time, no data is available that's why ACCTAG is appeared in the truth table and according to the Multi-Valued DNA Shift register principle one data molecular sequence shifted once on the right side. After n+1 operation, final data is ACCTAG, TGGATC, TGGATC, TGGATC. These data purely proved that

TABLE 4.7

Data Shifting Process in Multi-Valued DNA-Quantum Shift Register

clk	|x>	|Q>	|Q1>	|Q2>	|Q3>
ACCTAG	|0>	|0>	|0>	|0>	|0>
TGGATC	|0>	|0>	|0>	|0>	|0>
TGGATC	|1>	|1>	|0>	|0>	|0>
TGGATC	|1>	|1>	|1>	|0>	|0>
TGGATC	|1>	|1>	|1>	|1>	|0>
TGGATC		|1>	|1>	|1>	|1>

the Multi-Valued DNA shift register gives the correct required output by maintaining the Multi-Valued DNA principle. Multi-Valued DNA shift register can produce some garbage value but in this research, this is not the part of the study.

4.8   Multi-Valued DNA-Quantum Ripple Counter

A Multi-Valued DNA-Quantum counter is basically used to count the number of clock pulses applied to a Multi-Valued DNA-Quantum flip-flop. It can also be used for Multi-Valued DNA-Quantum Frequency divider, Multi-Valued DNA-Quantum time measurement, Multi-Valued DNA-Quantum frequency measurement, Multi-Valued DNA-Quantum distance measurement and also for generating square wave-forms. In this, the Multi-Valued DNA-Quantum flip-flops are Multi-Valued DNA-Quantum asynchronous counters and are supplied with different clock signals, there may be a delay in producing output. Also, a few numbers of Multi-Valued DNA-Quantum logic gates are needed to design asynchronous counters. So they are elementary in design and also are less expensive.

A n-bit ripple counter can count up to 2n states. It is also known as MOD n counter. It is known as a ripple counter because of the way the clock pulse ripples its way through the flip-flops. It is an asynchronous counter. Different Ripple Counters are used with a different clock pulse. All the flip-flops are used in toggle mode. Only one flip-flop is applied with an external clock pulse and another flip-flop clock is obtained from the output of the previous flip-flop. The flip-flop applied with an external clock pulse acts as LSB (Least Significant Bit) in the counting sequence. A counter may be an up counter that counts upwards or can be a down counter that counts downwards or can do both i.e. count up as well as count downwards depending on the input control. The sequence of counting usually gets repeated after a limit

Multi-Valued DNA-Quantum Ripple Counter is made out of four JK flip flops. Using these Multi-Valued DNA JK Flip flops, the Multi-Valued DNA-Quantum Ripple Counter creates four molecular sequence outputs. Here in JK flip flop, J and K are not shortened abbreviated letters of other words, such as "S" for Set and "R" for Reset, but are autonomous letters chosen by its inventor Jack Kilby to distinguish the flip-flop design from other types. Though Jack Kilby invented the digital electronics JK flip flop. Multi-Valued DNA-Quantum Ripple Counter is an asynchronous counter. It is created using Multi-Valued DNA JK flip flops and these flip flops are only controlled by clock pulse input.

Multi-Valued DNA-Quantum Ripple Counter produces much heat to produce the molecule's superposition state and also produces some garbage value.

4.8.1   General Organization of Multi-Valued DNA-Quantum Ripple Counter

Multi-Valued DNA-Quantum Ripple Counter uses three Multi-Valued DNA JK flip flops to operate two molecular sequence inputs. Multi-Valued DNA JK flip flop has the two-molecular sequence input named J and K.

Multi-Valued DNA JK flip-flop has the molecular sequence input J and K. This Multi-Valued DNA JK flip flop consists of many Ternary DNA NAND operations. At first basic Ternary DNA NAND operation performs a couple of then this operation output is entered into the SR flip flop as well as got the output Q and $\overline{Q}$. First of all J and clk input performs the Ternary DNA NAND operation. The output of the Ternary DNA NAND operation and output of the Multi-Valued DNA JK flip flop $\overline{Q}$ performs another Ternary DNA NAND operation and produces the output named as clk. clk input is shared so K and clk input also performs the Ternary DNA NAND operation. This Ternary DNA NAND operations output and Q output of the Multi-Valued DNA JK flip flop performs another Ternary DNA NAND operation as well as produces the output named as R.

These clk and R entered into the Multi-Valued DNA SR flip flop and produce two outputs Q and $\overline{Q}$. In Multi-Valued DNA SR flip flop, it has four Ternary DNA NAND operations. clk input and clk input performs Ternary DNA NAND operation as well as R input and shared clk input also performs the Ternary DNA NAND operation. These two NAND operations output entered the Multi-Valued DNA SR latch as input. In Multi-Valued DNA SR latches two Q and one input as well as $\overline{Q}$ and other inputs perform the Ternary DNA NAND operation. Finally, after all of this Multi-Valued DNA operation, the Multi-Valued DNA JK flip flops final output Q and $\overline{Q}$ are obtained.

After processing the molecular sequence inputs in three JK flip flops like this, the output is stored in a Multi-Valued DNA Cache Memory, which is made out of Multi-Valued DNA Shift Register. Then this Multi-Valued DNA cache memory serves the molecular sequence into the Multi-Valued Quantum computing circuit's portion. These molecular sequences need to be qubit first of all so molecular sequences are performed here NMR process. In the NMR process, EMR is emitted. After the NMR process molecular sequences are relaxing their state and are converted into the qubit.

In Multi-Valued Quantum computing molecular sequence needs to go through many processes during the operation. In Multi-Valued Quantum computing, melting and annealing are very important and these steps require a vast amount of heat. That's why in Multi-Valued DNA-Quantum T flip-flop Multi-Valued Quantum portion transfers a small amount of heat produced from the Multi-Valued Quantum circuits into Multi-Valued Quantum circuits. Though this amount of heat is not enough, this heat or temperature will help to perform Multi-Valued Quantum computation. In Multi-Valued Quantum computing Multi-Valued Quantum NOR SR latch performs where two Multi-Valued Quantum NOR gates perform parallelly and input comes from Multi-Valued Quantum circuit portion. After that operation, the output molecular sequence are found from Multi-Valued DNA-Quantum T flip-flop.

4.8.2 Circuit Architecture of Multi-Valued DNA-Quantum Ripple Counter

Multi-Valued DNA-Quantum Ripple Counter (Figure 4.12) uses three Multi-Valued DNA JK flip flops and one Multi-Valued Quantum JK flip flop to create qubit outputs.

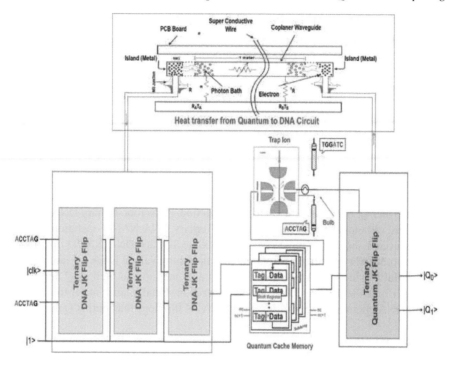

FIGURE 4.12

Block Diagram of Multi-Valued DNA-Quantum Ripple Counter

Multi-Valued DNA JK flip flop has a two-molecular sequence input and one clock shared input. Multi-Valued DNA JK flip flops also depend on the clock. If the clock is enabled then the circuit will enable, otherwise not.

In Multi-Valued DNA-Quantum Ripple Counter there's a one clock input and one logic input which is shared in both J and K input ports. Input J and input K both the value will perform the Ternary DNA NAND operation with the shared clk input differently. J and clk input performs the Ternary DNA NAND operation. NAND operation made by using Multi-Valued DNA basic gates. Basic gates in mainly Multi-Valued DNA computing are V, V+, and CNOT gates. For error correction here used an ancillary bit. In Quant NAND operation the value of an ancillary bit in TGGATC. After J and clk perform the Ternary DNA NAND operation produced an output molecular sequence and this output molecular sequence and the output of the flip flop |Q> perform the Ternary DNA NAND operation and produce the output clk. Like in the same procedure K, clk and |Q> input produces the R output. These operation circuits are mainly Ternary DNA NAND operations. This clk and R input is performed the Multi-Valued DNA SR flip-flop operation. In Multi-Valued DNA SR flip flop operational circuit architecture also made by the proposed basic component of Multi-Valued DNA computing is Ternary DNA NAND operation.

After processing the inputs in a Multi-Valued DNA JK flip flop, the output of the flip flop is going to be stored as an output of the Multi-Valued DNA-Quantum

Ripple Counter. Multi-Valued DNA-Quantum Ripple Counter intermediate architectures four Multi-Valued DNA JK flip-flops connected in serial connection using the logical molecular sequence input. Every clock input as clk enters in every molecular sequence and from 2nd Multi-Valued DNA JK flip-flop clk input is previous Multi-Valued DNA Jk flip-flops first output |Q> . With the same architecture, Multi-Valued DNA-Quantum Ripple Counter can be performed as an up counter or as a down counter but clock pulse as clk need to sometimes have a positive edge-triggered and sometimes a negative edge triggered.

The JK flip flop's circuit output lines are connected into the Multi-Valued DNA cache memory. This is the first portion of the circuit and is fully designed by using the principle of Multi-Valued DNA computing. Multi-Valued DNA cache memory is made by using some Multi-Valued Quantum shift register and in this proposed circuit it saves data in the Multi-Valued DNA array. When the time arrives Multi-Valued DNA cache memory supplies the data into the NMR process. Multi-Valued DNA cache memory supply line fully connected to NMR process as well as here EMR is emitted. Then the NMR process output line is connected to the Ternary Quantum NAND gate. Two outputs of the NMR process are connected to two different Multi-Valued Quantum NOR gates. Multi-Valued Quantum NOR gates another input come from the output which is Q and $\overline{Q}$ qubits. Hence this Ternary Quantum NAND gates' final output line is the final output of Multi-Valued DNA-Quantum Ripple Counter.

Multi-Valued Quantum Circuit produces more heat and the Multi-Valued Quantum circuit needs to process the input. For that reason, it would be best to transfer the heat into Multi-Valued Quantum circuits. Multi-Valued Quantum portion Heat transfers using the heat transfer circuit into Multi-Valued Quantum circuit's portion. In the heat transfer circuit, two junctions are connected to Multi-Valued Quantum Circuits and Multi-Valued Quantum circuits. Maximum one meter can transfer heat to this circuit. This heat transfer circuit has superconductive ware, photon batch, PCB board to make the architecture. This circuit mainly transfers the heat using the photon bath. With this circuit proposed Multi-Valued DNA-Quantum T flip-flop circuit is constructed fully.

4.8.3 Working Principle of Multi-Valued DNA-Quantum Ripple Counter

In the Multi-Valued DNA-Quantum Ripple Counter there are four Multi-Valued DNA JK flip-flop operational circuits. These Multi-Valued DNA JK flip-flop operational circuits are connected in serial connection. In this Multi-Valued DNA-Quantum Ripple Counter there is one clock input and a logic input which shared into the port J and K of Multi-Valued DNA JK flip-flop operational circuit. The inputs J and K of a Multi-Valued DNA JK flip-flop conducts two Multi-Valued DNA processes in parallel. The Ternary DNA NAND operation is performed using J and shared input clk. The Ternary DNA NAND operations are then completed, and one of the Multi-Valued DNA JK flip flop's outputs executes the Ternary DNA NAND operation, producing clk. The K and |clk > inputs are performed first in the Ternary DNA NAND operation. The output of the Multi-Valued DNA JK flip-flop |Q>, as

TABLE 4.8

Truth Table of Multi-Valued DNA-Quantum Ripple
Counter

clk	$|Q_0>$	$|Q_1>$	$|Q_2>$	$|Q_3>$
ACCTAG	ACCTAG	ACCTAG	ACCTAG	$|0>$
TGGATC	TGGATC	TGGATC	TGGATC	$|1>$
TGGATC	ACCTAG	ACCTAG	ACCTAG	$|0>$

well as the result of the Multi-Valued DNA first NAND operation, are then used to perform another Ternary DNA NAND operation, yielding R. The steps for creating clk and R are carried out simultaneously. It is known that one of the distinctive properties of Multi-Valued DNA operations is that they may do several operations at the same time, and this is exactly what is happening. The Multi-Valued DNA SR flip flop operation is then conducted on these clk and R inputs. The Ternary DNA NAND operation, which is employed here, is also used to make Multi-Valued DNA SR flip flops. Two outputs are discovered after conducting the Multi-Valued DNA SR flip-flop operation. The opposite of one output is the opposite of the other.

Here if the clk is ACCTAG then the Multi-Valued DNA JK flip-flop will not be triggered but if clk is TGGATC the Multi-Valued DNA JK flip-flop will be triggered and it will toggle the output. First of all Multi-Valued DNA JK flip-flop operational circuit getting clk value is TGGATC and toggles the output value from the previous state value. Then the output of the initial Multi-Valued DNA JK flip-flop will be clk input of next Multi-Valued DNA JK flip-flop. If the clk value is TGGATC then the output value will be toggled, otherwise the output will be the previous state output. Maintaining the same procedure every Multi-Valued DNA JK flip-flop operated in the Multi-Valued DNA-Quantum Ripple Counter. Multi-Valued DNA JK flip-flop is toggled very much, that's why in the Multi-Valued DNA-Quantum Ripple Counter Multi-Valued DNA JK flip-flop operational circuit here is used as a basic component. Table 4.8 shows the truth table of multi-valued DNA-Quantum Ripple Counter.

The JK flip flop's circuit output lines are connected into the Multi-Valued DNA cache memory. This is the first portion of the circuit and is fully designed by using the principle of Multi-Valued DNA computing. Multi-Valued DNA cache memory is made by using some Multi-Valued DNA shift register and in this proposed circuit it saves data in the Multi-Valued DNA array. When the time arrives Multi-Valued DNA cache memory supplies the data into the NMR process. Multi-Valued DNA cache memory supply line fully connected to NMR process as well as here EMR is emitted. Then the NMR process output line is connected to the Ternary Quantum NAND gate. Two outputs of the NMR process are connected to two different Multi-Valued Quantum NOR gates. Multi-Valued Quantum NOR gates another input come from the output which is Q and $\overline{Q}$ qubits. Hence this Ternary Quantum NAND gates' final output line is the final output of Multi-Valued DNA-Quantum Ripple Counter.

Multi-Valued Quantum Cache memory works here as the intermediate process. Multi-Valued Quantum circuit produces much heat and here two Ternary Quantum AND operation performs so it produces much heat. But Multi-Valued Quantum circuit needs to be in close to zero Kelvin temperature to perform the operation. So from the Multi-Valued Quantum circuit, it is needed to reduce the temperature to maintain the superposition state of the qubit. Hence in Multi-Valued Quantum circuit operation, it needs much heat in several steps. Mainly in melting and annealing require much heat. This topic is briefly described in the first chapter named Multi-Valued DNA-Quantum circuit operation. For that reason, the excessive heat can be transferred from the Multi-Valued Quantum circuit portion to the Multi-Valued Quantum circuit portion using a heat transfer circuit. This heat transfer circuit using the junction captures the heat from the Multi-Valued Quantum circuit and using photon bath heat flows through the circuit and gives this into the Multi-Valued Quantum circuit. This circuit can transfer heat maximum in the one-meter distance and in Multi-Valued DNA-Quantum flip-flop distance is less than one meter between Ternary Quantum AND Multi-Valued Quantum circuit. This heat transfer circuit cannot transfer full excessive heat produced from the Multi-Valued Quantum circuit and this heat is not enough to perform the Multi-Valued Quantum circuit operation. But this heat transfer can optimize the cost of heat based on the needs. After all of this operation and architecture fully then Multi-Valued DNA-Quantum T flip-flop can produce two output molecular sequences.

4.8.4 Example

To check the Multi-Valued DNA-Quantum Ripple Counter proposed circuit, assume a clock signal is TGGATC and the logical molecular sequence is high. So, if initially clock signal TGGATC is delivered then for the first Multi-Valued DNA JK flip-flop operational circuit, output will be same.

According to the work principle of Multi-Valued DNA-Quantum Ripple Counter if clock input value is TGGATC then the previous state value will be toggled. Then now from the first Multi-Valued DNA JK flip-flop got one output which is TGGATC. This TGGATC will be the clock input for the second Multi-Valued DNA JK flip-flop circuit according to the architecture of the Multi-Valued DNA-Quantum Ripple Counter.

As like the previous Multi-Valued DNA JK flip-flop for thirds principle is the same. Fourth flip-flop is a Multi-Valued Quantum JK flip-flop operational circuit. This circuit got qubit by NMR process. The last output will be |0>.

Hence from the truth table, it is seen that the proposed Multi-Valued DNA-Quantum Ripple Counter is correct theoretically. Multi-Valued DNA JK flip-flop used here for toggling. This Multi-Valued DNA-Quantum Ripple Counter produces much heat when the molecule's normal state becomes a superposition state. Multi-Valued DNA-Quantum Ripple Counters full operation needs to be an environment where other particles are totally prohibited to maintain the coherence state.

4.9 Multi-Valued DNA-Quantum Synchronous Counter

A Multi-Valued DNA-Quantum counter is a Multi-Valued DNA-Quantum device which can count any particular event on the basis of how many times the particular event(s) has occurred. In a Multi-Valued DNA-Quantum logic system or computers, this Multi-Valued DNA-quantum counter can count and store the number of times any particular event or process has occurred, depending on a Multi-Valued DNA-Quantum clock signal. Most common type of Multi-Valued DNA-Quantum counter is a sequential Multi-Valued DNA logic circuit with a single clock input and multiple molecular sequence outputs. The molecular sequence outputs represent two-valued decimal numbers. Each clock pulse either increases the number or decreases the number.

Multi-Valued DNA-Quantum Synchronous Circuit generally refers to something which is coordinated with others based on time. Multi-Valued DNA-Quantum Synchronous Signals occur at same clock rate and all the clocks follow the same reference clock. Multi-Valued DNA asynchronous Counter have shown that the molecular sequence output of that Multi-Valued DNA counter is directly connected to the input of next subsequent counter and making a chain system, and due to this chain system propagation delay appears during counting stage and create counting delays. In a Multi-Valued DNA-Quantum Synchronous Counter, the clock molecular sequence input across all the Multi-Valued DNA flip-flops use the same source and create the same clock signal at the same time. So, a Multi-Valued DNA counter which is using the same clock signal from the same source at the same time is called a Multi-Valued DNA-Quantum Synchronous Counter.

Multi-Valued DNA-Quantum Synchronous Counter is made out of four JK flip flops and two Ternary DNA AND Gates. Using these Multi-Valued DNA JK Flip flops and Ternary DNA AND Gates, the Multi-Valued DNA-Quantum Synchronous Counter creates four molecular sequence outputs. A Multi-Valued DNA-Quantum Synchronous Counter produces much heat and this circuit operation needs to happen in the required environment of Multi-Valued DNA computing.

4.9.1 General Organization of Multi-Valued DNA-Quantum Synchronous Counter

Multi-Valued DNA-Quantum Synchronous Counter uses three Multi-Valued DNA JK flip flops and one Multi-Valued Quantum JK flip flop to create qubit outputs from molecular sequence outputs. Multi-Valued DNA JK flip flop has the two-molecular sequence input named as J and K.

Multi-Valued DNA JK flip-flop has the molecular sequence input J and K. This Multi-Valued DNA JK flip flop consists of many Ternary DNA NAND operations. At first basic Ternary DNA NAND operation performs a couple of then this operation output is entered into the SR flip flop as well as got the output Q and Q. First of all J and clk input performs the Ternary DNA NAND operation. The output of the Ternary DNA NAND operation and output of the Multi-Valued DNA JK flip flop $\overline{Q}$ performs another Ternary DNA NAND operation and produces the output named as S. clk input is shared so K and clk input also performs the Ternary DNA NAND operation. This Ternary DNA NAND operations output and Q output of the Multi-Valued DNA JK flip flop performs another Ternary DNA NAND operation as well as produces the output named as R.

These S and R entered into the Multi-Valued DNA SR flip flop and produce two outputs Q and $\overline{Q}$. In Multi-Valued DNA SR flip flop, it has four Ternary DNA NAND operations. S input and clk input performs Ternary DNA NAND operation as well as R input and shared clk input also performs the Ternary DNA NAND operation. These two NAND operations output entered the Multi-Valued DNA SR latch as input. In Multi-Valued DNA SR latches two Q and one input as well as $\overline{Q}$ and other inputs perform the Ternary DNA NAND operation. Finally, after all of these Multi-Valued DNA operations, the Multi-Valued DNA JK flip flops final output Q and $\overline{Q}$ are obtained.

After processing the inputs in a Multi-Valued DNA JK flip flop, the output of the flip flop is going to be stored as an output of the Multi-Valued DNA-Quantum Synchronous Counter. Thus, Multi-Valued DNA-Quantum Synchronous Counter creates four molecular sequence outputs using four Multi-Valued DNA JK flip flops. The clock inputs for all of the four Multi-Valued DNA Jk flip flops come from the same source. For this, all of the flip flops work synchronously. One output of each of the second and third flip flops go through Ternary DNA AND Gates.

Then the output of the Multi-Valued DNA portion is stored in the Multi-Valued DNA Cache Memory. Whenever needed the output then goes through the "NMR" process in which the EMR is emitted. Thus the molecular sequences of the superstrate become the qubit and used as inputs for the Multi-Valued Quantum JK flip flop.

In Multi-Valued Quantum computing molecular sequence needs to go through many processes during the operation. In Multi-Valued Quantum computing, melting and annealing are very important and these steps require a vast amount of heat. That's why in Multi-Valued DNA-Quantum T flip-flop Multi-Valued Quantum portion transfers a small amount of heat produced from the Multi-Valued Quantum circuits into Multi-Valued Quantum circuits. Though this amount of heat is not enough, this heat or temperature will help to perform Multi-Valued Quantum computation. In Multi-Valued Quantum computing Ternary Quantum NAND SR latch performs where two Ternary Quantum NAND gates perform parallelly and input comes from Multi-Valued Quantum circuit portion. After that operation, the output molecular sequence are found from Multi-Valued DNA-Quantum T flip-flop.

4.9.2 Circuit Architecture of Multi-Valued DNA-Quantum Synchronous Counter

Multi-Valued DNA-Quantum Synchronous Counter uses three Multi-Valued DNA JK flip flops and one Multi-Valued Quantum JK flip flops to create qubit outputs. Multi-Valued DNA JK flip flop has a two-molecular sequence input and one clock shared input. Multi-Valued DNA JK flip flops also depend on the clock. If the clock is enabled then the circuit will enable, otherwise not. Figure 4.13 shows the block diagram of multi-valued DNA-Quantum Synchronous Counter.

Input J and input K both the value will perform the Ternary DNA NAND operation with the shared clk input differently. J and clk input performs the Ternary DNA NAND operation. NAND operation made by using Multi-Valued DNA basic gates.

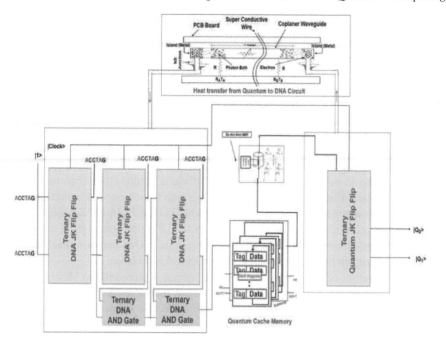

FIGURE 4.13
Block Diagram of Multi-Valued DNA-Quantum Synchronous Counter

Basic gates in mainly Multi-Valued DNA computing are v, V+, and CNOT gates. For error correction here used an ancillary bit. In Quant NAND operation the value of an ancillary bit in TGGATC. After J and clk perform the Ternary DNA NAND operation and an output molecular sequence is obtained and this output molecular sequence and the output of the flip flop $|\overline{Q}\rangle$ perform the Ternary DNA NAND operation and produce the output S. Like in the same procedure K, clk and $|Q\rangle$ input produces the R output. These operational circuits are mainly Ternary DNA NAND operations. This S and R input is performed the Multi-Valued DNA SR flip-flop operation. In Multi-Valued DNA SR flip flop operation circuit architecture also made by the proposed basic component of Multi-Valued DNA computing is Ternary DNA NAND operation.

After processing the inputs in a Multi-Valued DNA JK flip flop, the output of the flip flop is going to be stored as an output of the Multi-Valued DNA-Quantum Synchronous Counter. Multi-Valued DNA-Quantum Synchronous Counter intermediate architectures three Multi-Valued DNA JK flip-flops connected in serial connection using the logical molecular sequence input. Every clock input as clk enters in every molecular sequence and from 2nd Multi-Valued DNA JK flip-flop clk input is previous Multi-Valued DNA Jk flip-flops first output $|Q\rangle$. With the same architecture Multi-Valued DNA-Quantum Synchronous Counter can be performed as a up counter or as a down counter but clock pulse as clk need to sometimes have a positive edge triggered and sometimes a negative edge triggered.

The JK flip flop's circuit output lines are connected into the Multi-Valued DNA cache memory. This is the first portion of the circuit and is fully designed by using the principle of Multi-Valued DNA computing. Multi-Valued DNA cache memory is made by using some Multi-Valued DNA shift register and in this proposed circuit it saves data in the Multi-Valued DNA array. When the time arrives Multi-Valued DNA cache memory supplies the data into the NMR process. Multi-Valued DNA cache memory supply line fully connected to NMR process as well as here EMR is emitted. Then the NMR process output line is connected to the Ternary Quantum NAND gate. Two outputs of the NMR process are connected to two different Ternary Quantum NAND gates. Ternary Quantum NAND gates another input come from the output which is Q and $\bar{Q}$ qubits. Hence this Ternary Quantum NAND gates' final output line is the final output of Multi-Valued DNA-Quantum Synchronous Counter.

Multi-Valued Quantum Circuit produces more heat and the Multi-Valued Quantum circuit needs to process the input. For that reason, it would be best to transfer the heat into Multi-Valued Quantum circuits. Multi-Valued Quantum portion Heat transfers using the heat transfer circuit into Multi-Valued Quantum circuit's portion. In the heat transfer circuit, two junctions are connected to Multi-Valued Quantum Circuits and Multi-Valued Quantum circuits. Maximum one meter can transfer heat to this circuit. This heat transfer circuit has superconductive ware, photon batch, PCB board to make the architecture. This circuit mainly transfers the heat using the photon bath. With this circuit proposed Multi-Valued DNA-Quantum T flip-flop circuit is constructed fully.

4.9.3 Working Principle of Multi-Valued DNA-Quantum Synchronous Counter

Multi-Valued DNA-Quantum Synchronous counter has one logical molecular sequence input which is shared and one clock input. Multi-Valued DNA-Quantum Synchronous counter is constructed by the basic component as a Multi-Valued DNA JK flip-flop operational circuit. Multi-Valued DNA JK flip flop is the two-molecular sequence circuit. Multi-Valued DNA JK flip flop is mostly used flip-flop in Multi-Valued DNA computing. Multi-Valued DNA Jk flip-flop's input J and K perform two Multi-Valued DNA operations parallelly. J and shared input clk performs the Ternary DNA NAND operation. Then this Ternary DNA NAND operations result and one output of the Multi-Valued DNA JK flip flop performs the Ternary DNA NAND operation and produce S. Like Ternary DNA NAND operation performs K and |clk > input first. Then the output of Multi-Valued DNA JK flip-flop |Q> and the result of Multi-Valued DNA first NAND operation executes again another Ternary DNA NAND operation and produce R. The procedure of producing S and R are executed in parallel. It is known that Multi-Valued DNA operations have one of the iconic characteristics that perform multiple operations parallelly and here it is happening. These S and R inputs are performed then Multi-Valued DNA SR flip flop operation. Multi-Valued DNA SR flip flop is also basically made by the Ternary DNA NAND operation that is used here. After performing the Multi-Valued DNA SR flip-flop operation, two outputs are found. One output is the opposite of another. This

TABLE 4.9

Truth Table of Multi-Valued DNA-Quantum
Synchronous Counter

clk	$|Q_0>$	$|Q_1>$	$|Q_2>$	$|Q_3>$
ACCTAG	$|0>$	$|0>$	$|0>$	$|0>$
TGGATC	$|0>$	$|0>$	$|0>$	$|1>$
TGGATC	$|0>$	$|0>$	$|1>$	$|0>$
TGGATC	$|0>$	$|0>$	$|1>$	$|1>$
TGGATC	$|0>$	$|1>$	$|0>$	$|0>$
TGGATC	$|0>$	$|1>$	$|0>$	$|1>$
TGGATC	$|0>$	$|1>$	$|1>$	$|0>$
TGGATC	$|0>$	$|1>$	$|1>$	$|1>$
TGGATC	$|1>$	$|0>$	$|0>$	$|0>$
TGGATC	$|1>$	$|0>$	$|0>$	$|1>$
TGGATC	$|1>$	$|0>$	$|1>$	$|0>$
TGGATC	$|1>$	$|0>$	$|1>$	$|1>$
TGGATC	$|1>$	$|1>$	$|0>$	$|0>$
TGGATC	$|1>$	$|1>$	$|0>$	$|1>$
TGGATC	$|1>$	$|1>$	$|1>$	$|0>$
TGGATC	$|1>$	$|1>$	$|1>$	$|1>$
TGGATC	$|0>$	$|0>$	$|0>$	$|0>$

Multi-Valued DNA JK flip flop actually removes the problem of Multi-Valued DNA SR flip flop. Multi-Valued DNA SR flip-flop describes briefly in the 5th chapter.

Multi-Valued DNA JK flip-flop is triggered when the value of clock is TGGATC. So, in Multi-Valued DNA-Quantum Synchronous the counter clock needs to be always high. According to the principle counters need to be toggled, that's why Multi-Valued DNA JK flip-flop is perfect for Multi-Valued DNA-Quantum Synchronous counters. Multi-Valued DNA JK flip-flop operational circuit triggered then output of Multi-Valued DNA JK flip-flop will be shared input of second Multi-Valued DNA JK flip-flop operational circuit. Then first Multi-Valued DNA JK flip-flops output and second Multi-Valued DNA JK flip-flops output enters in the Ternary DNA AND operation circuit and performs the Ternary DNA AND operation. The produced output from Ternary DNA AND operation performs the Multi-Valued DNA JK flip-flop operation. Then again produced output from the first Ternary DNA AND operation and the third Multi-Valued DNA JK flip-flop operation's output performs another Ternary DNA AND operation. Hence the previous Ternary DNA AND operations output is shared into the Multi-Valued DNA JK flip-flops as two inputs as well as performs the Multi-Valued DNA JK flip-flop operational circuit. Multi-Valued DNA-Quantum Synchronous counter circuit mainly performs as a finite counter. Table 4.9 shows the truth table of multi-valued DNA-Quantum Synchronous Counter.

Between the second and the third Multi-Valued DNA JK flip flop, Ternary DNA AND Gate is used. The JK flip flop's circuit output lines are connected into the Multi-

Valued DNA cache memory. This is the first portion of the circuit and is fully designed by using the principle of Multi-Valued DNA computing. Multi-Valued DNA cache memory is made by using some Multi-Valued DNA shift register and in this proposed circuit it saves data in the Multi-Valued DNA array. When the time arrives Multi-Valued DNA cache memory supplies the data into the NMR process. Multi-Valued DNA cache memory supply line fully connected to NMR process as well as here EMR is emitted. Then the NMR process output line is connected to the Ternary Quantum NAND gate. Two outputs of the NMR process are connected to two different Ternary Quantum NAND gates. Ternary Quantum NAND gates another input come from the output which is Q and $\overline{Q}$ qubits. Hence this Ternary Quantum NAND gates' final output line is the final output of the Multi-Valued DNA-Quantum Ripple Counter.

Multi-Valued DNA Cache memory works here as the intermediate process. Multi-Valued DNA circuit produces much heat and here two Ternary DNA AND operations perform so it produces much heat. But Multi-Valued DNA circuit needs to be close to zero Kelvin temperature to perform the operation. So from the Multi-Valued DNA circuit, the temperature is needed to be reduced to maintain the superposition state of the molecular sequence. Hence in Multi-Valued Quantum circuit operation, it needs much heat in several steps. Mainly in melting and annealing require much heat. This topic is briefly described in the first chapter named Multi-Valued DNA-Quantum circuit operation. For that reason, the excessive heat can be transferred from the Multi-Valued DNA circuit portion to the Multi-Valued Quantum circuit portion using a heat transfer circuit. This heat transfer circuit using the junction captures the heat from the Multi-Valued DNA circuit and using photon bath heat flows through the circuit and gives this into the Multi-Valued Quantum circuit. This circuit can transfer heat maximum in the one-meter distance and in Multi-Valued DNA-Quantum flip-flop distance is less than one meter between Ternary DNA AND Multi-Valued Quantum circuit. This heat transfer circuit cannot transfer full excessive heat produced from the Multi-Valued DNA circuit and this heat is not enough to perform the Multi-Valued Quantum circuit operation. But this heat transfer can optimize the cost of heat based on the needs. After all of this operation and architecture fully then Multi-Valued DNA-Quantum Ripple Counter can produce two output qubits.

4.9.4 Example

Suppose the input molecular sequence TGGATC enters into the Multi-Valued DNA-Quantum Synchronous counter. Multi-Valued DNA-Quantum Synchronous counters basic component is Multi-Valued DNA Jk flip-flop operational circuit so for triggered the circuit clock value is also TGGATC. Now first Multi-Valued DNA JK flip-flop both J input and K input is TGGATC. If two value is TGGATC the flip-flop triggered and value toggles. Now Q_0 is TGGATC from previous state ACCTAG.

But after the first Multi-Valued DNA JK flip-flop operation $\overline{Q}_0$'s value became ACCTAG. Then $\overline{Q}_0$ entered as an input in the second Multi-Valued DNA JK flip-flop operation and this circuit won't be triggered. Then again $\overline{Q}_1$ and $\overline{Q}_0$ performs the Ternary DNA AND operation. As Ternary DNA AND operation depicts that if any input is ACCTAG then the produced output will ACCTAG. Then ACCTAG will be input for the third Multi-Valued DNA JK flip-flop operational circuit. As this circuit Multi-Valued DNA JK flip-flop depicts that if two input is ACCTAG then Multi-Valued DNA Jk flip-flop won't be triggered. Again third Multi-Valued DNA flip-flop's output $\overline{Q_2}$ and previous Ternary DNA AND operations produced output performs Ternary DNA AND operation again. As this propose circuit depicts previous Ternary DNA AND operation output is ACCTAG so, this Ternary DNA AND operations output will be ACCTAG. Hence ACCTAG will be input of J and K for the fourth Multi-Valued DNA JK flip-flop operation as well as finally it produces ACCTAG.

So, this proposed circuit illustrates the exact required output of a Multi-Valued DNA-Quantum Synchronous counter theoretically.

4.10   Summary

The multi-valued DNA-quantum SR latch circuits can be utilized as storage devices in power gating circuits and clocks since the multi-valued DNA-quantum SR latch is a single-bit storage element. Sequential circuits in multi-valued DNA-quantum computing are a new thing in the modern world. This is the form of quantum molecular biology. This chapter has presented the detailed working principle and circuit architecture of all sequential circuits in multi-valued DNA-quantum computing with example. Necessary figures are also shown here.

Bibliography

[1] Adleman, L. M. (1994). Molecular computation of solutions to combinatorial problems. Science, 266(5187), 1021-1024.

[2] Smith, L. M., Sanders, J. Z., Kaiser, R. J., Hughes, P., Dodd, C., & Connell, C. R. (1986). C. Heiner, S. B. Kent. and L. E. Hood. Nature, 321(674), 7679.

[3] Zheng, X., Yang, J., Zhou, C., Zhang, C., Zhang, Q., & Wei, X. (2019). Allosteric DNAzyme-based DNA logic circuit: Operations and dynamic analysis. Nucleic Acids Research, 47(3), 1097-1109.

[4] Thaker, D. D., Metodi, T. S., Cross, A. W., Chuang, I. L., & Chong, F. T. (2006, June). Quantum memory hierarchies: Efficient designs to match available parallelism in quantum computing. In 33rd International Symposium on Computer Architecture (ISCA'06) (pp. 378-390). IEEE.

Part II

Multiple-Valued Memory Devices in Quantum Molecular Biology

Overview

It is broadly established that quantum mechanics has attributes which is not present in classical physics. These unusual features – like entanglement – can be used as a resource to construct technologies not historically understood to be possible. Quantum sensors, quantum computers, and quantum cryptography all have specific enhancements over their classical counterparts. The construction of these new quantum-enabled technologies, however, remains extremely challenging. Out of the efforts to build quantum technologies has emerged the understanding that various quantum components will be essential, or very beneficial. In the realm of photonics-based quantum technologies , key quantum components include quantum memories, photon sources, frequency converters, quantum random number generators, and single-photon detectors. The focus of this review is quantum memory, a device that can store a single photon and recreate the quantum state. Quantum memories are under development by many groups around the world. Approaches to quantum memory encompass the full gamut of our understanding of electromagnetic interactions, and as such these research programs represent the most advanced techniques for the quantum control of optical signals. For example, resonant (first-order) and Raman (second-order) interactions in warm, cold, trapped, or Bose-condensed atoms, in amorphous and crystalline solids, molecular gases, structured media, and metamaterials are actively pursued, with engineered couplings to electronic, magnetic, vibrational and hybrid degrees of freedom. Photonics is a unique platform for quantum technologies because it can support broadband signals over long distances in ambient conditions, enabling, for example, the first commercial quantum cryptographic devices. However, to extend the range of quantum cryptography systems, quantum memories are needed for repeaters. More generally for quantum technologies, mechanisms are needed to generate and guide photons, mediate non-linearity, and detect photons. Quantum memories would enable the development of large photonic quantum processing systems, by providing the capability to coherently manipulate, buffer and retime photonic signals. While the primary focus of quantum memory research has been the synchronization of entanglement swapping in quantum repeater protocols for long-distance quantum communication, the ability to interconvert material and optical quanta, to prepare non-classical states and read them out optically, and to generate and distribute long-lived entanglement, has led to a range of other applications, so that quantum memories is now a catch-all term for a broad class of research, linked by the common theme of coherent interfacing between light and matter. As the field has developed, there have been outstanding review articles on quantum memories.

5

Multiple-Valued Quantum Memory Devices

5.1 Introduction

Quantum computing is a type of computation that harnesses the collective properties of quantum states, such as superposition, interference, and entanglement, to perform calculations. The devices that perform quantum computations are known as quantum computers. Quantum computing harnesses the phenomena of quantum mechanics to deliver a huge leap forward in computation to solve certain problems. IBM designed quantum computers to solve complex problems that today's most powerful supercomputers cannot solve, and never will. Any computational problem that can be solved by a classical computer can also be solved by a quantum computer. Conversely, any problem that can be solved by a quantum computer can also be solved by a classical computer, at least in principle given enough time. Quantum computers provide no additional advantages over classical computers in terms of computability, quantum algorithms for certain problems have significantly lower time complexities than corresponding known classical algorithms. Quantum algorithms take a new approach to these sorts of complex problems – creating multidimensional spaces where the patterns linking individual data points emerge. Classical computers cannot create these computational spaces, so they cannot find these patterns. In the case of proteins, there are already early quantum algorithms that can find folding patterns in entirely new, more efficient ways, without the laborious checking procedures of classical computers. As quantum hardware scales and these algorithms advance, they could tackle protein folding problems too complex for any supercomputer . In quantum computing, operations instead use the quantum state of an object to produce what's known as a qubit. These states are the undefined properties of an object before they've been detected, such as the spin of an electron or the polarization of a photon. Rather than having a clear position, unmeasured quantum states occur in a mixed 'superposition', not unlike a coin spinning through the air before it lands in hand. These superposition can be entangled with those of other objects, meaning their final outcomes will be mathematically related. The complex mathematics behind these unsettled states of entangled 'spinning coins' can be plugged into special algorithms to make short work of problems that would take a classical computer a long time to work out... if they could ever calculate them at all. Such algorithms would be useful in solving complex mathematical problems, producing hard-to-break security codes, or predicting multiple particle interactions in chemical reactions. Short for memory device, a generic term for an integrated circuit that can be programmed in a

DOI: 10.1201/9781003381921-5

laboratory to perform complex functions. Memory device consists of arrays of AND and OR gates. A system designer implements a logic design with a device programmer that blows fuses on the PROM to control gate operation. A programmable read only memory (PROM) has a fixed AND gate array, which links to a programmable OR gate array, which can then be conditionally complemented to produce an output. A PROM is similar to a ROM concept; however, a PROM does not provide full decoding of a variable and does not generate all the minterms as in ROM. PROM devices have arrays of transistor cells arranged in a "fixed-OR, programmable-AND" plane used to implement "sum-of-products" binary logic equations for each of the outputs in terms of the inputs and either synchronous or asynchronous feedback from the outputs. System designers can use development software that converts basic code into instructions a device programmer needs to implement a design. Quantum logic used to design RAM and Cache memory devices make the device more powerful.

5.2   Multiple-Valued Quantum Random Access Memory

A ternary computer (also called a trinary computer) is one that uses ternary logic (i.e., base 3) instead of the more common binary system (i.e., base 2) in its calculations. This means it uses trits instead of bits, as most computers do. The significant concern is to provide low-cost, robust, high-density, reliable, and energy-efficient memory technologies through designing multi-valued quantum-based RAM. One early calculating machine, built entirely from wood by Thomas Fowler in 1840, operated in balanced ternary. The first modern, electronic ternary computer, Setun, was built in 1958 in the Soviet Union at the Moscow State University by Nikolay Brusentsov, and it had notable advantages over the binary computers that eventually replaced it, such as lower electricity consumption and lower production cost. In 1970 Brusentsov built an enhanced version of the computer, which he called Setun-70. In the United States, the ternary computing emulator Ternac working on a binary machine was developed in 1973. The ternary computer QTC-1 was developed in Canada. Multi-Valued Quantum computing has the features of parallel processing and fast processing capabilities that make it special from other conventional computing systems. Unlike conventional memory, which stores information as ternary states (represented by "$|2>$"s, "$|1>$"s and "$|0>$"s).

MQRAM is an abbreviation for Multi-Valued Quantum Random Access Memory. It refers to computer memory chips that hold permanent or semi-permanent data and incorporate both the multi-valued quantum decoder and multi-valued quantum OR operations onto a single integrated circuit (IC). The contents of multi-valued quantum ROM are non-volatile; even if the computer is turned off, the contents of multi-valued quantum ROM persist. To update the programming in multi-valued Quantum ROM, these chips have to be physically removed and replaced. Data saved in quantum ROM cannot be electrically changed once the memory device is manufactured. A block diagram of a multi-valued ROM is shown in Figure 5.1.

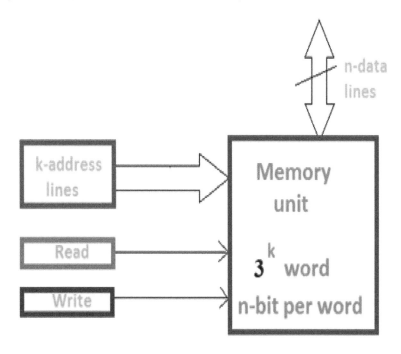

FIGURE 5.1
3^k-to-n RAM Block Diagram

It consists of k input lines and n output lines. Each bit combinations of the input variables is called an address. Each bit combination that comes out of the output lines is called a word. The number of bits per word is equal to the number of output lines n. An address is essentially a ternary number that denotes one of the minterms of k variables.

Initially, the multi-valued quantum RAM is a combinational circuit with multi-valued quantum AND operations connected as a multi-valued quantum decoder, multi-valued qubit cells and a number of multi-valued quantum OR operations to the outputs in the unit. With k address lines and n output lines in RAM, the output functions will calculate through the sum of minterms. The number of distinct addresses possible with k input variables is 3^k. An output word can be selected by a unique address, and since there are 3^k distinct addresses in a multi-valued RAM, there are 3^k distinct words that are said to be stored in the unit. The word available at the output lines at any given time depends on the address value applied to the input lines. Therefore, A RAM is characterized by the number of words 3^k and the number of bits per word n. For input, k=2 and output, n=1 the RAM circuit will be called 9-to-1 multi-valued quantum RAM.

5.2.1 Basic Definitions

For memory size, this is referred to as a $3 \wedge$k-to-n memory. There are k address lines, which can specify one of the $3 \wedge$k addresses. Each address contains an n-bit word. For example, a $3 \wedge 2$ x 1 RAM contains $3 \wedge 2 = 9$ words, each 1 bit long.
 – The RAM would need 2 address lines.
 – The total storage capacity is $3 \wedge 2$ x $3 \wedge 0 = 3 \wedge 2$ bits.

Many operating systems implement virtual memory, which makes the memory seem larger than it really is. – Most systems allow up to 32-bit addresses. This works out to 232, or about four billion, different possible addresses. – With a data size of one byte, the result is a 4GB memory! – The operating system uses hard disk space as a substitute for "real" memory.

Reading RAM

To read from this RAM, the controlling circuit must: – Enable the chip by ensuring CS = 1. – Select the read operation, by setting WR = 0. – Send the desired address to the ADRS input. – The contents of that address appear on OUT after a little while.

50 MHz CPU – 20 ns clock cycle time Memory access time= 65 ns Maximum time from the application of the address to the appearance of the data at the Data Output.

Writing RAM

To write to this RAM, it needs to: – Enable the chip by setting CS = 1. – Select the write operation, by setting WR = 1. – Send the desired address to the ADRS input. – Send the word to store to the DATA input. The output OUT is not needed for memory write operations.

5.2.2 Block Diagram

Consider a multi-valued quantum 9-to-1 RAM general organization of block diagram (Figure 5.2), the unit consists of 9 words of 2 input sequence ($|A> = |0>$ **or** $|1>$ **or** $|2>$ and $|B> = |0>$ **or** $|1>$ **or** $|2>$) each.

This implies there are a single output line ($|Z1>$) and 9 distinct word sequences stored in the unit, each of which may be applied to the output lines. The particular word selected that is presently available on the output line is determined from the 1 input line. There are only 3 input sequences in multi-valued DNA 9-to-1 RAM because $3 \wedge 2 = 9$ and with 3 sequence variables, 9 addresses or minterms can be specified.

5.2.3 Architecture of Basic Components

RAM consists of three basic components- Decoder, Molecular qubit Cells, and OR gates. To execute DNA 9-to-1 RAM operation, one 2-to-9 quantum Decoder, and

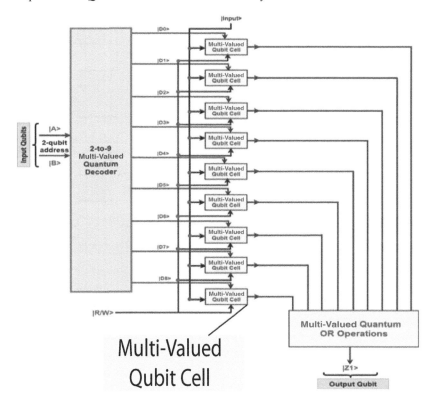

FIGURE 5.2

Block Diagram of Multi-Valued Quantum 9-to-1 RAM

Multi-Valued Qubit Cells and Multi-Valued Quantum OR operations for corresponding minterms are required.

5.2.3.1 Multi-Valued Quantum 2-to-9 Decoder

Multi-Valued Quantum 2-to-9 decoder is a combinational logic circuit which is designed by using two Multi-Valued 1-to-3 Quantum decoders and nine multi-valued Quantum AND (D0-D8) operations (Figure 5.3).

The multi-valued 2-to-9 Quantum decoder operation is performed the two input DNA sequences and each of three states, the combinations are shown in Table 5.1.

From these combinations, the corresponding multi-valued AND operation outputs are the minimum of input sequence states. Ternary logic functions are those functions which have significance if a third value is acquainted with the binary logic. Thus $|0>$, $|1>$ and $|2>$ denote the ternary levels for basic logic gates to represent sequence as "0", "1" and "2" respectively.

For multi-valued Quantum 2-to-9 decoder 9 output sequences, $|D0>$ to $|D8>$ and two input sequences $|A>$, $|B>$.

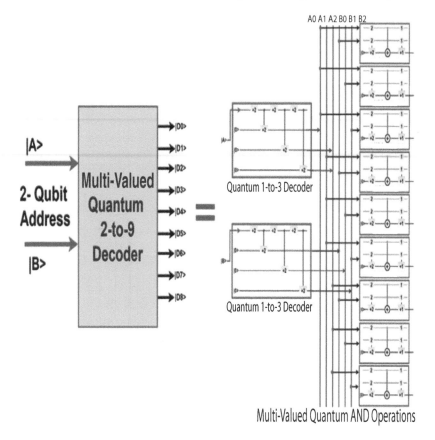

Multi-Valued Quantum AND Operations

FIGURE 5.3
Multi-Valued Quantum 2-to-9 Decoder

TABLE 5.1
Truth Table of Quantum 2-to-9 Decoder

$|A>$	$|B>$	$|D8>$	$|D7>$	$|D6>$	$|D5>$	$|D4>$	$|D3>$	$|D2>$	$|D1>$	$|D0>$
$|0>$	$|0>$	$|0>$	$|0>$	$|0>$	$|0>$	$|0>$	$|0>$	$|0>$	$|0>$	$|2>$
$|0>$	$|1>$	$|0>$	$|0>$	$|0>$	$|0>$	$|0>$	$|0>$	$|0>$	$|2>$	$|0>$
$|0>$	$|2>$	$|0>$	$|0>$	$|0>$	$|0>$	$|0>$	$|0>$	$|2>$	$|0>$	$|0>$
$|1>$	$|0>$	$|0>$	$|0>$	$|0>$	$|0>$	$|0>$	$|2>$	$|0>$	$|0>$	$|0>$
$|1>$	$|1>$	$|0>$	$|0>$	$|0>$	$|0>$	$|2>$	$|0>$	$|0>$	$|0>$	$|0>$
$|1>$	$|2>$	$|0>$	$|0>$	$|0>$	$|2>$	$|0>$	$|0>$	$|0>$	$|0>$	$|0>$
$|2>$	$|0>$	$|0>$	$|0>$	$|2>$	$|0>$	$|0>$	$|0>$	$|0>$	$|0>$	$|0>$
$|2>$	$|1>$	$|0>$	$|2>$	$|0>$	$|0>$	$|0>$	$|0>$	$|0>$	$|0>$	$|0>$
$|2>$	$|2>$	$|2>$	$|0>$	$|0>$	$|0>$	$|0>$	$|0>$	$|0>$	$|0>$	$|0>$

[i]

For input sequences |A>, |B> = **|0>**, **|0>**, the multi-valued 1-to-3 decoder will perform (|A0> && |B0>) and the |D0> line will be open. So the output qubit of |D0> = **|2>** and rest of the lines |D1> to |D8> will remain closed **|0>**.

[ii]

For input sequences |A>, |B> = **|0>**, **|1>**, the multi-valued 1-to-3 decoder will perform (|A0> && |B1>) and the |D1> line will be open. So the output sequence of |D1> = **|2>** and rest of the lines |D0>, |D2> to |D8> will remain closed**|0>**.

[iii]

For input sequences |A>, |B> = **|0>**, **|2>**, the multi-valued 1-to-3 decoder will perform (|A0> && |B2>) and the |D2> line will be open. So the output sequence of |D2> = **|2>** and rest of the lines |D0>, |D1> and |D3> to |D8> will remain closed **|0>**.

[iv]

For input sequences |A>, |B> = **|1>**, **|0>**, the multi-valued 1-to-3 decoder will perform (|A1> && |B0>) and the |D3> line will be open. So the output sequence of |D3> = **|2>** and rest of the lines |D0> to |D2> and |D4> to |D8> will remain closed **|0>**.

[v]

For input sequences |A>, |B> = **|1>**, **|1>** the multi-valued 1-to-3 decoder will perform (|A1> && |B1>) and the |D4> line will be open. So the output sequence of |D4> = **|2>** and rest of the lines |D0> to |D3> and |D5> to |D8> will remain closed **|0>**.

[vi]

For input sequences |A>, |B> = **|1>**, **|2>**, the multi-valued 1-to-3 decoder will perform (|A1> && |B2>) and the |D5> line will be open. So the output sequence of |D5> = **|2>** and rest of the lines |D0> to |D4> and |D6> to |D8> will remain closed **|0>**.

[vii]

For input sequences A, B= **|2>**, **|0>**, the multi-valued 1-to-3 decoder will perform (|A2> && |B0>) and the |D6> line will be open. So the output sequence of |D6> = **|2>** and rest of the lines |D0> to |D5> and |D7> to |D8> will remain closed **|0>**.

[viii]

For input sequences |A>, |B> = **|2>**, **|1>**, the multi-valued 1-to-3 decoder will perform (|A2> && |B1>) and the |D7> line will be open. So the output sequence of |D7> = **|2>** and rest of the lines |D0> to |D6> and |D8> will remain closed **|0>**.

[ix]

For input sequences |A>, |B> = **|2>**, **|2>**, the multi-valued 1-to-3 decoder will perform (|A2> && |B2>) and the |D8> line will be open. So the output sequence of |D8> = **|2>** and rest of the lines |D0> to |D7> will remain closed **|0>**.

5.2.3.2 Multi-Valued Qubit Cell

The fundamental design of this Molecular cell is based on the R-S flip-flop (Figure 5.4). To begin with, the molecular cell has three inputs and a single output. The

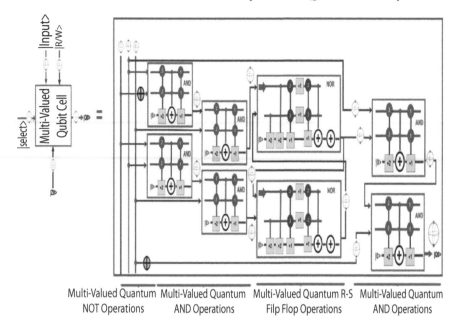

Multi-Valued Quantum   Multi-Valued Quantum   Multi-Valued Quantum R-S   Multi-Valued Quantum
NOT Operations   AND Operations   Filp Flop Operations   AND Operations

FIGURE 5.4
Single Multi-Valued Qubit Cell

inputs are labeled "Select", "R/W", and "Input". The output line is labeled "|Q0>".
To perform the qubit cell output, two quantum NOT, six quantum AND and two
quantum NOR operations are needed to perform.

Step 1: First draw three input sequences Input, R/W and Select. Two possible
states for a sequence are the states "|0>" false, and "|2>" true.

Step 2: Draw Quantum NOT operation with the Input and R/W sequences.

Step 3: Each Quantum AND gate is with three inputs. So two inputs are taken
(NOT Input and Select) to one AND gate input and the output of this gate will go to
another AND gate with input sequence R/W as input.

Step 4: Again, each Quantum AND gate is with three inputs. So two inputs (Input
and Select) are taken to one AND gate input and the output of this gate will go to
another AND gate with input sequence R/W as input.

Step 5: The outputs of step 3 and 4 will go to the R-S flip-flop as input.

Step 6: Finally, the output of RS flip-flop and select will go to a Quantum AND
gate, then this output with NOT of R/W will go through another Quantum AND gate
to produce desired qubit cell output.

In a sequential device as simple as an R-S flip-flop, it could be used to remember
one bit of data. To develop a complete memory cell, called a sequence cell, it needs
to be based on the flip-flop. The number of total qubit cells per word will be m × n
where m represents words with n bits. The "select" input is used to access the cell,

either for reading or writing also used to access any one sequence cell when there is more than one sequence cell. When the select line is high or |0> then the cell performs the memory operation. But when the select line of the DNA cell is low or |2> then the cell is not interested to perform a read from or written to.

The next input sequence is "R/W" where a system clock will conduct this input. If the clock value on the read/write line is |2>, this will signify "read" and when it is |0>, it will perform the "write" phase. When such a cell is selected and in "read" mode, the current value of its underlying flip-flop will be transferred to the cell's output line.

When the cell is selected and in "write" mode, an input data signal will determine the value remembered by the flip-flop. Let's consider the cell that has been selected. In that case, if the clock value is |2> then the cell contents are to be read and this time the output value will depend only on the P value of the flip flop. But if P is low, the cell output will be |2> and if P is high, the cell output will be |0>. It occurs because the DNA AND gate added to the cell's output which has three input-negated R/W, Select, and P; and both "negated R/W" and "Select are currently high |0>.

To perform the quantum 9-to-1 RAM, four selection lines are needed to select four quantum qubit cells. The output of quantum 2-to-9 decoder with nine output qubits will perform as selection input qubit of qubit cells as |Select>.

Figure 5.5 shows the multi-valued 9-qubit cells to perform qubit cell output qubits from |Q0> to |Q8> for further minterms OR operation.

5.2.3.3 Multi-Valued Quantum OR operations

In a Ternary Quantum OR operation, using two inputs and one output is obtained. The higher logical sequence value from the input sequences is considered as the output of Ternary quantum OR operations and when the input sequences have the same logical value then one of them is considered as the output. Table 5.2 shows the ternary quantum OR operation truth table and Figure 5.6 shows the Ternary quantum OR operation.

Ternary OR function is also similar to the Binary "MAXIMUM OF" function. The following equation helps to get the OR gate output for the ternary inputs.

$$y_{OR} = \max(x, y)$$

TABLE 5.2
Truth Table of Multi-Valued Quantum OR Operation

|A>+|B>		|B>	
		|0>	|1>	|2>
A	|0>	|0>	|1>	|2>
	|1>	|1>	|1>	|2>
	|2>	|2>	|2>	|2>

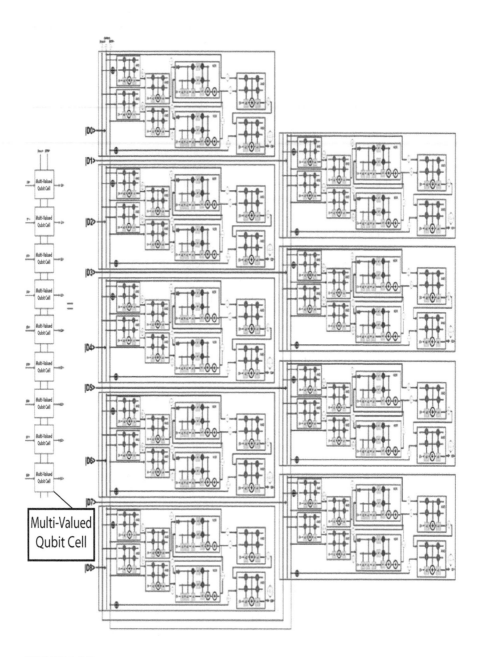

FIGURE 5.5
Multi-Valued Qubit Cells

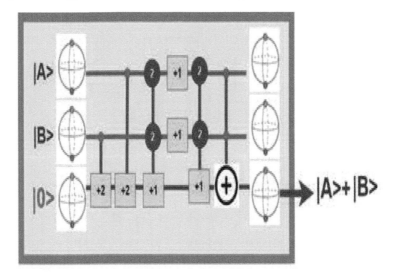

FIGURE 5.6
Multi-Valued Quantum OR operations

[i]
A= |0> and B = |0>, though the logical value of A and B are the same so after doing ternary DNA OR operation the output is |0>. In this way, when the value of A and B is |2> then the output is also |2>, when the value of A and B is |1> then the output is also |1>.

[ii]
A= |2> and B= |1>, then the logical value of B is less than the logical value of A, So the output result is |2>.

When A= |1> and B= |0>, then the logical value of B is less than the logical value of A, So the result is |1>.

A= |2> and B= |0>, then the logical value of B is less than the logical value of A, so the output result is |2>.

[iii]
B= |2> and A= |1>, then the logical value of A is less than the logical value of B, So the output result is |2>.

When B= |1> and A= |0>, then the logical value of A is less than the logical value of B, So the result is |1>.

B= |2> and A= |0>, then the logical value of A is less than the logical value of B, So the output result is |2>.

Figure 5.7 depicts the the multi-valued quantum OR operations in quantum 4-to-1 RAM.

To perform output Z_1 of multi-valued quantum 9-to-1 RAM, the quantum decoder output sequences will go through qubit cells as selection sequences to operate

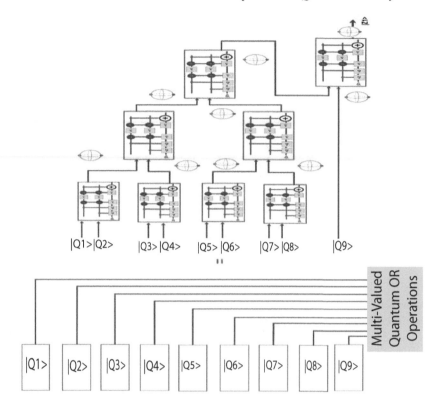

FIGURE 5.7
Multi-Valued Quantum OR Operations in Quantum 4-to-1 RAM

in the sum of minterms form, $|Z1\rangle = \sum (|Q0\rangle, |Q1\rangle, |Q2\rangle, |Q3\rangle, |Q4\rangle, |Q5\rangle, |Q6\rangle, |Q7\rangle, |Q8\rangle)$.

5.2.4 Circuit Architecture of Multiple-Valued Quantum RAM Memory

For the circuit architecture of Multi-Valued Quantum 9-to-1 RAM (Figure 5.8), a 2-to-9 ternary decoder is needed which is designed by using two Multi-Valued 1-to-3 Quantum decoders to get the output of nine Quantum ternary RAM cells operations. Finally, the multi-valued 9-ternary RAM cells perform qubit cell output qubits from $|Q0\rangle$ to $|Q8\rangle$ for further minterms ternary OR operation to perform desired outputs.

Figure 5.8 represents the implementation of 9-to-1 multi-valued quantum RAM. This quantum RAM consists of 9 separate "Words" of memory and each is 1 se-quence wide. The quantum RAM Cell has three inputs and one output. Nine words of memory need two address lines. A and B are the two-sequence address lines input that goes through a multi-valued quantum 2-to-9 decoder that selects one of the nine words. The memory-enabled input enables the decoder. If the memory enables is $|2\rangle$,

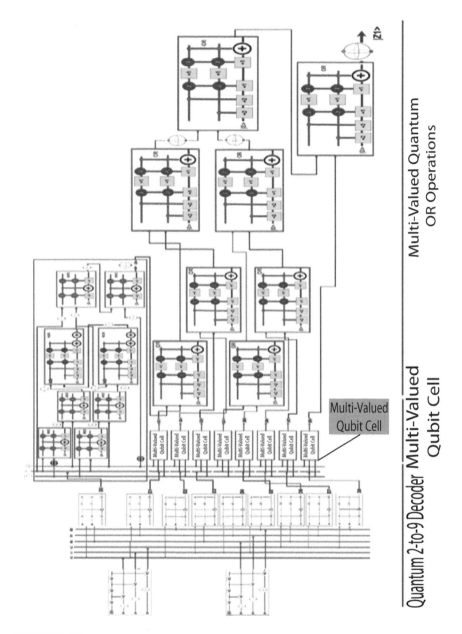

FIGURE 5.8

Circuit Architecture of Multi-Valued Quantum 9-to-1 RAM

all output of the decoder will be |2> and in that case, none of the memory addresses will be selected. But when the memory enables is |0> then one of the nine words is

TABLE 5.3

Control Input to Memory Chip

R/W	Memory Operation
X	*None*
|0>	*Write to the selected word*
|2>	*Read from the selected word*

selected. The word is selected by the value in the two address lines. When a word has been selected, the read/write input determines the operation. During the read operation, the nine sequences of the selected word pass to the multi-valued quantum OR gates to the output |Z1> terminals. But during the write operation, the data which is available in the input lines are transferred into the multi-valued quantum RAM cells of the selected word. The quantum RAM cells that are not selected become disabled and their previous sequence never changes. But when the memory enable input that passes into the decoder is equal to |2>, none of the words are selected, and then all RAM cells remain unchanged regardless of the value of the read/ write input. Table 5.3 shows the control input to memory chip.

5.3   Multiple-Valued Quantum Read-Only Memory

A ternary computer (also called a trinary computer) is one that uses ternary logic (i.e., base 3) instead of the more common binary system (i.e., base 2) in its calculations. This means it uses trits instead of bits, as most computers do. The significant concern is to provide low-cost, robust, high-density, reliable, and energy-efficient memory technologies through designing multi-valued quantum-based ROM. One early calculating machine, built entirely from wood by Thomas Fowler in 1840, operated in balanced ternary. The first modern, electronic ternary computer, Setun, was built in 1958 in the Soviet Union at the Moscow State University by Nikolay Brusentsov, and it had notable advantages over the binary computers that eventually replaced it, such as lower electricity consumption and lower production cost. In 1970 Brusentsov built an enhanced version of the computer, which he called Setun-70. In the United States, the ternary computing emulator Ternac working on a binary machine was developed in 1973. The ternary computer QTC-1 was developed in Canada. Multi-Valued Quantum computing has the features of parallel processing and fast processing capabilities that make it special from other conventional computing systems. Unlike conventional memory, which stores information as ternary states (represented by "|2>"s, "|1>"s and "|0>"s).

MQROM is an abbreviation for Multi-Valued Quantum Read-Only Memory. It refers to computer memory chips that hold permanent or semi-permanent data and incorporate both the multi-valued quantum decoder and multi-valued quantum OR operations onto a single integrated circuit (IC). The contents of multi-valued quantum

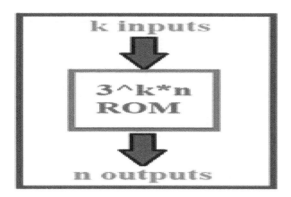

FIGURE 5.9
Block Diagram of 3^k-to-n ROM

ROM are non-volatile; even if the computer is turned off, the contents of multi-valued quantum ROM persist. To update the programming in multi-valued Quantum ROM, these chips have to be physically removed and replaced. Data saved in quantum ROM cannot be electrically changed once the memory device is manufactured. A block diagram of a multi-valued ROM is shown in Figure 5.9. It consists of k input lines and n output lines. Each bit combination of the input variables is called an address. Each bit combination that comes out of the output lines is called a word. The number of bits per word is equal to the number of output lines n. An address is essentially a ternary number that denotes one of the minterms of k variables.

Initially, the multi-valued quantum ROM is a combinational circuit with multi-valued quantum AND operations connected as a multi-valued quantum decoder and a number of multi-valued quantum OR gates equal to the outputs in the unit. With k input lines and n output lines in ROM, the output functions will calculate through the sum of minterms form. The number of distinct addresses possible with k input variables is 3^k. An output word can be selected by a unique address, and since there are 3^k distinct addresses in a multi-valued ROM, there are 3^k distinct words that are said to be stored in the unit. The word available at the output lines at any given time depends on the address value applied to the input lines. Therefore, a ROM is characterized by the number of words 3^k and the number of bits per word n. For input, k=2 and output, n=2 the ROM circuit will be called 9-to-2 multi-valued quantum ROM and the function output $|F_{1>}$ and $|F_{2>}$ in sum of minterms form, $\sum$ (0, 1, 2, 3, 4, 5, 6, 7, 8).

5.3.1 Basic Definitions

It is easy to see that the following rules, concerning memory parameters, hold on growing memory volume makes the memory cost per bit of stored information decreased, lower access time makes the memory cost per bit increased, growing memory volume frequently corresponds to larger access time. The most important parameters of memory are:

1. **Memory capacity or memory volume** is the number of locations that exist in a given memory. Memory capacity is measured in bits, bytes, or words. When words are used, the length of a word in bits or bytes has to be given.

2. **Memory access time** is the time that separates sending a memory access request and the reception of the requested information. The access time determines the unitary speed of memory (the reception time of unitary data). The access time is small for fast memories.

3. **Memory cycle time** is the shortest time that has to elapse between consecutive access requests to the same memory location. The memory cycle time is another parameter that characterizes the overall speed of the memory. The speed is big when the cycle time is small.

4. **Memory transfer rate** is the speed of reading or writing data in the given memory, measured in bits/sec or bytes/sec.

5.3.2   Block Diagram of Multiple-Valued Quantum Read-Only Memory

Consider a multi-valued quantum 9-to-2 ROM general organization of block diagram (Figure 14.2), the unit consists of 9 words of 2 input sequence ($|A> = |0>$ **or** $|1>$ **or** $|2>$ and $|B> = |0>$ **or** $|1>$ **or** $|2>$) each. This implies there are 2 output lines ($|F1>$ and $|F2>$) and 9 distinct words sequences stored in the unit, each of which may be applied to the output lines. Figure 5.10 shows the block diagram of multi-valued Quantum 9-to-2 ROM.

The particular word selected that is presently available on the output line is determined from the 2 input lines. There are only 3 input sequences in multi-valued DNA 9-to-2 ROM because $3 \wedge 2 = 9$ and with 3 sequence variables, 9 addresses or minterms can be specified. To perform minterms of nine addresses, a multi-valued quantum 2-to-9 decoder and multi-valued quantum OR operations are required.

5.3.3   Circuit Architecture of Multiple-Valued Quantum ROM

For the circuit architecture of Multi-Valued Quantum 9-to-2 ROM (Figure 5.11), a 2-to-9 ternary decoder is needed which is designed by using two Multi-Valued 1-to-3 Quantum decoders to get the output of nine Quantum ternary AND operations. Finally, the multi-valued 2-to-9-ternary decoder performs further minterms ternary OR operation to perform desired outputs.

In this 9-to-2 multi-valued DNA based ROM, two inputs DNA sequences A and B with three states $|A>$ ($|0>, |1>, |2>$) and $|B>$ ($|0>, |1>, |2>$) and two output functions $|F1>$ and $|F2>$. Input lines are 2 so with the ternary logic, it will produce $3 \wedge 2 = 9$ address lines and two output function lines are data lines of multi-valued ROM.

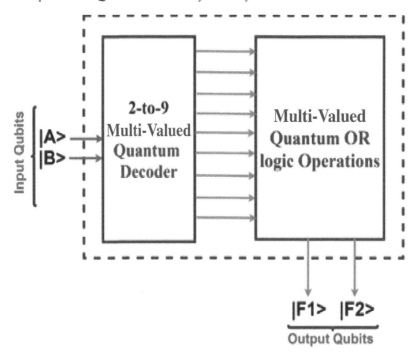

FIGURE 5.10
Block Diagram of Multi-Valued Quantum 9-to-2 ROM

That's why it is called 9-to-2 multi-valued ROM. Thus it will perform with quantum qubits so it is called Quantum multi-valued ROM.

5.3.3.1 Design Procedure of Multiple-Valued Quantum ROM

Step 1:
First draw two input qubits |A> and |B>. Three possible states for input qubits are the states |0>, |1>, |2>. These 3 will produce 9 combinations of 2 input qubits.

Step 2:
For each input qubit, a ternary 1-to-3 quantum decoder is needed for the selection of input qubit values combinations. For example, if any input qubit is |A> then three values can be performed as |A0>, |A1> and |A2>. So the ternary decoder will select combinations from the values of inputs qubits.

Step 3:
After performing the two ternary 1-to-3 quantum decoder operations for input A and B, |A0>, |A1>, |A2> and |B0>, |B1>, |B2> will be found.

Step 4:
With each input value of step 3, it is needed to perform AND operations that will produce 9 combinations of 2 input sequences and enable anyone output line.

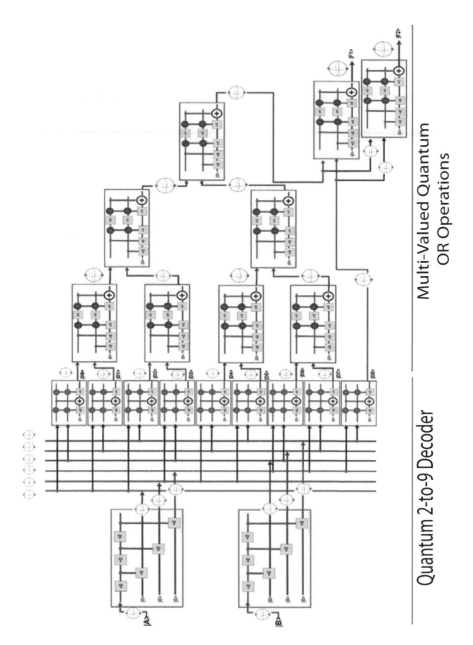

FIGURE 5.11
Multi-Valued 9-to-2 Quantum ROM

Step 5:

From each combination of ternary AND operations in Step 4, perform multi-valued quantum OR operations with (|D0>, |D1>), (|D2>, |D3>), (|D4>, |D5>), (|D6>, |D7>).

Step 6:

From each outputs sequences of Step 5, again combine them with multi-valued quantum OR operations, (|D0>, |D1>, |D2>, |D3>), (|D4>, |D5>, |D6>, |D7>).

Step 7:

From each outputs sequences of Step 6, again combine them with multi-valued quantum OR operations as (|D0>, |D1>, |D2>, |D3>, |D4>, |D5>, |D6>, |D7>).

Step 8:

Finally, perform multi-valued quantum OR operations with |D8> and the output of step 7 to produce desired multi-valued 9-to-2 quantum ROM for output functions |F1> and |F2>.

5.3.3.2 Working Principle

Table 5.4 shows the truth table of DNA 9-to-2 ROM where inputs are |A> and |B>. |F1> and |F2> are the outputs. The working principle are described a according to the truth table.

[1]

For input sequences |A>, |B>= **|0>, |0>**

|A>= |A0>, |A1>, |A2> =|2>, |0>, |0> and |B> = |B0>, |B1>, |B2>= |2>, |0>, |0>. With this input combinations, only input |A0> and |B0> is true. Thus the |A0> and |B0> are connected to output line |D0> so only the |D0> will **|2>** and rest of the gates will produce **|0>**.

To perform 9-to-2 multi-valued DNA ROM functions outputs,

[i] For input qubits |D0>, |D1> = **|2>, |0>**, the multi-valued OR gates will perform maximum operations of input qubits. So the output qubit is **|2>.**

TABLE 5.4

Truth Table of Quantum 9-to-2 ROM

|A>	|B>	|F1>	|F2>
|0>	|0>	|2>	|2>
|0>	|1>	|2>	|2>
|0>	|2>	|2>	|2>
|1>	|0>	|2>	|2>
|1>	|1>	|2>	|2>
|1>	|2>	|2>	|2>
|2>	|0>	|2>	|2>
|2>	|1>	|2>	|2>
|2>	|2>	|2>	|2>

[ii] For input qubits |D2>, |D3> = **|0>, |0>**, the multi-valued OR gates will perform maximum operations of input qubits. So the output qubit is **|0>**.

[iii] For input qubits |D4>, |D5> = **|0>, |0>**, the multi-valued OR gates will perform maximum operations of input qubits. So the output qubit is **|0>**.

[iv] For input qubits |D6>, |D7> = **|0>, |0>** the multi-valued OR gates will perform maximum operations of input qubits. So the output qubit is **|0>**.

[v] Combine the output qubits **|2>, |0>** from [i] and [ii] respectively as input qubits, the multi-valued OR gates will perform maximum operations of input qubits. So the output qubit is **|2>**.

[vi] Combine the output qubits **|0>, |0>** from [iii] and [iv] respectively as input qubits, the multi-valued OR gates will perform maximum operations of input qubits. So the output qubit is **|0>**.

[vii] Combine the output qubits **|2>, |0>** from [v] and [vi] respectively as input qubits, the multi-valued OR gates will perform maximum operations of input qubits. So the output qubit is **|2>**.

[viii] Finally, Combine the output qubits **|2>** from [vii] and |D8> (**|0>**) as input qubit, the multi-valued OR gates will perform maximum operations of input qubits. So the output qubit is **|2>** for the both functions output |F1> and |F2> multi-valued 9-to-2 Quantum ROM.

[2]
For input sequences |A>, |B>= **|0>, |1>**

|A>= |A0>, |A1>, |A2> = |2>, |0>, |0> and |B> = |B0>, |B1>, |B2> = |0>, |2>, |0>. With this input combinations, only input |A0> and |B1> is true. Thus the |A0> and |B1> are connected to output line |D1> so only the |D1> will **|2>** and rest of the gates will produce **|0>**.

For Functions |F1> and |F2>, perform OR operations among all decoder (|D0>-|D8>) outputs. So, max (|D0>, |D1>, |D2>, |D3>, |D4>, |D5>, |D6>, |D7>, |D8>) = max (|0>, |2>, |0>, |0>, |0>, |0>, |0>, |0>, |0>) = **|2>** will produce as the functions outputs |F1> and |F2> of multi-valued 9-to-2 Quantum ROM.

[3]
For input sequences |A>, |B>= **|0>, |2>**

|A>= |A0>, |A1>, |A2> = |2>, |0>, |0> and |B0>, |B1>, |B2> = |0>, |0>, |2>. With this input combinations, only input |A0> and |B2> is true. Thus the |A0> and

|B2> are connected to output line |D2> so only the D2 will **|2>** and rest of the gates will produce **|0>**.

For Functions |F1> and |F2>, perform OR operations among all decoder (|D0>-|D8>) outputs. So, max (|D0>, |D1>, |D2>, |D3>, |D4>, |D5>, |D6>, |D7>, |D8>) = max (|0>, |0>, |2>, |0>, |0>, |0>, |0>, |0>, |0>) = **|2>** will produce as the functions outputs |F1> and |F2> of multi-valued 9-to-2 Quantum ROM.

[4]
For input sequences |A>, |B>= **|1>, |0>**

|A>= |A0>, |A1>, |A2> = |0>, |2>, |0> and |B0>, |B1>, |B2> = |2>, |0>, |0>. With this input combinations, only input |A1> and |B0> is true. Thus the |A1> and |B0> are connected to output line |D3> so only the |D3> will **|2>** and rest of the gates will produce **|0>**.

For Functions |F1> and |F2>, perform OR operations among all decoder (|D0>-|D8>) outputs. So, max (|D0>, |D1>, |D2>, |D3>, |D4>, |D5>, |D6>, |D7>, |D8>) = max (|0>, |0>, |0>, |2>, |0>, |0>, |0>, |0>, |0>) = **|2>** will produce as the functions outputs |F1> and |F2> of multi-valued 9-to-2 Quantum ROM.

[5]
For input sequences |A>, |B>= **|1>, |1>**

|A>= |A0>, |A1>, |A2> = |0>, |2>, |0> and |B0>, |B1>, |B2> = |0>, |2>, |0>. With this input combinations, only input |A1> and |B1> is true. Thus the |A1> and |B1> are connected to output line |D4> so only the |D4> will **|2>** and rest of the gates will produce **|0>**.

For Functions |F1> and |F2>, perform OR operations among all decoder (|D0>-|D8>) outputs. So, max (|D0>, |D1>, |D2>, |D3>, |D4>, |D5>, |D6>, |D7>, |D8>) = max (|0>, |0>, |0>, |0>, |2>, |0>, |0>, |0>, |0>) = **|2>** will produce as the functions outputs |F1> and |F2> of multi-valued 9-to-2 Quantum ROM.

[6]
For input sequences |A>, |B>= **|1>, |2>**

|A>= |A0>, |A1>, |A2> = |0>, |2>, |0> and |B0>, |B1>, |B2> = |0>, |0>, |2>. With this input combinations, only input |A1> and |B2> is true. Thus the |A1> and |B2> are connected to output line |D5> so only the |D5> will **|2>** and rest of the gates will produce **|0>**.

For Functions |F1> and |F2>, perform OR operations among all decoder (|D0>-|D8>) outputs. So, max (|D0>, |D1>, |D2>, |D3>, |D4>, |D5>, |D6>, |D7>, |D8>) = max (|0>, |0>, |0>, |0>, |0>, |2>, |0>, |0>, |0>) = **|2>** will produce as the functions outputs |F1> and |F2> of multi-valued 9-to-2 Quantum ROM.

[7]

For input sequences |A>, |B>= **|2>, |0>**

|A>= |A0>, |A1>, |A2> = |0>, |0>, |2> and |B0>, |B1>, |B2> = |2>, |0>, |0>. With this input combinations, only input |A2> and |B0> is true. Thus the |A2> and |B0> are connected to output line |D6> so only the |D6> will **|2>** and rest of the gates will produce **|0>**.

For Functions |F1> and |F2>, perform OR operations among all decoder (|D0>-|D8>) outputs. So, max (|D0>, |D1>, |D2>, |D3>, |D4>, |D5>, |D6>, |D7>, |D8>) = max (|0>, |0>, |0>, |0>, |0>, |0>, |2>, |0>, |0>) = **|2>** will produce as the functions outputs |F1> and |F2>of multi-valued 9-to-2 Quantum ROM.

[8]

For input sequences |A>, |B>= **|2>, |1>**

|A>= |A0>, |A1>, |A2> = |0>, |0>, |2> and |B0>, |B1>, |B2> = |0>, |2>, |0>. With this input combinations, only input |A2> and |B1> is true. Thus the |A2> and |B1> are connected to output line |D7> so only the |D7> will **|2>** and rest of the gates will produce **|0>**.

For Functions |F1> and |F2>, perform OR operations among all decoder (|D0>-|D8>) outputs. So, max (|D0>, |D1>, |D2>, |D3>, |D4>, |D5>, |D6>, |D7>, |D8>) = max (|0>, |0>, |0>, |0>, |0>, |0>, |0>, |2>, |0>) = **|2>** will produce as the functions outputs |F1> and |F2> of multi-valued 9-to-2 Quantum ROM.

[9]

For input sequences |A>, |B>= **|2>, |2>,**

|A>= |A0>, |A1>, |A2> = |0>, |0>, |2> and |B0>, |B1>, |B2> = |0>, |0>, |2>. With this input combinations, only input |A2> and |B2> is true. Thus the |A2> and |B2> are connected to output line |D8> so only the |D8> will **|2>** and rest of the gates will produce **|0>**.

For Functions |F1> and |F2>, perform OR operations among all decoder (|D0>-|D8>) outputs. So, max (|D0>, |D1>, |D2>, |D3>, |D4>, |D5>, |D6>, |D7>, |D8>) = max (|0>, |0>, |0>, |0>, |0>, |0>, |0>, |0>, |2>) = **|2>** will produce as the function's outputs |F1> and |F2> of multi-valued 9-to-2 Quantum ROM.

5.4 Multiple-Valued Quantum Programmable Read-Only Memory

Multi-Valued Quantum computing has the features of parallel processing and fast processing capabilities that make it special from other conventional computing systems. Unlike conventional memory, which stores information as ternary states (represented by "$|2>$"s, "$|1>$"s and "$|0>$"s). MQPROM is an abbreviation for Multi-Valued Quantum Programmable Read-Only Memory. It refers to computer memory chips that hold permanent or semi-permanent data and incorporate both the multi-valued quantum decoder and multi-valued quantum OR operations onto a single integrated circuit (IC). The contents of multi-valued quantum PROM are non-volatile; even if the computer is turned off, the contents of multi-valued quantum PROM persist. A block diagram of a multi-valued PROM is shown in Figure 5.12. It consists of k input lines and n output lines. Each bit combination of the input variables is called an address.

Initially, the multi-valued quantum PROM is a combinational circuit with multi-valued quantum AND operations connected as a multi-valued quantum decoder and several multi-valued quantum OR gates equal to the outputs in the unit. With k input lines and n output lines in PROM, the output functions will calculate through the sum of minterms form. The number of distinct addresses possible with k input variables is $3 \wedge k$. An output word can be selected by a unique address, and since there are $3 \wedge k$ distinct addresses in a multi-valued PROM, there are $3 \wedge k$ distinct words that are said to be stored in the unit. The word available at the output lines at any given

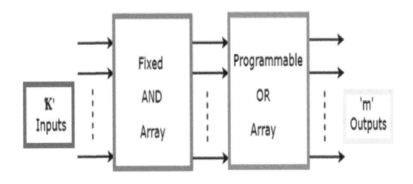

FIGURE 5.12
Block Diagram of 3^k-to-n PROM

time depends on the address value applied to the input lines. Therefore, A ROM is characterized by the number of words 3∧k and the number of bits per word n. For input, k=2 and output, n=2 the PROM circuit will be called 9-to-2 multi-valued quantum PROM, and the function output $|F_{1>}$ is in the sum of minterms form, $\sum$ (0, 1, 2, 3) and $|F_{2>}$ is in the sum of minterms form, $\sum$ (4, 5, 6, 7, 8).

5.4.1 Basic Definitions

PROM is sometimes considered in the same category of the circuit as programmable logic, although in this text, PROM is discussed only in the memory category. Since the PROM architecture reaches its limits when many inputs are linked to many outputs, the PLD architecture, Programmable Logic Device (PLD), provides a more flexible concept.

1. Possible to write data or program only once. However, once it is written, it can be read any number of times

2. A PROM chip is used mainly in the start-up process of a modern computer

3. A PROM, non-volatile memory stores only several megabytes (MB) of data, up to 4 MB or more per chip.

A random memory enables unrestricted space and time access to any location at any address in the address space. The access is possible independently of the order of all previous accesses. The access can take place to addresses in any order. Each location in a random access memory has independent hardware circuits that provide the access. These circuits are activated as a result of address decoding. To such memories belong semiconductor memories of the ROM types. It is possible to distinguish different parameters which determine the properties of different kinds of memories. The most important parameters are Memory capacity, Memory access time, Memory cycle time, and Memory transfer rate.

5.4.2 Block Diagram

Consider a multi-valued quantum 9-to-2 PROM, general organization of block diagram (Figure 5.13), the unit consists of 9 words of 2 input sequence ($|A> = |0>$ or $|1>$ or $|2>$ and $|B> = |0>$ or $|1>$ or $|2>$) each. This implies there are 2 output lines ($|F1>$ and $|F2>$) and 9 distinct word sequences stored in the unit, each of which may be applied to the output lines.

5.4.3 Circuit Architecture of Multiple-Valued Quantum PROM

For the circuit architecture of Multi-Valued Quantum 9-to-2 PROM , a 2-to-9 ternary decoder is needed which is designed by using two Multi-Valued 1-to-3 Quantum decoders to get the output of nine Quantum ternary AND operations. Finally, the multi-valued 2-to-9-ternary decoder performs further minterms ternary OR operation to perform desired outputs. Following given the design algorithms of multiple-valued

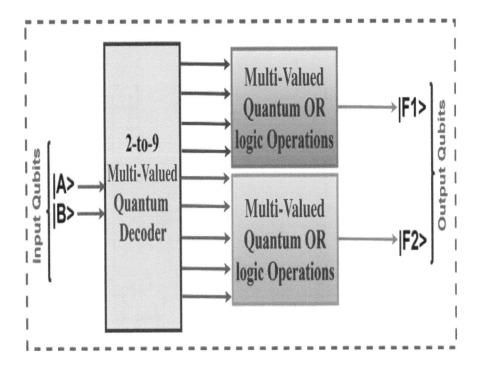

FIGURE 5.13

Block Diagram of Multi-Valued Quantum 9-to-2 PROM

Quantum 9-to-2 PROM Memory. The circuit architecture of multi-valued quantum PROM is shown in Figure 5.14.

5.4.3.1 Design Procedure of Multiple-Valued Quantum PROM

Step 1:

First, draw two input qubits |A> and |B>. Three possible states for input qubits are the states **|0>, |1>, |2>**. These 3 will produce 9 combinations of 2 input qubits.

Step 2:

For each input qubit, a ternary 1-to-3 quantum decoder is needed for the selection of input qubit values combinations. For example, if any input qubit is |A> then three values can be performed as |A0>, |A1>, and |A2>. So the ternary decoder will select combinations from the values of inputs qubits.

Step 3:

After performing the two ternary 1-to-3 quantum decoder operations for input A and B, then |A0>, |A1>, |A2> and |B0>, |B1>, |B2> will be found.

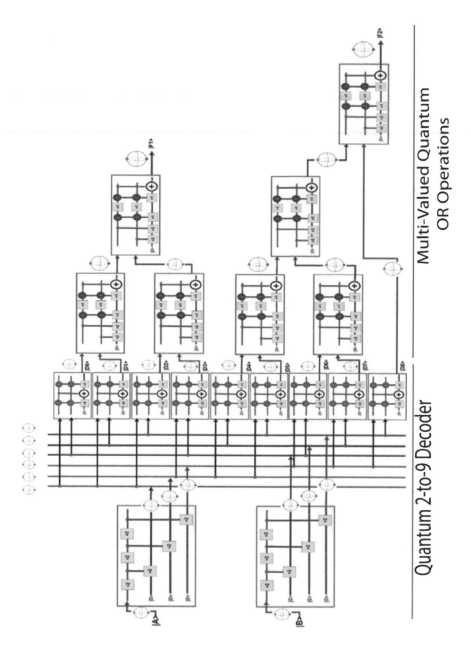

FIGURE 5.14
Multi-Valued 9-to-2 Quantum PROM

Step 4:

With each input value of step 3, it is needed to perform AND operations that will produce 9 combinations of 2 input sequences and enable anyone output line.

Step 5:

From each combination of ternary AND operations in Step 4, perform multi-valued quantum OR operations with (|D0>, |D1>), (|D2>, |D3>), (|D4>, |D5>), (|D6>, |D7>).

Step 6:

From each output sequence of Step 5, again combine them with multi-valued DNA OR gates, (D0, D1, D2, and D3) to generate output function F1 of multi-valued 9-to-2 DNA PROM.

Step 7:

Again combine (D4, D5, D6, D7, and D8) to generate output function F2 of multi-valued 9-to-2 DNA PROM.

5.4.3.2 Working Principle of Multiple-Valued Quantum PROM

The Truth Table of Quantum 9-to-2 ROM is shown in Table 5.5. Here the inputs are |A> and |B>, outputs are |F1> and |F2>.

[1] For input sequences |A>, |B>= |0>, |0>

|A>= |A0>, |A1>, |A2> =|2>, |0>, |0> and |B> = |B0>, |B1>, |B2>= |2>, |0>, |0>. With this input combinations, only input |A0> and |B0> is true. Thus the |A0> and |B0> are connected to output line |D0> so only the |D0> will **|2>** and rest of the gates will produce **|0>**.

To perform 9-to-2 multi-valued DNA ROM functions outputs,

TABLE 5.5

Truth Table of Quantum 9-to-2 ROM

|A>	|B>	|F1>	|F2>
. |0>	|0>	|2>	|0>
|0>	|1>	|2>	|0>
|0>	|2>	|2>	|0>
|1>	|0>	|2>	|0>
|1>	|1>	|0>	|2>
|1>	|2>	|0>	|2>
|2>	|0>	|0>	|2>
|2>	|1>	|0>	|2>
|2>	|2>	|0>	|2>

[i] For input qubits |D0>, |D1> = **|2>, |0>**, the multi-valued OR gates will perform maximum operations of input qubits. So the output qubit is **|2>.**

[ii] For input qubits |D2>, |D3> = **|0>, |0>**, the multi-valued OR gates will perform maximum operations of input qubits. So the output qubit is **|0>.**

[iii] For input qubits |D4>, |D5> = **|0>, |0>**, the multi-valued OR gates will perform maximum operations of input qubits. So the output qubit is **|0>.**

[iv] For input qubits |D6>, |D7> = **|0>, |0>** the multi-valued OR gates will perform maximum operations of input qubits. So the output qubit is **|0>.**

[v] Combine the output sequences **|2>, |0>** from [i] and [ii] respectively as input sequences, the multi-valued OR gates will perform maximum operations of input sequences. So the output sequence of **|2>.**

[vi] Combine the output sequences **|0>** from [iii] and [iv] respectively as input sequences, the multi-valued OR gates will perform maximum operations of input sequences. So the output sequence of **|0>.**

[vii] Finally, Combine the output sequence **|0>** from [vii] and D8 (**|0>**) as input sequences, the multi-valued OR gates will perform maximum operations of input sequences. So the output sequence of **|2> (v)** for the function output |F1> and **|0> (vii)** for |F2> multi-valued 9-to-2 Quantum PROM.

[2] For input sequences |A>, |B> = **|0>, |1>**

|A>= |A0>, |A1>, |A2> = |2>, |0>, |0> and |B> = |B0>, |B1>, |B2> = |0>, |2>, |0>. With this input combinations, only input |A0> and |B1> is true. Thus the |A0> and |B1> are connected to output line |D1> so only the |D1> will **|2>** and rest of the gates will produce **|0>.**

For output sequence |F1>, perform OR operations among (|D0> - |D3>) outputs. So, max (|D0>, |D1>, |D2>, |D3>) = max (|0>, |2>, |0>, |0) = **|2>** and for |F2> perform OR operations among (|D4>, |D5>, |D6>, |D7>, |D8>) = max (|0>, |0>, |0>, |0>, |0>) = **|0>** of multi-valued 9-to-2 Quantum PROM.

[3]For input sequences |A>, |B>= **|0>, |2>**

|A>= |A0>, |A1>, |A2> = |2>, |0>, |0> and |B0>, |B1>, |B2> = |0>, |0>, |2>. With this input combinations, only input |A0> and |B2> is true. Thus the |A0> and |B2> are connected to output line |D2> so only the D2 will **|2>** and rest of the gates will produce **|0>.**

For output sequence |F1>, perform OR operations among (|D0> - |D3>) outputs. So, max (|D0>, |D1>, |D2>, |D3>) = max (|0>, |0>, |2>, |0) = **|2>** and for |F2> perform OR operations among (|D4>, |D5>, |D6>, |D7>, |D8>) = max (|0>, |0>, |0>, |0>, |0>) = **|0>** of multi-valued 9-to-2 Quantum PROM.

[4] For input sequences |A>, |B>= **|1>, |0>**

|A>= |A0>, |A1>, |A2> = |0>, |2>, |0> and |B0>, |B1>, |B2> = |2>, |0>, |0>. With this input combinations, only input |A1> and |B0> is true. Thus the |A1> and |B0> are connected to output line |D3> so only the |D3> will **|2>** and rest of the gates will produce **|0>**.

For output sequence |F1>, perform OR operations among (|D0> - |D3>) outputs. So, max (|D0>, |D1>, |D2>, |D3>) = max (|0>, |0>, |0>, |2) = **|2>** and for |F2> perform OR operations among (|D4>, |D5>, |D6>, |D7>, |D8>) = max (|0>, |0>, |0>, |0>, |0>) = **|0>** of multi-valued 9-to-2 Quantum PROM.

[5]
For input sequences |A>, |B> = **|1>, |1>**

|A>= |A0>, |A1>, |A2> = |0>, |2>, |0> and |B0>, |B1>, |B2> = |0>, |2>, |0>. With this input combinations, only input |A1> and |B1> is true. Thus the |A1> and |B1> are connected to output line |D4> so only the |D4> will **|2>** and rest of the gates will produce **|0>**.

For output sequence |F1>, perform OR operations among (|D0> - |D3>) outputs. So, max (|D0>, |D1>, |D2>, |D3>) = max (|0>, |0>, |0>, |0) = **|0>** and for |F2> perform OR operations among (|D4>, |D5>, |D6>, |D7>, |D8>) = max (|2>, |0>, |0>, |0>, |0>) = **|2>** of multi-valued 9-to-2 Quantum PROM.

[6]
For input sequences |A>, |B>= **|1>, |2>**

|A>= |A0>, |A1>, |A2> = |0>, |2>, |0> and |B0>, |B1>, |B2> = |0>, |0>, |2>. With this input combinations, only input |A1> and |B2> is true. Thus the |A1> and |B2> are connected to output line |D5> so only the |D5> will **|2>** and rest of the gates will produce **|0>**.

For output sequence |F1>, perform OR operations among (|D0> - |D3>) outputs. So, max (|D0>, |D1>, |D2>, |D3>) = max (|0>, |0>, |0>, |0) = **|0>** and for |F2> perform OR operations among (|D4>, |D5>, |D6>, |D7>, |D8>) = max (|0>, |2>, |0>, |0>, |0>) = **|2>** of multi-valued 9-to-2 Quantum PROM.

[7]
For input sequences |A>, |B>= **|2>, |0>**

|A>= |A0>, |A1>, |A2> = |0>, |0>, |2> and |B0>, |B1>, |B2> = |2>, |0>, |0>. With this input combinations, only input |A2> and |B0> is true. Thus the |A2> and

|B0> are connected to output line |D6> so only the |D6> will **|2>** and rest of the gates will produce **|0>**.

For output sequence |F1>, perform OR operations among (|D0> - |D3>) outputs. So, max (|D0>, |D1>, |D2>, |D3>) = max (|0>, |0>, |0>, |0) = **|0>** and for |F2> perform OR operations among (|D4>, |D5>, |D6>, |D7>, |D8>) = max (|0>, |0>, |2>, |0>, |0>) = **|2>** of multi-valued 9-to-2 Quantum PROM.

[8]
For input sequences |A>, |B>= **|2>, |1>**

|A>= |A0>, |A1>, |A2> = |0>, |0>, |2> and |B0>, |B1>, |B2> = |0>, |2>, |0>. With this input combinations, only input |A2> and |B1> is true. Thus the |A2> and |B1> are connected to output line |D7> so only the |D7> will **|2>** and rest of the gates will produce **|0>**.

For output sequence |F1>, perform OR operations among (|D0> - |D3>) outputs. So, max (|D0>, |D1>, |D2>, |D3>) = max (|0>, |0>, |0>, |0) = **|0>** and for |F2> perform OR operations among (|D4>, |D5>, |D6>, |D7>, |D8>) = max (|0>, |0>, |0>, |2>, |0>) = **|2>** of multi-valued 9-to-2 Quantum PROM.

[9]
For input sequences |A>, |B> = **|2>, |2>,**

|A>= |A0>, |A1>, |A2> = |0>, |0>, |2> and |B0>, |B1>, |B2> = |0>, |0>, |2>. With this input combinations, only input |A2> and |B2> is true. Thus the |A2> and |B2> are connected to output line |D8> so only the |D8> will **|2>** and rest of the gates will produce **|0>**.

For output sequence |F1>, perform OR operations among (|D0> - |D3>) outputs. So, max (|D0>, |D1>, |D2>, |D3>) = max (|0>, |0>, |0>, |0) = **|0>** and for |F2> perform OR operations among (|D4>, |D5>, |D6>, |D7>, |D8>) = max (|0>, |0>, |0>, |0>, |2>) = **|2>** of multi-valued 9-to-2 Quantum PROM.

5.5   Multiple-Valued Quantum Cache Memory

The Cache memory is one of the fastest memories. Though it is costlier than the main memory but more useful than the registers. The Cache memory basically acts as a buffer between the main memory and the CPU. Moreover, it synchronizes with the speed of the CPU. Besides, it stores the data and instructions that the CPU uses more frequently so that it does not have to access the main memory again and again. Therefore, the average time to access the main memory decreases. It is placed between the main memory and the CPU. Moreover, for any data, the CPU first checks the Cache and then the main memory. There can be various levels of Cache memory, they are as follows:

Level 1 (L1) or Registers

It stores and accepts the data which is immediately stored in the CPU. For example, instruction register, Cache memory, Cache counter, accumulator, address register, etc.

Level 2 (L2) or Cache Memory

It is the fastest memory that stores data temporarily for fast access by the CPU. Moreover, it has the fastest access time.

Level 3 (L3) or Main Memory

It is the main memory where the computer stores all the current data. It is a volatile memory which means that it loses data on power OFF.

Level 4 (L4) or Secondary Memory

It is slow in terms of access time. But the data stays permanently in this memory.

Hardware implements Cache as a block of memory for the temporary storage of data likely to be used again. Central processing units (CPUs) and hard disk drives (HDDs) frequently use a Cache, as do web browsers and web servers [18]. A Cache is made up of a pool of entries. Each entry has associated data, which is a copy of the same data in some backing store. Each entry also has a tag, which specifies the identity of the data in the backing store of which the entry is a copy. Tagging allows simultaneous Cache-oriented algorithms to function in a multilayered fashion without differential relay interference.

5.5.1 Basic Definitions

1. The CPU first checks any required data in the Cache. Furthermore, it does not access the main memory if that data is present in the Cache.

2. On the other hand, if the data is not present in the Cache then it accesses the main memory.

3. The block of words that the CPU accesses currently is transferred from the main memory to the Cache for quick access in the future.

4. The hit ratio defines the performance of the Cache memory.

5.5.2 Block Diagram

A 4-to-1 Cache memory chip has a memory capacity of 4 words of one bit per word. This requires a 9-bit address and a 1-bit bidirectional data bus. The 1-bit bidirectional data bus allows the transfer of data either from memory to CPU during a **read** operation or from CPU to memory during a **write** operation. The **read** and **write** inputs specify the memory operation, and the two chip select (CS) control inputs are for enabling the chip only when the microprocessor selects it.

Figure 5.15 represents the Block Diagram of Quantum 9-to-1 Cache Memory. This quantum Cache memory consists of 9 separate "Words" of memory and each is single qubits wide.

5.5.3 Architecture of Basic Components

Cache memory consists of three basic components. To execute Quantum 9-to-1 Cache memory following operations are required.

1. A Quantum 2-to-9 Decoder
2. Multi-Valued RAM cells and
3. Quantum ternary OR operations for corresponding minterms.

Quantum 2-to-9 Decoder and quantum ternary OR operation are discussed before.

5.5.3.1 Multi-Valued Quantum RAM Cell

Figure 5.16 depicts the quantum multi-valued single RAM Cell . The fundamental design of this qubit cell is based on the D flip-flop. To begin with, the cell has three inputs and a single output. The inputs are labeled "|Select>", "|R/W>", and "|Input>". The output line is labeled "|output>". To perform the Quantum Cache memory cell output, two quantum NOT, three Quantum AND and four Quantum NAND operations are required.

Step 1:
First draw three input qubits |Input>, |R/W> and |Select>. Two possible states for a qubit are the states |0 > false, and |1> true.

Step 2:
Draw Quantum NOT operation with the |Input> and |R/W> qubits.

Step 3:
Two input qubits (|R/W> and |select>) will go through quantum AND operations.

Step 4:
Again, Two input qubits (NOT of |R/W> and |select>) will go through quantum AND operations.

Step 5:
The outputs of step 4 and NOT of qubit |input> will go to quantum NAND operations. Also, the outputs of step 4 and qubit |input> will go to another quantum NAND operations.

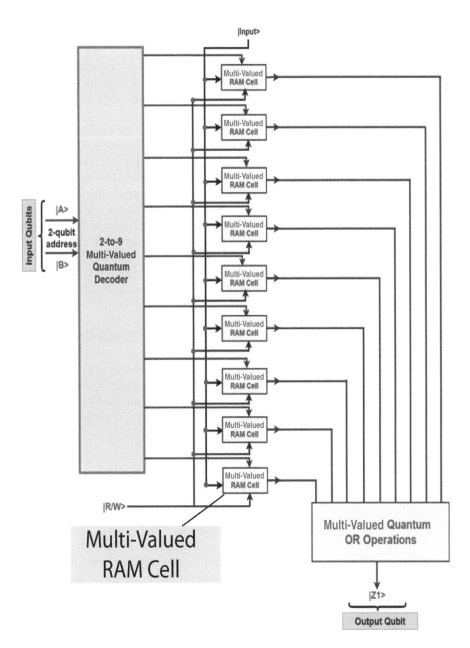

FIGURE 5.15
Block Diagram of Quantum 9-to-1 Cache Memory

Step 6:
The outputs of step 5 will go to the D flipflop as input.

Step 7:
Finally, D flipflop output and step 3 output qubit will go to a quantum AND operation, then this output with NOT of |R/W will go through another quantum AND operation to produce desired quantum cache memory |output> qubit.

In a sequential device as simple as a D flip-flop it could be used to remember one bit of data. To develop a complete memory cell, called a qubit cell, it needs to be based on the flip-flop. The number of total quantum cells per word will be m × n where m represents words with n bits. The "|select>" input is used to access the cell, either for reading or writing also used to access anyone quantum RAM cell when there is more than one quantum RAM cell. When the select line is high or |1> then the cell performs the memory operation. But when the select line of the quantum RAM cell is low or |0> then the cell is not interested to perform a read from or written to. The next input qubit is "|R/W>" where a system clock will conduct this input. If the clock value on the read/write line is |0>, this will signify "read" and when it is |1>, it will perform the "write" phase. When such a cell is selected and in "read" mode, the current value of its underlying flip-flop will be transferred to the cell's output line. When the cell is selected and in "write" mode, an input data signal will determine the value remembered by the flip-flop. Figure 5.17 shows the quantum multi-valued RAM cells.

To perform the quantum 9-to-1 Cache memory, four selection lines are needed to design nine quantum RAM cells. The output of quantum 2-to-9 decoder with nine output qubits will perform as selection input qubit of quantum RAM cells. Figure 5.17 shows the 9 quantum RAM cells to perform |Q0> to |Q8> for further minterms operation.

5.5.4   Circuit Architecture of Multiple-Valued Quantum Cache Memory

For the circuit architecture of Multi-Valued Quantum 9-to-1 Cache, need a 2-to-9 ternary decoder which is designed by using two Multi-Valued 1-to-3 Quantum decoders to get the output of nine Quantum ternary RAM cells operations. Finally the multi-valued 9-ternary quantum RAM cells perform qubit cell output qubits from |Q0> to |Q8> for further minterms ternary OR operation to perform desired output. Following given the design algorithms of multiple-valued Quantum 9-to-1 Cache Memory.

5.5.4.1   Working Principle

Figure 5.18 represents the implementation of 9-to-1 qubit Cache memory. This quantum Cache memory consists of 9 separate "Words" of memory and each is

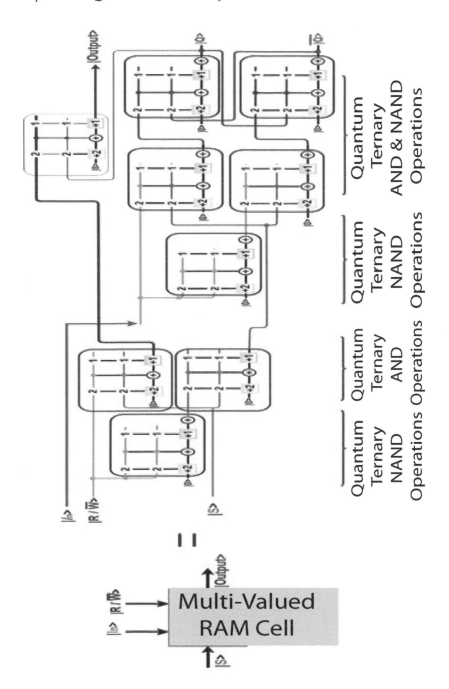

FIGURE 5.16
Quantum Multi-Valued Single RAM Cell

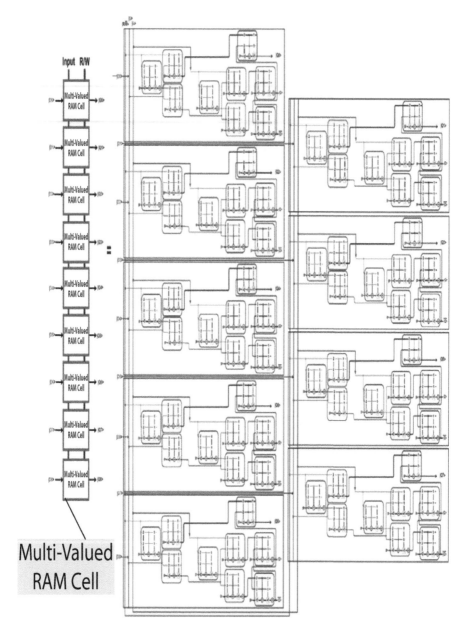

FIGURE 5.17
Quantum Multi-Valued RAM Cells

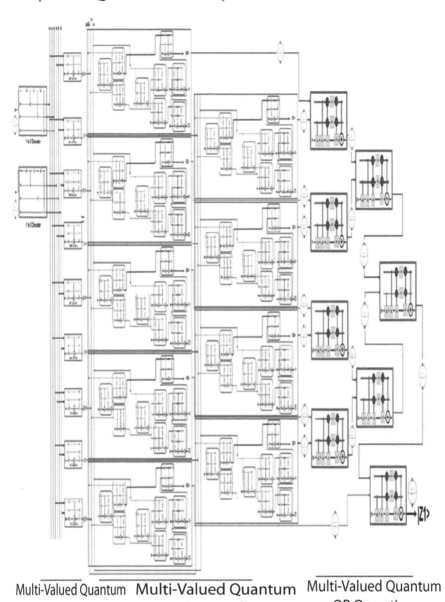

Multi-Valued Quantum 2-to-9 Decoder

Multi-Valued Quantum RAM Cell

Multi-Valued Quantum OR Operations

FIGURE 5.18
Circuit Architecture of Quantum 9-to-1 Cache Memory

1 qubit wide. The quantum Cache memory Cell has three inputs and one output. The complete circuit of a quantum Cache memory cell is described in figure 8 with proper explanation. A word consists of two quantum Cache memory cells arranged in such a way so that both qubits can be accessed simultaneously. Four words of memory need two address lines. $|A>$ and $|B>$ are the two-qubit address lines input that goes through a 2-to-9 decoder that selects one of the nine words. The memory-enabled input enables the decoder. If the memory enables is $|0>$, all output of the decoder will be $|0>$ and in that case, none of the memory addresses will be selected. But when the memory enables is $|2>$ then one of the nine words is selected. The word is selected by the value in the two address lines. When a word has been selected, the read/write input determines the operation. During the read operation, the nine qubits of the selected word pass to the quantum OR gates to the output $|Z1>$ terminals. But during the write operation, the data which is available in the input lines are transferred into the four quantum cells of the selected word. The quantum Cache memory cells that are not selected have become disabled and their previous qubit never changes. But when the memory enable input that passes into the decoder is equal to $|0>$, none of the words are selected, and then all quantum cells remain unchanged regardless of the value of the read/ write input.

5.6   Summary

This chapter has presented four types of memory devices in multi-valued quantum, DNA, quantum-DNA and DNA-quantum computing. The combination of quantum and DNA computing can be expressed as quantum molecular biology. DNA computing represents information by using DNA molecules, where a sequence of DNA molecules is data. Quantum molecular biology is two types of combinations; quantum-DNA and DNA-quantum computing. This chapter shows multi-valued memory devices in quantum molecular biology with a detailed description and their working principles. Necessary figures are also shown.

Bibliography

[1] Gyongyosi, L., & Imre, S. (2019). A survey on quantum computing technology. Computer Science Review, 31, 51-71.

[2] De Vos, A. (2011). Reversible Computing: Fundamentals, Quantum Computing, and Applications. John Wiley & Sons.

[3] Steane, A. (1998). Quantum computing. Reports on Progress in Physics, 61(2), 117.

[4] Acín, A., Bloch, I., Buhrman, H., Calarco, T., Eichler, C., Eisert, J., ... & Wilhelm, F. K. (2018). The quantum technologies roadmap: a European community view. New Journal of Physics, 20(8), 080201.

[5] Huang, H. L., Wu, D., Fan, D., & Zhu, X. (2020). Superconducting quantum computing: a review. Science China Information Sciences, 63(8), 1-32.

[6] Häffner, H., Roos, C. F., & Blatt, R. (2008). Quantum computing with trapped ions. Physics Reports, 469(4), 155-203.

[7] Munro, J. I., & Raman, V. (1996). Selection from read-only memory and sorting with minimum data movement. Theoretical Computer Science, 165(2), 311-323.

[8] Gill, S. S., Kumar, A., Singh, H., Singh, M., Kaur, K., Usman, M., & Buyya, R. (2022). Quantum computing: A taxonomy, systematic review and future directions. Software: Practice and Experience, 52(1), 66-114.

[9] Perkowski, M. A. (2005). Multiple-valued quantum circuits and research challenges for logic design and computational intelligence communities. IEEE Connections, 3(4), 6-12.

[10] Tymchenko, L., Petrovskiy, M., Kokryatskaya, N., Gubernatorov, V., & Kutaev, Y. (2013). Modeling of the high-performance PLD-based sectioning method for classification of the shape of optical object images. SpringerPlus, 2(1), 1-12.

[11] Suh, G. E., Rudolph, L., & Devadas, S. (2004). Dynamic partitioning of shared cache memory. The Journal of Supercomputing, 28(1), 7-26.

[12] Hill, M. D. (1987). Aspects of Cache Memory and Instruction Buffer Performance. University of California, Berkeley.

[13] Hidaka, H., Matsuda, Y., Asakura, M., & Fujishima, K. (1990). The cache DRAM architecture: A DRAM with an on-chip cache memory. IEEE Micro, 10(2), 14-25.

6

Multiple-Valued DNA Memory Devices

6.1 Introduction

Human civilization went through paradigm shifts with new ways of storing and disseminating information. To survive in the complex and ever-changing environment, our ancestors created utensils out of wood, bone and stone, and used them as media for recording information. This was the beginning of human history. With the development of computer technology, the information age has revolutionized the global scene. Digital information stored in magnetic (floppy disks), optical (CDs) and electronic media (USB sticks) and transmitted through the internet promoted the explosion of next-generation science, technology and arts. With the total amount of worldwide data skyrocketing, traditional storage methods face daunting challenges. International Data Corporation forecasts that the global data storage demand will grow to 175 ZB or 1.75×10^{14} GB by 2025 (in this review, 'B' refers to Byte while 'b' refers to base pair). With the current storage media having a maximal density of 10^3 GB/mm^3, this will far exceed the storage capacity of any currently available storage method. Meanwhile, the costs of maintaining and transferring data, as well as limited lifespans and significant data losses, also call for novel solutions for information storage. On the other hand, since the very beginning of life on Earth, nature has solved this problem in its own way: it stores the information that defines the organism in unique orders of four bases (A, T, C, G) located in tiny molecules called deoxyribonucleic acid (DNA), and this way of storing information has continued for 3 billion years. DNA molecules as information carriers have many advantages over traditional storage media. Its high storage density, potentially low maintenance cost and other excellent characteristics make it an ideal alternative for information storage, and it is expected to provide wide practicality in the future.

DOI: 10.1201/9781003381921-6

6.2   Multiple-Valued DNA Random Access Memory

Sequential circuits all depend upon the presence of memory. – A flip-flop can store one bit of information. – A register can store a single "word," typically 32-64 bits. Memory allows us to store even larger amounts of data. – Random Access Memory (RAM - volatile). Random-access memory, or RAM, provides large quantities of temporary storage in a computer system. Memory cells can be accessed to transfer information to or from any desired location, with the access taking the same time regardless of the location A RAM can store many values. An address will specify which memory value is needed and each value can be a multiple-bit word (e.g., 32 bits).

A RAM should be able to:
- Store many words, one per address
- Read the word that was saved at a particular address
- Change the word that's saved at a particular address

In multi-valued memory, the ternary logic function has inputs that can assume three states (say 0, 1 and 2) and generates one output signal that can have one of these three states. In ternary DNA computing, two DNA sequences are used as inputs and one DNA sequence is used as output. During ternary DNA computing, sequence **ACCTAG** is considered as "0", sequence **CAAGCT** strands as "1" and **TGGATC** stands as "2". In ternary DNA computing, the fluorescent level is used to detect the DNA sequence. Fluorescence is the temporary absorption of electromagnetic wave-lengths from the visible light spectrum by fluorescent molecules, and the subsequent emission of light at a lower energy level. The block diagram of main interface to RAM is shown in Figure 6.1.

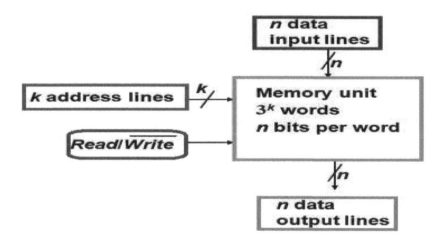

FIGURE 6.1
Block Diagram of Main Interface to RAM

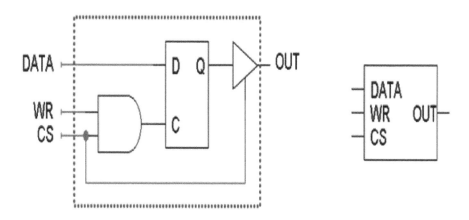

FIGURE 6.2
One-Bit RAM Cell

This block diagram introduces the main interface to RAM. A Chip Select, CS, enables or disables the RAM. ADRS specifies the address or location to read from or write to. WR selects between reading from or writing to the memory. To read from memory, WR should be set to 0. OUT will be the n-bit value stored at ADRS. To write to memory, WR = 1 is set. DATA is the n-bit value to save in memory.

To start, one latch is used to store each bit. A one-bit RAM cell is shown in Figure 6.2.

Since this is just a one-bit memory, an ADRS input is not needed. Writing to the RAM cell:

– When CS = 1 and WR = 1, the latch control input will be 1.

– The DATA input is thus saved in the D latch.

Reading from the RAM cell and maintaining the current contents:

– When CS = 0 or when WR = 0, the latch control input is also 0, so the latch just maintains its present state.

– The current latch content will appear on OUT when CS= 1.

To make a 4 × 1 RAM, these cells can be used.

Since there are four words, ADRS is two bits. Each word is only one bit, so DATA and OUT are one bit each. Word selection is done with a decoder attached to the CS inputs of the RAM cells. Only one cell can be read or written at a time. Notice that the outputs are connected together with a single line!

6.2.1 Block Diagram

Figure 6.3 shows the block diagram of a multi-valued DNA 9-to-1 RAM, the unit consists of 9 words of 2 input sequence (A = **ACCTAG (0) CAAGCT (1) or TG-GATC (2)** and B = **ACCTAG (0) CAAGCT (1) or TGGATC (2)**) each.

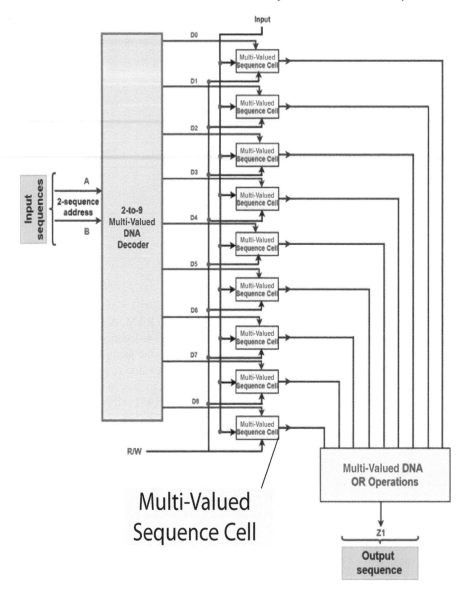

FIGURE 6.3
Block Diagram of Multi-Valued DNA 9-to-1 RAM

This implies there are 2 output lines (F1 and F2) and 9 distinct word sequences stored in the unit, each of which may be applied to the output lines. The particular word selected that is presently available on the output line is determined from the 2 input lines. There are only 3 input sequences in multi-valued DNA 9-to-1 RAM because $3^2=9$ and with 3 sequence variables can specify 9 addresses or minterms.

To perform minterms of nine addresses, a multi-valued DNA 2-to-9 decoder and multi-valued DNA OR operations are required.

6.2.2 Architecture of Basic Components

RAM consists of three basic components- Decoder, Molecular Cells and OR gates. To execute DNA 9-to-1 RAM operation:

1. A 2-to-9 DNA Decoder
2. Multi-Valued Sequence Cells and
3. DNA OR operations for corresponding minterms are required.

6.2.2.1 DNA 2-to-9 Decoder

Three DNA molecular sequences,
  TGGATC represents **2,**
  CAAGCT represents **1 and**
  ACCTAG represents **0.**

To perform the ternary 2-to-9 decoder, at first it is needed to know about the 1-to-3 decoder.

The 1-to-3 decoder in the circuit generates unary functions for input variable x as x0, x1 and x2 which is used for ternary function implementation. Table 6.1 shows the truth table of ternary 1-to-3 Decoder and Figure 6.4 shows the circuit diagram of DNA Ternary Decoder.

6.2.2.1.1 Circuit Design of DNA 2-to-9 Decoder
Multi-Valued DNA 2-**to**-9 decoder is a combinational logic circuit which is designed by using two Multi-Valued 1-to-3 DNA decoders and nine DNA AND (D0-D8) operations (Figure 6.5).

6.2.2.1.2 Working Procedure of DNA 2-to-9 Decoder
The multi-valued 2-to-9 DNA decoder operation is performed the two input DNA sequences and each of three states, the combination will be according to the Table 6.2. From these combinations, the corresponding multi-valued AND operation outputs are the minimum of input sequence states. Ternary logic functions are those

TABLE 6.1

Ternary 1-**to**-3 Decoder

A	A2	A1	A0
0	0	0	2
1	0	2	0
2	2	0	0

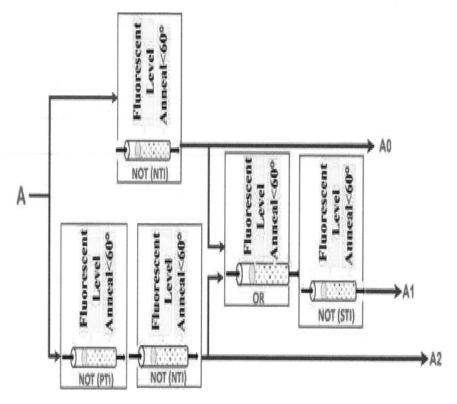

FIGURE 6.4
DNA Ternary Decoder

functions which have significance if a third value is acquainted with the binary logic.
Thus **ACCTAG**, **CAAGCT** and **TGGATC** denote the ternary levels for basic logic
gates to represent sequence as "0", "1" and "2" respectively.

For multi-valued 2-**to**-9 decoder 9 outputs. D0, D1, D2, D3, D4, D5, D6, D7, D8,
and two inputs A, B.

[i]
For input sequences, A, B= **ACCTAG, ACCTAG**, the multi-valued 1-to-3 decoder will perform A0 and B0 and the D0 line will be open. So the output sequence
of D0 = **TGGATC** and the rest of the lines D1 to D8 will remain closed.

[ii]
For input sequences, A, B= **ACCTAG, CAAGCT**, the multi-valued 1-to-3 decoder will perform A0 and B1 and the D1 line will be open. So the output sequence
of D1 = **TGGATC** and the rest of the lines D0, D2 to D8 will remain closed.

TABLE 6.2
DNA 2-to-9 Decoder

A	B	D8	D7	D6	D5	D4	D3	D2	D1	D0
ACCTAG	ACCTAG	ACCTAG	ACCTAG	ACCTAG	ACCTAG	ACCTAG	ACCTAG	ACCTAG	ACCTAG	TGGATC
ACCTAG	CAAGCT	ACCTAG	ACCTAG	ACCTAG	ACCTAG	ACCTAG	ACCTAG	ACCTAG	TGGATC	ACCTAG
ACCTAG	TGGATC	ACCTAG	ACCTAG	ACCTAG	ACCTAG	ACCTAG	ACCTAG	TGGATC	ACCTAG	ACCTAG
CAAGCT	ACCTAG	ACCTAG	ACCTAG	ACCTAG	ACCTAG	ACCTAG	TGGATC	ACCTAG	ACCTAG	ACCTAG
CAAGCT	CAAGCT	ACCTAG	ACCTAG	ACCTAG	ACCTAG	TGGATC	ACCTAG	ACCTAG	ACCTAG	ACCTAG
CAAGCT	TGGATC	ACCTAG	ACCTAG	ACCTAG	TGGATC	ACCTAG	ACCTAG	ACCTAG	ACCTAG	ACCTAG
TGGATC	ACCTAG	ACCTAG	ACCTAG	TGGATC	ACCTAG	ACCTAG	ACCTAG	ACCTAG	ACCTAG	ACCTAG
TGGATC	CAAGCT	ACCTAG	TGGATC	ACCTAG	ACCTAG	ACCTAG	ACCTAG	ACCTAG	ACCTAG	ACCTAG
TGGATC	TGGATC	TGGATC	ACCTAG	ACCTAG	ACCTAG	ACCTAG	ACCTAG	ACCTAG	ACCTAG	ACCTAG

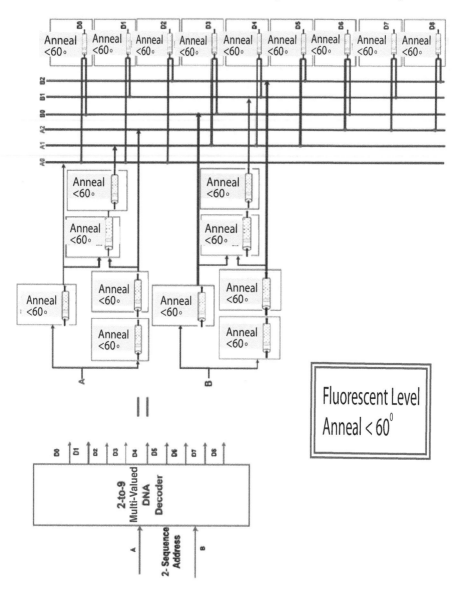

FIGURE 6.5
Multi-Valued DNA 2-to-9 Decoder

[iii]

For input sequences, A, B= **ACCTAG**, **TGGATC**, the multi-valued 1-to-3 decoder will perform A0 and B2 and the D2 line will be open. So the output sequence of D2 = **TGGATC** and the rest of the lines D0, D1 and D3 to D8 will remain closed.

[iv]

For input sequences, A, B= **CAAGCT**, **CAAGCT**, the multi-valued 1-to-3 decoder will perform A1 and B0 and the D3 line will be open. So the output sequence of D3 = **TGGATC** and the rest of the lines D0 to D2 and D4 to D8 will remain closed.

[v]

For input sequences, A, B= **CAAGCT**, **CAAGCT**, the multi-valued 1-to-3 decoder will perform A1 and B1 and the D4 line will be open. So the output sequence of D4 = **TGGATC** and the rest of the lines D0 to D3 and D5 to D8 will remain closed.

[vi]

For input sequences, A, B= **CAAGCT**, **TGGATC**, the multi-valued 1-to-3 decoder will perform A1 and B2 and the D5 line will be open. So the output sequence of D5 = **TGGATC** and the rest of the lines D0 to D4 and D6 to D8 will remain closed.

[vii]

For input sequences, A, B= **TGGATC**, **ACCTAG**, the multi-valued 1-to-3 decoder will perform A2 and B0 and the D6 line will be open. So the output sequence of D6 = **TGGATC** and the rest of the lines D0 to D5 and D7 to D8 will remain closed.

[viii]

For input sequences, A, B= **TGGATC**, **CAAGCT**, the multi-valued 1-to-3 decoder will perform A2 and B1 and the D7 line will be open. So the output sequence of D7 = **TGGATC** and the rest of the lines D0 to D6 and D8 will remain closed.

[ix]

For input sequences, A, B= **TGGATC**, **TGGATC**, the multi-valued 1-to-3 decoder will perform A2 and B2 and the D8 line will be open. So the output sequence of D8 = **TGGATC** and the rest of the lines D0 to D7 will remain closed.

All the corresponding values from nine DNA multi-valued AND gates sequences will go through multi-valued DNA OR gates and the multi-valued OR gates will perform according to the maximum of the input sequences.

6.2.2.2 Multi-Valued DNA Sequence Cells

The fundamental design of this Molecular cell is based on the R-S flip-flop (Figure 6.6). To begin with, the molecular cell has three inputs and a single output. The inputs are labeled "Select", "R/W", and "Input". The output line is labeled "M0". To perform the Molecular cell output, two DNA NOT, six DNA AND and two DNA NOR operations are needed to perform.

25.2.2.2.1 Circuit Design of Molecular Cell

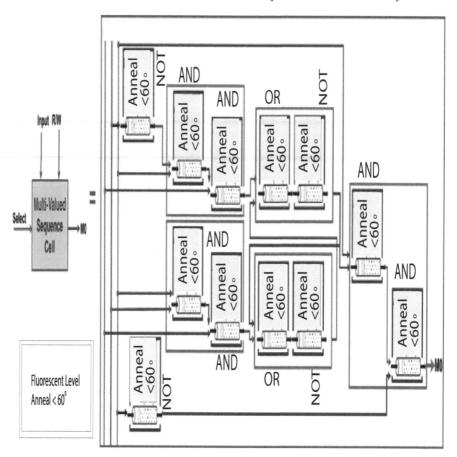

FIGURE 6.6
DNA Multi-Valued Sequence Cell

Step 1:
First draw three input sequences Input, R/W and Select. Two possible states for a sequence are the states **"TGGATC"** false, and **"ACCTAG"** true.

Step 2:
Draw DNA NOT operation with the Input and R/W sequences.

Step 3:
Each DNA AND gate is with three inputs. So two inputs (NOT Input and Select) are taken to one AND gate input and the output of this gate will go to another AND gate with input sequence R/W as input.

Step 4:

Again, each DNA AND gate is with three inputs. So two inputs (Input and Select) are taken to one AND gate input and the output of this gate will go to another AND gate with input sequence R/W as input.

Step 5:

The outputs of step 3 and 4 will go to the R-S flip-flop as input.

Step 6:

Finally, the output of R − S flip flop and select will go to a DNA AND gate, then this output with NOT of R/W will go through another DNA AND gate to produce desired Molecular cell output.

25.2.2.2.2 Working Procedure of Molecular Cell

In sequential devices as simple as an R-S flip-flop, it could be used to remember one bit of data. To develop a complete DNA memory cell based on the architecture of a DNA flip-flop which is called a DNA sequence cell. The number of total DNA cells per word will be m × n where m represents words with n bits. The "select" input is used to access the cell, either for reading or writing also used to access any one sequence cell when there is more than one sequence cell. When the select line is high or **ACCTAG** then the cell performs the memory operation. But when the select line of the DNA cell is low or **TGGATC** then the cell is not interested to perform a read from or written to.

The next input sequence is "R/W" where a system clock will conduct this input. If the clock value on the read/write line is **TGGATC**, this will signify "read" and when it is **ACCTAG**, it will perform the "write" phase. When such a cell is selected and in "read" mode, the current value of its underlying flip-flop will be transferred to the cell's output line. When the cell is selected and in "write" mode, an input data signal will determine the value remembered by the flip-flop. Let's consider the cell that has been selected. In that case, if the clock value is **TGGATC** then the cell contents are to be read and this time the output value will depend only on the P-value of the flip flop. But if P is low, the cell output will be **TGGATC** and if P is high, the cell output will be **ACCTAG**. It occurs because the DNA AND gate added to the cell's output which has three input-negated R/W, Select, and P; and both "negated R/W" and "Select is currently high **ACCTAG**.

To perform the DNA 9-to-1 RAM, four selection lines are needed to design nine DNA sequence cells. The output of DNA 2-to-9 decoder with nine output sequences will perform as selection input sequence of sequence cell. Figure 6.7 shows the 9-sequence cells to perform M0 to M8 for further minterms operation.

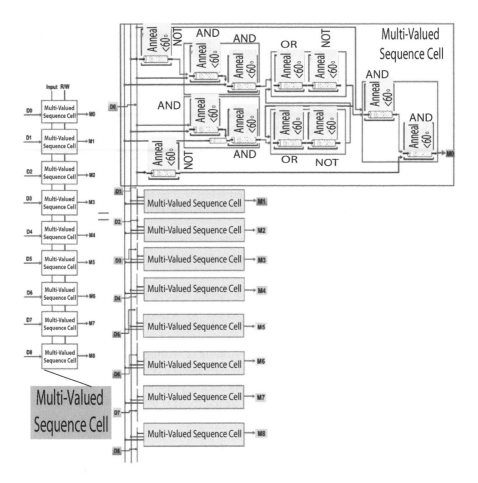

FIGURE 6.7
DNA Multi-Valued Sequence Cells of RAM

6.2.3   Circuit Architecture of Multiple-Valued DNA RAM

For the design algorithm of Multi-Valued DNA 9-to-1 RAM, it needs a 2-to-9 ternary decoder which is designed by using two Multi-Valued 1-to-3 DNA decoders to get the output of nine DNA ternary RAM cells operations. Finally, the multi-valued 9-ternary RAM cells perform qubit cell output sequences from Q0 to Q8 for further minterms ternary OR operation to perform desired output. Following the design algorithms of multiple-valued DNA 9-to-1 RAM Memory. Figure 6.8 shows the circuit architecture of DNA 9-to-1 RAM.

In DNA 9-to-1 RAM architecture (Figure 6.8), two address lines with 9-to-1 sequence RAM and each address line needs to be DNA NOT form as well. These address lines combination will be the input of DNA 2-to-9 decoders which consist of

TABLE 6.3
Control Input to Memory Chip

R/W	Memory Operation
X	*None*
TGGATC	*Write to the selected word*
ACCTAG	*Read from the selected word*

nine DNA AND gates and this decoder has one enable input. Four select lines from this decoder are got and each select line will go through each molecular cell. Note that the word calculation of RAM will be $3\wedge k$ where k is the address line and $3\wedge k$ is the total words of n bit and the decoder combination will be $k\times 3\wedge k$. This single sequence RAM consists of 9 separate Molecular cells and each cell has 3 inputs-D0, anyone selects a line and read/ write inputs. The obtained output from 9 DNA molecular RAM cells will be the input of a DNA OR gate which produces the final output. This is the whole design procedure of 9-to-1 sequence RAM. Table 6.3 shows the Control Input to Memory Chip

6.2.4 Working Principle of Multi-Valued DNA RAM

This Figure 6.8 represents the implementation of 9-to-1 sequence RAM. This DNA RAM consists of 9 separate "Words" of memory and each one is 1 sequence wide. The DNA RAM Cell has three inputs and one output. The complete circuit of a DNA RAM cell is described in Figure 6.8 with proper explanation. A word consists of two DNA RAM cells arranged in such a way so that both sequences can be accessed simultaneously. Four words of memory need two address lines. A and B are the two-sequence address lines input that goes through a DNA 2-to-9 decoder that selects one of the nine words. The memory-enabled input enables the decoder. If the memory enables is **TGGATC**, all output of the decoder will be **TGGATC** and in that case, none of the memory addresses will be selected. But when the memory enables is **ACCTAG** then one of the nine words is selected. The word is selected by the value in the two address lines.

When a word has been selected, the read/write input determines the operation. During the read operation, the nine sequences of the selected word pass to the DNA OR gates to the output Z1 terminals. But during the write operation, the data which is available in the input lines are transferred into the nine DNA cells of the selected word. The DNA RAM cells that are not selected have become disabled and their previous sequence never changes. But when the memory enable input that passes into the decoder is equal to **TGGATC**, none of the words are selected, and then all DNA cells remain unchanged regardless of the value of the read/ write input. This is the working procedure of 9-to-1 DNA RAM.

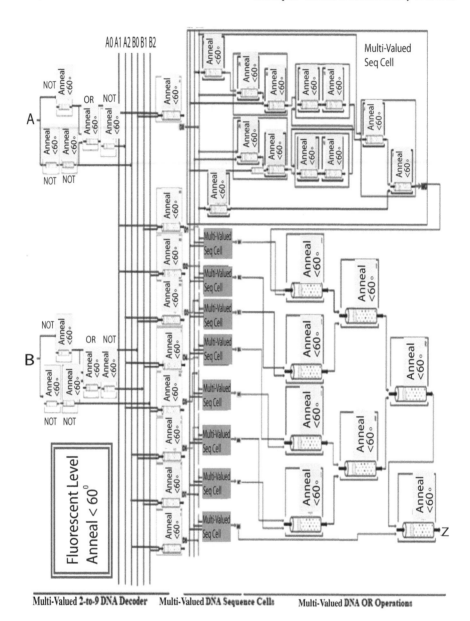

FIGURE 6.8
Circuit Architecture of DNA 9-to-1 RAM

6.3 Multiple-Valued DNA Read-Only Memory

A ternary computer (also called a trinary computer) is one that uses ternary logic (i.e., base 3) instead of the more common binary system (i.e., base 2) in its calculations. This means it uses trits instead of bits, as most computers do. Continuous technological advancement – driven by Moore's Law - electronic gadgets and their linked memory systems have been able toward becoming simultaneously smaller and more efficient. Current storage capacity and other cognitive mechanisms, on the other hand, consume more energy. Furthermore, internal thermal resistance and other restrictions may hinder further memory storage advancements, even as worldwide demand for enhanced data storage and retrieval capabilities continues to expand. So, the main concern is to provide low-cost, robust, high-density, reliable, and energy-efficient memory technologies through designing multi-valued DNA-based ROM. This technology has the ability to be written on, read from, and erased rapidly, and not deteriorate with time. Though traditional ROM is a slower memory so multi-valued DNA computing enables the creation of new types of computers capable of operating with multiple sequences as input states, therefore increasing storage capacity. Security experts have identified a bug in Intel's read-only memory that they believe is unfixable, leaving all Intel devices. The fact is that once an attacker has a static circuit, it is just a matter of time before they can reverse engineer its configuration. As a result, dynamic ROM configurations are required, and biological multi-valued methods have been invented to imitate existing vulnerable static technologies in order to ensure reliable storage. Multi-Valued DNA computing introduces an approach for generating information that can be stored and retrieved reliably within such a DNA sequence. To process any circuit with more than two values for each input, called Multi-Valued. It is possible to minimize the complexity of interconnecting circuits by using multi-valued logic rather than binary logic. Since each wire can transmit more multi-valued logic information so results in smaller chip size.

DNA (Deoxyribose Nucleic Acid) computing has the features of parallel processing and large storage capability that make it special from other conventional computing systems. It is a type of bimolecular programming where different types of reactions are used to perform basic operations and the processing information is stored in nucleic acids and proteins. The traditional ROM (Read Only Memory) is a slower memory. Thus, DNA computing enables the creation of new types of computers which is capable of operating sequences as input states by increasing storage capacity. DNA memory is required for the formation of a synchronization tool that can match the multiple procedures in a DNA-based computer, a DNA gate that retains the identity of any state, and a method for turning preset molecular sequences into on-demand sequences, among other DNA information processing devices. DNA-based memory may be utilized in a variety of applications, including DNA computing and DNA communication. Continuous research and experimentation have enabled the storing of sequences in DNA memory. DNA memory is the DNA-mechanical equivalent of conventional computer memory in DNA computing. Unlike conventional memory, which stores information as binary states (represented by "2"s, "1"s, and "0"s), DNA memory saves a molecular state for subsequent retrieval. Sequences (represented by **ACCTAG as 0"**, **CAAGCT** as **1 and "TGGATC as 2**) which provide important computing information, are stored in these states. Unlike traditional computer memory, the states saved in DNA memory can be in a sequence, providing far more practical flexibility in DNA algorithms than traditional information storage.

One early calculating machine, built entirely from wood by Thomas Fowler in 1840, operated in balanced ternary. The first modern, electronic ternary computer, Setun, was built in 1958 in the Soviet Union at the Moscow State University by Nikolay Brusentsov, and it had notable advantages over the binary computers that eventually replaced it, such as lower electricity consumption and lower production cost. In 1970 Brusentsov built an enhanced version of the computer, which he called Setun-70. In the United States, the ternary computing emulator Ternac working on a binary machine was developed in 1973. The ternary computer QTC-1 was developed in Canada.

6.3.1   Basic Definition

MDROM is an abbreviation for Multi-Valued DNA Read-Only Memory. It refers to computer memory chips that hold permanent or semi-permanent data and incorporate both the multi-valued DNA decoder and multi-valued DNA OR operations onto a single integrated circuit (IC). The contents of multi-valued DNA ROM are nonvolatile; even if the computer is turned off, the contents of multi-valued DNA ROM persist. It is used to hold a computer's boot-up instructions. Almost every computer has a tiny sequence of multi-valued DNA ROM that contains the boot software. This is made up of a few kilobytes of code that instructs the computer on what to do when it boots up, such as conducting hardware diagnostics and loading the operating system into multi-valued DNA RAM. The BIOS is the boot firmware on a computer. To update the programming in multi-valued DNA ROM, these chips have to be physically removed and replaced. Data saved in DNA ROM cannot be electrically changed once the memory device is manufactured. A block diagram of a multi-valued ROM is shown in Figure 6.9.

It consists of k input lines and n output lines. Each bit combination of the input variables is called an address. Each bit combination that comes out of the output lines

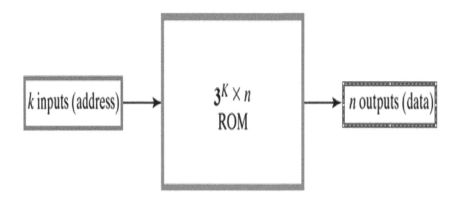

FIGURE 6.9
Block Diagram of 3^k-to-n ROM

is called a word. The number of bits per word is equal to the number of output lines n. An address is essentially a ternary number that denotes one of the minterms of k variables.

Initially, the multi-valued DNA ROM is a combinational circuit with multi-valued DNA AND gates connected as a multi-valued DNA decoder and a number of multi-valued DNA OR gates equal to the outputs in the unit. Therefore it is a two-level implementation in the sum of minterms form. With k input lines and n output lines in ROM, the output functions will calculate through the sum of minterms form. The number of distinct addresses possible with k input variables is $3 \wedge k$. An output word can be selected by a unique address, and since there are $3 \wedge k$ distinct addresses in a multi-valued ROM, there are $3 \wedge k$ distinct words which are said to be stored in the unit. The word available at the output lines at any given time depends on the address value applied to the input lines. Therefore, A ROM is characterized by the number of words $3 \wedge k$ and the number of bits per word n. For input, k=2 and output, n=2 the ROM circuit will be called 9-to-2 multi-valued ROM and the function output F_1 and F_2 in the sum of minterms form, $\sum (0, 1, 2, 3, 4, 5, 6, 7, 8)$. It does not have to be a multi-valued DNA AND-OR implementation but it can be any other possible two-level minterms implementation.

Multi-Valued DNA Read-Only Memory has some parameters to the memory processing capabilities. Some important parameters are:

1. Memory capacity, often known as memory volume, refers to the number of places in memory. Bits, bytes, and words are used to describe memory capacity. When using words, the length of each word in bits or bytes must be specified.

2. The time between sending a memory access request and receiving the desired information is known as memory access time. The access time of a memory defines its unitary speed (the reception time of unitary data). For rapid memory, the access time is minimal.

3. The shortest time between successive access requests to the same memory region is known as memory cycle time. Another measure that describes the memory's overall speed is the memory cycle time. When the cycle period is short, the speed is high.

4. Memory transfer rate is the rate at which data is read or written in a specific memory, measured in bits per second or bytes per second.

6.3.2 Block Diagram

Consider a multi-valued DNA 9-to-2 ROM, a general organization of block diagram (Figure 6.10), the unit consists of 9 words of 2 input sequences (A = **ACCTAG or CAAGCT or TGGATC** and B = **ACCTAG or CAAGCT or TGGATC**) each. This implies there are 2 output lines (F1 and F2) and 9 distinct word sequences stored in the unit, each of which may be applied to the output lines.

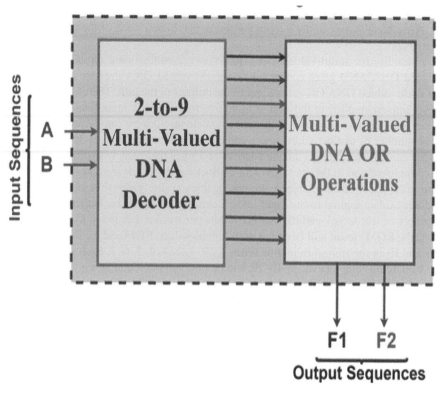

FIGURE 6.10
Block Diagram of Multi-Valued DNA 9-to-2 ROM

The particular word selected that is presently available on the output line is determined from the 2 input lines. There are only 3 input sequences in multi-valued DNA 9-to-2 ROM because $3\wedge 2=9$ and with 3 sequence variables can specify 9 addresses or minterms. To perform minterms of nine addresses, a multi-valued DNA 2-to-9 decoder and multi-valued DNA OR operations are required.

6.3.3   Circuit Architecture of Multiple-Valued DNA ROM

For the design algorithm of Multi-Valued DNA 9-to-2 ROM, a 2-to-9 ternary decoder is needed which is designed by using two Multi-Valued 1-to-3 DNA decoders to get the output of nine DNA ternary AND operations. Finally, the multi-valued 2-to-9-ternary decoder performs further minterms ternary OR operation to perform desired outputs. Following given the design algorithms of multiple-valued DNA 9-to-2 ROM Memory. Figure 6.11 depicts the Multi-Valued 9-to-2 DNA ROM.

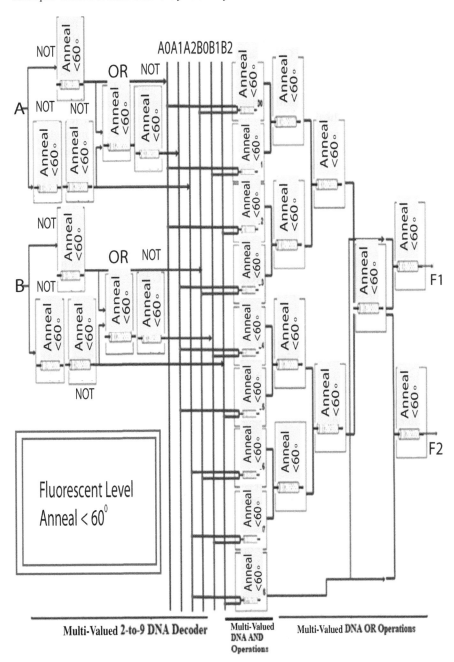

FIGURE 6.11
Multi-Valued 9-to-2 DNA ROM

In this 9-to-2 multi-valued DNA based ROM, two inputs DNA sequences A and B with three states A (**ACCTAG = 0, CAAGCT =1, TGGATC = 2**) and B (**ACCTAG = 0, CAAGCT =1, TGGATC = 2**) and two output functions F1 and F2. Input line are 2 so with the ternary logic, it will produce $3 \wedge 2 = 9$ address lines and two output function lines are data lines of multi-valued ROM. That's why it is called 9-to-2 multi-valued ROM. Thus, it will perform with DNA sequences so it is called DNA multi-valued ROM. In this ternary DNA computing, the fluorescent level is used to detect the DNA sequence. Fluorescence is defined as fluorescent molecules temporarily absorbing electromagnetic wavelengths from the visible light spectrum and then emitting light at a lower energy level.

6.3.4   Design Procedure

Step 1:
First, draw two input DNA sequences A and B. Three possible states for a DNA input sequence are the states **ACCTAG, CAAGCT, TGGATC**. These 3 will produce 9 combinations of 2 input DNA sequences.

Step 2:
For each input, a ternary decoder is needed for the selection of input value combinations. For example, if any input is A then three values can be performed A0, A1, and A2. So the ternary decoder will select combinations from the values of inputs.

Step 3:
After the ternary decoder operations for input A and B, A0, A1, A2 and B0, B1, B2 are found.

Step 4:
With each input value of step 3, it is needed to perform AND operations that will produce 9 combinations of 2 input sequences and enable anyone output line.

Step 5:
From each combination of AND operations in Step 4, draw multi-valued DNA OR gates with (D0, D1), (D2, D3), (D4, D5), (D6, D7).

Step 6:
From each outputs sequences of Step 5, again combine them with multi-valued DNA OR gates, (D0, D1, D2, D3), (D4, D5, D6, D7).

Step 7:
From each outputs sequences of Step 6, again combine them with multi-valued DNA OR gates as (D0, D1, D2, D3, D4, D5, D6, D7).

Step 8:
Finally, perform multi-valued DNA OR operations with D8 and the output of step 7 to produce desired multi-valued 9-to-2 DNA ROM for output functions F1 and F2.

TABLE 6.4
Truth Table of DNA 9-to-2 ROM

A	B	F1	F2
ACCTAG	ACCTAG	TGGATC	TGGATC
ACCTAG	CAAGCT	TGGATC	TGGATC
ACCTAG	TGGATC	TGGATC	TGGATC
CAAGCT	ACCTAG	TGGATC	TGGATC
CAAGCT	CAAGCT	TGGATC	TGGATC
CAAGCT	TGGATC	TGGATC	TGGATC
TGGATC	ACCTAG	TGGATC	TGGATC
TGGATC	CAAGCT	TGGATC	TGGATC
TGGATC	TGGATC	TGGATC	TGGATC

6.3.5 Working Principle

Table 6.4 shows the Truth Table of DNA 9-to-2 ROM. Here two inputs are A and B, two outputs are F1 and F2.

[1]
For input sequences A, B= **ACCTAG, ACCTAG**

1. To perform the value of input A0, A will go through the NOT (NTI), and produces TGGATC in the A0.

2. To perform the value of input A2, A will go through the NOT (PTI) and produce TGGATC then go through NOT (NTI) and produces ACCTAG in the A2.

3. To perform the value of input A1, the value of A0 and A2 go through DNA NOR (OR, NOT) Operations and produces ACCTAG in the A1.

4. To perform the value of input B0, B will go through the NOT (NTI), and produces TGGATC in the B0.

5. To perform the value of input B2, B will go through the NOT (PTI) and produce TGGATC then go through NOT (NTI) and produces ACCTAG in the B2.

6. To perform the value of input B1, the value of B0 and B2 go through DNA NOR (OR, NOT) Operations and produces ACCTAG in the B1.

ACCTAG = |0> **CAAGCT** =|1> **TGGATC** = |2>
A= A0, A1, A2 = 2, 0, 0 and B= B0, B1, B2= 2, 0, 0. With this input combinations, only input A0 and B0 are true. Thus the A0 and B0 are connected to output line D0 so only the D0 will **TGGATC** and the rest of the gates will produce **ACCTAG**.

To perform 9 to 2 multi-valued DNA ROM functions outputs,

[i] For input sequences D0, D1 = **TGGATC, ACCTAG**, the multi-valued OR gates will perform maximum operations of input sequences. So the output sequence

of **TGGATC.**

[ii] For input sequences D2, D3 = **ACCTAG, ACCTAG**, the multi-valued OR gates will perform maximum operations of input sequences. So the output sequence of **ACCTAG.**

[iii] For input sequences D4, D5 = **ACCTAG, ACCTAG**, the multi-valued OR gates will perform maximum operations of input sequences. So the output sequence of **ACCTAG.**

[iv] For input sequences D6, D7 = **ACCTAG, ACCTAG** the multi-valued OR gates will perform maximum operations of input sequences. So the output sequence of **ACCTAG.**

[v] Combine the output sequences **TGGATC, ACCTAG** from [i] and [ii] respectively as input sequences, the multi-valued OR gates will perform maximum operations of input sequences. So the output sequence of **TGGATC.**

[vi] Combine the output sequences ACCTAG, **ACCTAG** from [iii] and [iv] respectively as input sequences, the multi-valued OR gates will perform maximum operations of input sequences. So the output sequence of **ACCTAG.**

[vii] Combine the output sequences **TGGATC, ACCTAG** from [v] and [vi] respectively as input sequences, the multi-valued OR gates will perform maximum operations of input sequences. So the output sequence of **TGGATC.**

[viii] Finally, Combine the output sequence **TGGATC** from [vii] and D8 (**ACCTAG**) as input sequences, the multi-valued OR gates will perform maximum operations of input sequences. So the output sequence of **TGGATC** for both functions output F1 and F2 multi-valued 9-to-2 DNA ROM.

[2]
For input sequences A, B= **ACCTAG, CAAGCT**

A= A0, A1, A2 = 2, 0, 0 and B= B0, B1, B2= 0, 2, 0. With these input combinations, only input A0 and B1 are true. Thus the A0 and B1 are connected to output line D1 so only the D1 will **TGGATC** and the rest of the gates will produce **ACCTAG.**

For Functions F1 and F2, perform OR operations among all decoder (D0-D8) outputs. So, max (D0, D1, D2, D3, D4, D5, D6, D7, D8) = max (0,2,0,0,0,0,0,0,0) = 2 "**TGGATC**" will produce as the functions outputs F1 and F2 of multi-valued 9-to-2 DNA ROM.

[3]

For input sequences A, B= **ACCTAG, TGGATC**

A= A0, A1, A2 = 2, 0, 0 and B= B0, B1, B2= 0, 0, 2. With these input combinations, only input A0 and B2 are true. Thus the A0 and B2 are connected to output line D2 so only the D2 will **TGGATC** and the rest of the gates will produce **ACCTAG**.

For Functions F1 and F2, perform OR operations among all decoder (D0-D8) outputs. So, max (D0, D1, D2, D3, D4, D5, D6, D7, D8) = max (0,0,2,0,0,0,0,0,0) = 2 "**TGGATC**" will produce as the functions outputs F1 and F2 of multi-valued 9-to-2 DNA ROM.

[4]

For input sequences A, B= **CAAGCT, ACCTAG**

A= A0, A1, A2 = 0, 2, 0 and B= B0, B1, B2= 2, 0, 0. With these input combinations, only input A1 and B0 is true. Thus the A1 and B0 are connected to output line D3 so only the D3 will **TGGATC** and the rest of the gates will produce **ACCTAG**.

For Functions F1 and F2, perform OR operations among all decoder (D0-D8) outputs. So, max (D0, D1, D2, D3, D4, D5, D6, D7, D8) = max (0,0,0,2,0,0,0,0,0) = 2 "**TGGATC**" will produce as the functions outputs F1 and F2 of multi-valued 9-to-2 DNA ROM.

[5]

For input sequences A, B= **CAAGCT, CAAGCT**

A= A0, A1, A2 = 0, 2, 0 and B= B0, B1, B2= 0, 2, 0. With these input combinations, only input A1 and B1 is true. Thus the A1 and B1 are connected to output line D4 so only the D4 will **TGGATC** and the rest of the gates will produce **ACCTAG**.

For Functions F1 and F2, perform OR operations among all decoder (D0-D8) outputs. So, max (D0, D1, D2, D3, D4, D5, D6, D7, D8) = max (0,0,0,0,2,0,0,0,0) = 2 "**TGGATC**" will produce as the functions outputs F1 and F2 of multi-valued 9-to-2 DNA ROM.

[6]

For input sequences A, B= **CAAGCT, TGGATC**

A= A0, A1, A2 = 0, 2, 0 and B= B0, B1, B2= 0, 0, 2. With these input combinations, only input A1 and B2 is true. Thus the A1 and B2 are connected to output line D5 so only the D5 will **TGGATC** and the rest of the gates will produce **ACCTAG**.

For Functions F1 and F2, perform OR operations among all decoder (D0-D8) outputs. So, max (D0, D1, D2, D3, D4, D5, D6, D7, D8) = max (0,0,0,0,0,2,0,0,0) = 2 "**TGGATC**" will produce as the functions outputs F1 and F2 of multi-valued 9-to-2 DNA ROM.

[7]
For input sequences A, B= **TGGATC, ACCTAG**

A= A0, A1, A2 = 0, 0, 2 and B= B0, B1, B2= 2, 0, 0. With these input combinations, only input A2 and B0 are true. Thus the A2 and B0 are connected to output line D6 so only the D6 will **TGGATC** and the rest of the gates will produce **ACCTAG**.

For Functions F1 and F2, perform OR operations among all decoder (D0-D8) outputs. So, max (D0, D1, D2, D3, D4, D5, D6, D7, D8) = max (0,0,0,0,0,0,2,0,0) = 2 "**TGGATC**" will produce as the functions outputs F1 and F2 of multi-valued 9-to-2 DNA ROM.

[8]
For input sequences A, B= **TGGATC, CAAGCT**

A= A0, A1, A2 = 0, 0, 2 and B= B0, B1, B2= 0, 2, 0. With these input combinations, only input A2 and B1 is true. Thus the A2 and B1 are connected to output line D7 so only the D7 will **TGGATC** and the rest of the gates will produce **ACCTAG**.

For Functions F1 and F2, perform OR operations among all decoder (D0-D8) outputs. So, max (D0, D1, D2, D3, D4, D5, D6, D7, D8) = max (0,0,0,0,0,0,0,2,0) = 2 "**TGGATC**" will produce as the functions outputs F1 and F2 of multi-valued 9-to-2 DNA ROM.

[9]
For input sequences A, B= **TGGATC, TGGATC**

A= A0, A1, A2 = 0, 0, 2 and B= B0, B1, B2= 0, 0, 2. With these input combinations, only input A2 and B2 is true. Thus the A2 and B2 are connected to output line D8 so only the D8 will **TGGATC** and the rest of the gates will produce **ACCTAG**.

For Functions F1 and F2, perform OR operations among all decoder (D0-D8) outputs. So, max (D0, D1, D2, D3, D4, D5, D6, D7, D8) = max (0,0,0,0,0,0,0,0,2) = 2 "**TGGATC**" will produce as the functions outputs F1 and F2 of multi-valued 9-to-2 DNA ROM.

6.4   Multiple-Valued DNA Programmable Read-Only Memory

MDPROM is an abbreviation for Multi-Valued DNA Programmable Read-Only Memory. It refers to computer memory chips that hold permanent or semi-permanent data and incorporate both the multi-valued DNA decoder and multi-valued DNA OR operations onto a single integrated circuit (IC). The contents of multi-valued DNA PROM are non-volatile; even if the computer is turned off, the contents of multi-valued DNA PROM persist. It is used to hold a computer's boot-up instructions. Almost every computer has a tiny sequence of multi-valued DNA PROM that contains the boot software. This is made up of a few kilobytes of code that instructs the computer on what to do when it boots up, such as conducting hardware diagnostics and loading the operating system into multi-valued DNA RAM. The BIOS is the boot firmware on a computer. To update the programming in multi-valued DNA ROM, these chips have to be physically removed and replaced. Data saved in DNA PROM cannot be electrically changed once the memory device is manufactured. Unlike conventional memory, which stores information as binary states (represented by "2"s, "1"s, and "0"s), DNA memory saves a molecular state for subsequent retrieval. Sequences (represented by **"ACCTAG as 0", "CAAGCT** as **1", and "TGGATC as 2"**) which provide important computing information, are stored in these states. Unlike traditional computer memory, the states saved in DNA memory can be in a sequence, providing far more practical flexibility in DNA algorithms than traditional information storage. A block diagram of a multi-valued PROM is shown in Figure 6.12. It consists of k input lines and n output lines. Each bit combination of the input variables is called an address. Each bit combination that comes out of the output lines is called a word. The number of bits per word is equal to the number of output lines n. An address is essentially a binary number that denotes one of the minterms of k variables.

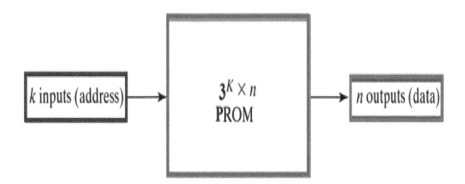

FIGURE 6.12
Block Diagram of 3^k-to-n PROM

Initially, the multi-valued DNA PROM is a combinational circuit with multi-valued DNA AND gates connected as a multi-valued DNA decoder and a number of multi-valued DNA OR gates equal to the outputs in the unit. Therefore it is a two-level implementation in the sum of minterms form. With k input lines and n output lines in PROM, the output functions will calculate through the sum of minterms form. The number of distinct addresses possible with k input variables is 3^k. An output word can be selected by a unique address, and since there are 3^k distinct addresses in a multi-valued ROM, there are 3^k distinct words that are said to be stored in the unit. The word available at the output lines at any given time depends on the address value applied to the input lines. Therefore, For input, k=2 and output, n=2 the ROM circuit will be called 9-to-2 multi-valued PROM and the function output F_1 and F_2 in the sum of minterms form, $\sum (0, 1, 2, 3)$ and $\sum (4, 5, 6, 7, 8)$, respectively.

6.4.1 Basic Definitions

Reading: Although the relative speed of RAM vs. PROM has varied over time, as of 2007 large RAM chips can be read faster than most PROMs. For this reason (and to allow uniform access), ROM content is sometimes copied to RAM or shadowed before its first use, and subsequently read from RAM.

Writing: For those types of PROMS that can be electrically modified, writing speed is always much slower than reading speed, and it may need unusually high voltage, the movement of jumper plugs to apply write-enable signals, and special lock/unlock command codes. Modern NAND Flash achieves the highest write speeds of any rewritable ROM technology, with speeds as high as 15 MB/s (or 70 ns/bit), by allowing (needing) large blocks of memory cells to be written simultaneously.

6.4.2 Block Diagram

Consider a multi-valued DNA 9-to-2 PROM, general organization of block diagram (Figure 6.13), the unit consists of 9 words of 2 input sequences (A = **ACCTAG or CAAGCT or TGGATC** and B = **ACCTAG or CAAGCT or TGGATC**) each. This implies there are 2 output lines (F1 and F2) and 9 distinct word sequences stored in the unit, each of which may be applied to the output lines.

6.4.3 Circuit Architecture of Multiple-Valued DNA PROM

For the design algorithm of Multi-Valued DNA 9-to-2 PROM, a 2-to-9 ternary decoder is needed which is designed by using two Multi-Valued 1-to-3 DNA decoders to get the output of nine DNA ternary AND operations. Finally, the multi-valued 2-to-9-ternary decoder performs further minterms ternary OR operation to perform desired outputs. Following the design algorithms of multiple-valued DNA 9-to-2 PROM Memory. Figure 6.14 shows the Multi-Valued 9-to-2 DNA PROM.

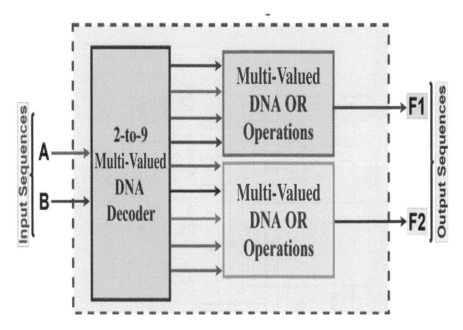

FIGURE 6.13
Block Diagram of DNA 9-to-2 PROM

6.4.4 Design Procedure

To perform multi-valued DNA ROM, a 2-to-9 decoder is needed and minterms of decoder output as the OR gates input to produce the desired multi-valued ROM output functions F1 and F2.

Step 1:
First, draw two input DNA sequences A and B. Three possible states for a DNA input sequence are the states **ACCTAG, CAAGCT, TGGATC**. These 3 will produce 9 combinations of 2 input DNA sequences.

Step 2:
For each input, a ternary decoder is needed for the selection of input value combinations. For example, if any input is A, then three values can be performed A0, A1 and A2. So the ternary decoder will select combinations from the values of inputs.

Step 3:
After the ternary decoder operations for input A and B, A0, then A1, A2 and B0, B1, B2 will be found.

Step 4:
With each input value of step 3, it is needed to perform AND operations that will produce 9 combinations of 2 input sequences and enable anyone output line.

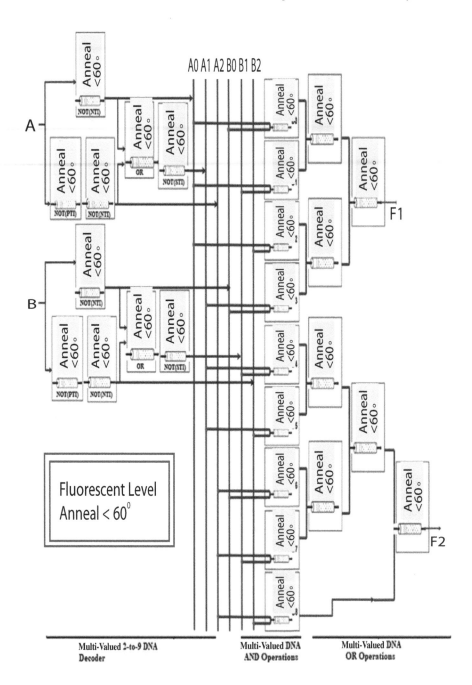

FIGURE 6.14
Multi-Valued 9-to-2 DNA PROM

TABLE 6.5
Truth Table of DNA 9-to-2 PROM

A	B	F1	F2
ACCTAG	ACCTAG	TGGATC	ACCTAG
ACCTAG	CAAGCT	TGGATC	ACCTAG
ACCTAG	TGGATC	TGGATC	ACCTAG
CAAGCT	ACCTAG	TGGATC	ACCTAG
CAAGCT	CAAGCT	ACCTAG	TGGATC
CAAGCT	TGGATC	ACCTAG	TGGATC
TGGATC	ACCTAG	ACCTAG	TGGATC
TGGATC	CAAGCT	ACCTAG	TGGATC
TGGATC	TGGATC	ACCTAG	TGGATC

Step 5:
From each combination of AND operations in Step 4, draw multi-valued DNA OR gates with (D0, D1), (D2, D3), (D4, D5), (D6, D7).

Step 6:
From each output sequence of Step 5, again combine them with multi-valued DNA OR gates, (D0, D1, D2, and D3) to generate output function F1 of multi-valued 9-to-2 DNA PROM.

Step 7:
Again combine (D4, D5, D6, D7, and D8) to generate output function F2 of multi-valued 9-to-2 DNA PROM.

6.4.5 Working Principle

Table 6.5 shows the truth table of DNA 9-to-2 PROM.

[1]
For input sequences A, B= ACCTAG, ACCTAG

1. To perform the value of input A0, A will go through the NOT (NTI) and produces TGGATC in the A0.

2. To perform the value of input A2, A will go through the NOT (PTI) and produce TGGATC then go through NOT (NTI) and produces ACCTAG in the A2.

3. To perform the value of input A1, the value of A0 and A2 go through DNA NOR (OR, NOT) Operations and produces ACCTAG in the A1.

4. To perform the value of input B0, B will go through the NOT (NTI) and produces TGGATC in the B0.

5. To perform value of input B2, B will go through the NOT (PTI) and produce TGGATC then go through NOT (NTI) and produces ACCTAG in the B2.

6. To perform the value of input B1, the value of B0 and B2 go through DNA NOR (OR, NOT) Operations and produces ACCTAG in the B1.

ACCTAG = |0> **CAAGCT** =|1> **TGGATC** = |2

A= A0, A1, A2 = 2, 0, 0 and B= B0, B1, B2= 2, 0, 0. With these input combinations, only input A0 and B0 are true. Thus the A0 and B0 are connected to output line D0 so only the D0 will **TGGATC** and the rest of the gates will produce **ACCTAG**.

To perform 9 to 2 multi-valued DNA ROM functions outputs,

[i] For input sequences D0, D1 = **TGGATC, ACCTAG**, the multi-valued OR gates will perform maximum operations of input sequences. So the output sequence of **TGGATC.**

[ii] For input sequences D2, D3 = **ACCTAG, ACCTAG**, the multi-valued OR gates will perform maximum operations of input sequences. So the output sequence of **ACCTAG.**

[iii] For input sequences D4, D5 = **ACCTAG, ACCTAG**, the multi-valued OR gates will perform maximum operations of input sequences. So the output sequence of **ACCTAG.**

[iv] For input sequences D6, D7 = **ACCTAG, ACCTAG** the multi-valued OR gates will perform maximum operations of input sequences. So the output sequence of **ACCTAG.**

[v] Combine the output sequences **TGGATC, ACCTAG** from [i] and [ii] respectively as input sequences, the multi-valued OR gates will perform maximum operations of input sequences. So the output sequence of **TGGATC.**

[vi] Combine the output sequences **ACCTAG** from [iii] and [iv] respectively as input sequences, the multi-valued OR gates will perform maximum operations of input sequences. So the output sequence of **ACCTAG.**

[vii] Finally, Combine the output sequence **ACCTAG** from [vii] and D8 (**ACCTAG**) as input sequences, the multi-valued OR gates will perform maximum operations of input sequences. So the output sequence of **TGGATC (v)** for the function output F1 and **ACCTAG (vii)** for F2 multi-valued 9 to 2 DNA PROM.

[2]

For input sequences A, B= ACCTAG, CAAGCT

A= A0, A1, A2 = 2, 0, 0 and B= B0, B1, B2= 0, 2, 0. With these input combinations, only input A0 and B1 are true. Thus the A0 and B1 are connected to output line D1 so only the D1 will **TGGATC** and the rest of the gates will produce **ACCTAG**.

For output sequence F1, perform OR operations among (D0-D3) outputs. So, max (D0, D1, D2, D3) = max (0,2,0,0) = 2 "**TGGATC**" and for F2 perform OR operations among (D4, D5, D6, D7, D8) = max (0,0,0,0,0)=0 "**ACCTAG**" of multi-valued 9-to-2 DNA PROM.

[3]
For input sequences A, B= ACCTAG, TGGATC

A= A0, A1, A2 = 2, 0, 0 and B= B0, B1, B2= 0, 0, 2. With these input combinations, only input A0 and B2 are true. Thus the A0 and B2 are connected to output line D2 so only the D2 will **TGGATC** and the rest of the gates will produce **ACCTAG**.

For output sequence F1, perform OR operations among (D0-D3) outputs. So, max (D0, D1, D2, D3) = max (0,0,2,0) = 2 "**TGGATC**" and for F2 perform OR operations among (D4, D5, D6, D7, D8) = max (0,0,0,0,0)=0 "**ACCTAG**" of multi-valued 9-to-2 DNA PROM.

[4]
For input sequences A, B= CAAGCT, ACCTAG

A= A0, A1, A2 = 0, 2, 0 and B= B0, B1, B2= 2, 0, 0. With these input combinations, only input A1 and B0 is true. Thus the A1 and B0 are connected to output line D3 so only the D3 will **TGGATC** and the rest of the gates will produce **ACCTAG**.

For output sequence F1, perform OR operations among (D0-D3) outputs. So, max (D0, D1, D2, D3) = max (0,0,0,2) = 2 "**TGGATC**" and for F2 perform OR operations among (D4, D5, D6, D7, D8) = max (0,0,0,0,0) =0 "**ACCTAG**" of multi-valued 9-to-2 DNA PROM.

[5] **For input sequences A, B= CAAGCT, CAAGCT**

A= A0, A1, A2 = 0, 2, 0 and B= B0, B1, B2= 0, 2, 0. With these input combinations, only input A1 and B1 is true. Thus the A1 and B1 are connected to output line D4 so only the D4 will **TGGATC** and the rest of the gates will produce **ACCTAG**.

For output sequence F1, perform OR operations among (D0-D3) outputs. So, max (D0, D1, D2, D3) = max (0,0,0,0) = 0 "**ACCTAG**" and for F2 perform OR operations among (D4, D5, D6, D7, D8) = max (2,0,0,0,0) = 2 "**TGGATC**" of multi-valued 9-to-2 DNA PROM.

[6]
For input sequences A, B= CAAGCT, TGGATC

A= A0, A1, A2 = 0, 2, 0 and B= B0, B1, B2= 0, 0, 2. With these input combinations, only input A1 and B2 is true. Thus the A1 and B2 are connected to output line D5 so only the D5 will **TGGATC** and the rest of the gates will produce **ACCTAG**.

For output sequence F1, perform OR operations among (D0-D3) outputs. So, max (D0, D1, D2, D3) = max (0,0,0,0) = 0 "**ACCTAG**" and for F2 perform OR operations among (D4, D5, D6, D7, D8) = max (0,2,0,0,0) = 2 "**TGGATC**" of multivalued 9-to-2 DNA PROM.

[7]
For input sequences A, B= TGGATC, ACCTAG

A= A0, A1, A2 = 0, 0, 2 and B= B0, B1, B2= 2, 0, 0. With these input combinations, only input A2 and B0 are true. Thus the A2 and B0 are connected to output line D6 so only the D6 will **TGGATC** and the rest of the gates will produce **ACCTAG**.

For output sequence F1, perform OR operations among (D0-D3) outputs. So, max (D0, D1, D2, D3) = max (0,0,0,0) = 0 "**ACCTAG**" and for F2 perform OR operations among (D4, D5, D6, D7, D8) = max (0,0,2,0,0) = 2 "**TGGATC**" of multivalued 9-to-2 DNA PROM.

[8]
For input sequences A, B= TGGATC, CAAGCT

A= A0, A1, A2 = 0, 0, 2 and B= B0, B1, B2= 0, 2, 0. With these input combinations, only input A2 and B1 are true. Thus the A2 and B1 are connected to output line D7 so only the D7 will **TGGATC** and the rest of the gates will produce **ACCTAG**.

For output sequence F1, perform OR operations among (D0-D3) outputs. So, max (D0, D1, D2, D3) = max (0,0,0,0) = 0 "**ACCTAG**" and for F2 perform OR operations among (D4, D5, D6, D7, D8) = max (0,0,0,2,0) = 2 "**TGGATC**" of multivalued 9-to-2 DNA PROM.

[9]
For input sequences A, B= TGGATC, TGGATC

A= A0, A1, A2 = 0, 0, 2 and B= B0, B1, B2= 0, 0, 2. With these input combinations, only input A2 and B2 is true. Thus the A2 and B2 are connected to output line D8 so only the D8 will **TGGATC** and the rest of the gates will produce **ACCTAG**.

For output sequence F1, perform OR operations among (D0-D3) outputs. So, max (D0, D1, D2, D3) = max (0,0,0,0) = 0 "**ACCTAG**" and for F2 perform OR operations among (D4, D5, D6, D7, D8) = max (0,0,0,0,2) = 2 "**TGGATC**" of multivalued 9-to-2 DNA PROM.

6.5 Multiple-Valued DNA Cache Memory

The main purpose of Cache memory in a computer is to read and write any data. Cache memory works with the computer's hard disk. Let us understand this with an example. When a Word file I opened, the Word file is stored in the computer's hard disk before it opens, and as soon as the Word file is opened the Word file is stored in the computer's cache memory.

There are many basic and main functions of a computer memory Cache memory which are as follows:

1. **Temporary Storage:** In addition to storing files read from the hard drive, Cache memory also stores data that program Caches are actively using but that doesn't need to be saved permanently. By keeping this data in Cache memory, program Caches can work with it quickly, improving speed and responsiveness.

2. **Loading Applications:** Loading a software application is also the main function of CACHE. Any software or application opens on the computer using Cache itself.

3. **Speed:** Cache memory speed is measured in Megahertz (MHz), millions of cycles per second so that it can be compared to other processor's clock speed.

6.5.1 Block Diagram

Consider a multi-valued DNA 9-to-1 Cache memory, a general organization of block diagram (Figure 6.15), the unit consists of 9 words of 1 input sequences (A = **ACCTAG or CAAGCT or TGGATC** and B = **ACCTAG or CAAGCT or TGGATC**) each. This implies there are a single output line (Z1) and 9 distinct word sequences stored in the unit, each of which may be applied to the output lines.

6.5.2 Architecture of Basic Components

Cache memory consists of three basic components- Decoder, Molecular RAM cells and OR gates. To execute DNA 9-to-1 Cache memory operation:

1. A 2-to-9 DNA Decoder
2. Molecular Sequence cell and
3. DNA OR operations for corresponding minterms are required.

6.5.2.1 Molecular RAM Cell

In a sequential device as simple as a D flip-flop could be used to remember one bit of data. To develop a complete memory cell, called a sequence cell, based on the flip-flop. The number of total DNA cells per word will be m × n where m represents words with n bits. The "select" input is used to access the cell, either for reading or writing also used to access any one sequence cell when there is more than one

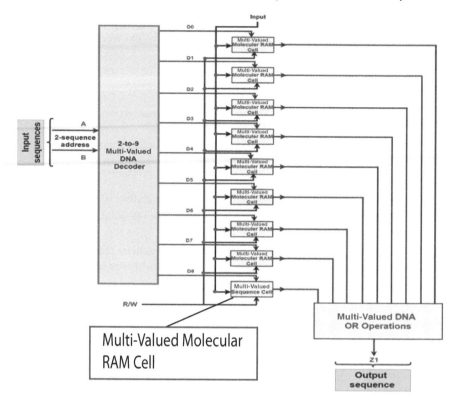

FIGURE 6.15
Block Diagram of DNA 9-to-1 Cache Memory

sequence cell. When the select line is high or **ACCTAG** then the cell performs the memory operation. But when the select line of the DNA cell is low or **TGGATC** then the cell is not interested to perform a read from or written to.

The next input sequence is "R/W" where a system clock will conduct this input. If the clock value on the read/write line is **TGGATC**, this will signify "read" and when it is **ACCTAG**, it will perform the "write" phase. When such a cell is selected and in "read" mode, the current value of its underlying flip-flop will be transferred to the cell's output line. When the cell is selected and in "write" mode, an input data signal will determine the value remembered by the flip-flop. Let's consider the cell that has been selected. In that case, if the clock value is **TGGATC** then the cell contents are to be read and this time the output value will depend only on the P-value of the flip flop. But if P is low, the cell output will be **TGGATC** and if P is high, the cell output will be **ACCTAG**. It occurs because the DNA AND gate added to the cell's output have three input-negated R/W, Select, and P; and both "negated R/W" and "Select is currently high **ACCTAG**. Figure 6.16 shows the diagram of DNA Single Molecular Cell.

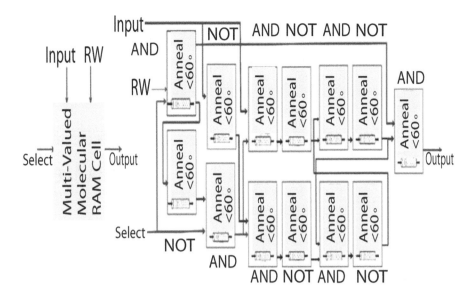

FIGURE 6.16
DNA Single Molecular Cell

To perform the DNA 9-to-1 Cache memory, four selection lines are needed to design four DNA sequence cells. The output of DNA 2-to-9 decoder with nine output sequences will perform as selection input sequence of sequence cell. Figure 6.17 depicts the Molecular RAM Cells .

6.5.3 Circuit Architecture of Multiple-Valued DNA Cache Memory

Figure 6.18 depicts the circuit architecture of Multi-Valued DNA 9-to-1 Cache memory. A 2-to-9 ternary decoder is needed which is designed by using two Multi-Valued 1-to-3 DNA decoders to get the output of nine DNA ternary RAM cells operations. Finally, the multi-valued 9-ternary DNA RAM cells perform qubit cell output sequences from Q0 to Q8 for further minterms ternary OR operation to perform desire output. Following given the design algorithms of multiple-valued DNA 9-to-1 Cache Memory.

6.5.4 Working Principle

Figure 6.18 represents the implementation of 9-to-1 sequence Cache. This DNA Cache consists of 9 separate "Words" of memory and each is 1 sequence wide. The DNA RAM Cell has three inputs and one output. The complete circuit of a DNA RAM cell is described in figure 8 with proper explanation. A word consists of two

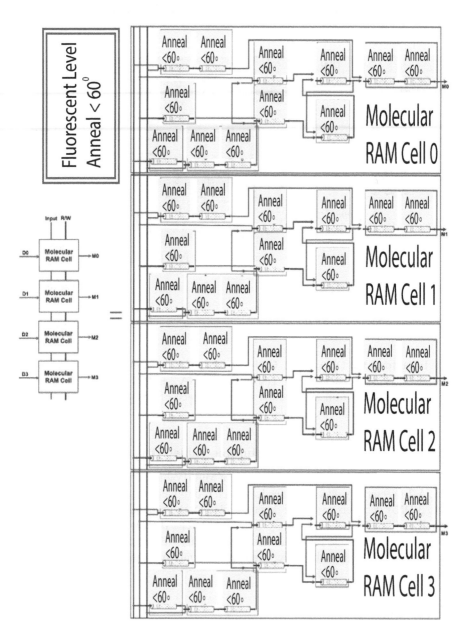

FIGURE 6.17
Molecular RAM Cells

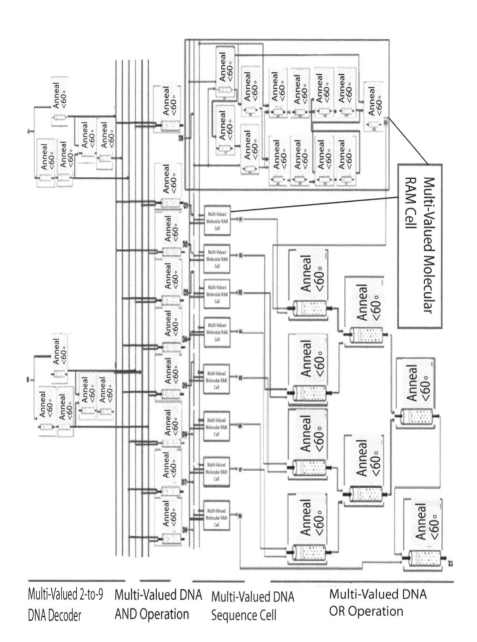

Multi-Valued 2-to-9 DNA Decoder   Multi-Valued DNA AND Operation   Multi-Valued DNA Sequence Cell   Multi-Valued DNA OR Operation

FIGURE 6.18
Circuit Architecture of DNA 9-to-1 Cache Memory

DNA RAM cells arranged in such a way so that both sequences can be accessed simultaneously. Four words of memory need two address lines. A and B are the two-sequence address lines input that goes through a DNA 2-to-9 decoder that selects one of the nine words. The memory-enabled input enables the decoder. If the memory enables is **TGGATC**, all output of the decoder will be **TGGATC** and in that case, none of the memory addresses will be selected. But when the memory enables is **ACC-TAG** then one of the nine words is selected. The word is selected by the value in the two address lines. When a word has been selected, the read/write input determines the operation. During the read operation, the nine sequences of the selected word pass to the DNA OR gates to the output Z1 terminals. But during the write operation, the data which is available in the input lines are transferred into the nine DNA cells of the selected word. The DNA RAM cells that are not selected are become disabled and their previous sequence never changes. But when the memory enable input that passes into the decoder is equal to **TGGATC**, none of the words are selected, and then all DNA cells remain unchanged regardless of the value of the read/ write input. This is the working procedure of 9-to-1 DNA Cache memory.

6.6   Summary

DNA computing is an emerging field and researchers are attracted to this field. Memory devices like RAM, ROM, PROM and Cache memory in DNA computing have been explained in this chapter for multi-valued logic. Memory devices are a mandatory part in computer system. In DNA computer, DNA memory devices are required. All topics are described here in details with necessary figures. Here ternary system is explained for every topic as multi-valued logic. For better understand, a related topic has been discussed also.

Bibliography

[1] Khoshkhahesh, A., Ebrahimi, S., & Sabbaghi-Nadooshan, R. (2018). Designing and optimizing DNA reversible adders and adder/subtractors. BioNanoScience, 8(1), 118-130.

[2] Clausen-Schaumann, H., Rief, M., Tolksdorf, C., & Gaub, H. E. (2000). Mechanical stability of single DNA molecules. Biophysical Journal, 78(4), 1997-2007.

[3] Sarker, A., Ahmed, T., Rashid, S. M., Anwar, S., Jaman, L., Tara, N., ... & Babu, H. M. H. (2011, October). Realization of reversible logic in DNA computing.

In 2011 IEEE 11th International Conference on Bioinformatics and Bioengineering (pp. 261-265). IEEE.

[4] Echols, H., & Goodman, M. F. (1991). Fidelity mechanisms in DNA replication. Annual Review of Biochemistry, 60(1), 477-511.

[5] Merindol, R., & Walther, A. (2017). Materials learning from life: concepts for active, adaptive and autonomous molecular systems. Chemical Society Reviews, 46(18), 5588-5619.

[6] Adleman, L. M. (1994). Molecular computation of solutions to combinatorial problems. science, 266(5187), 1021-1024.

[7] Seelig, G., Soloveichik, D., Zhang, D. Y., & Winfree, E. (2006). Enzyme-free nucleic acid logic circuits. Science, 314(5805), 1585-1588.

[8] Benenson, Y., Paz-Elizur, T., Adar, R., Keinan, E., Livneh, Z., & Shapiro, E. (2001). Programmable and autonomous computing machine made of biomolecules. Nature, 414(6862), 430-434.

[9] Pérez-Inestrosa, E., Montenegro, J. M., Collado, D., Suau, R., & Casado, J. (2007). Molecules with multiple light-emissive electronic excited states as a strategy toward molecular reversible logic gates. The Journal of Physical Chemistry C, 111(18), 6904-6909.

[10] Zoraida, B. S. E., Arock, M., Ronald, B. S. M., & Ponalagusamy, R. (2008, October). A novel generalized model for constructing reusable and reliable logic gates using DNA. In 2008 Fourth International Conference on Natural Computation (Vol. 7, pp. 533-537). IEEE.

[11] Gearheart, C. M., Rouchka, E. C., & Arazi, B. (2010, August). DNA-based dynamic logic circuitry. In 2010 53rd IEEE International Midwest Symposium on Circuits and Systems (pp. 248-251). IEEE.

[12] Pan, T. C., Misra, S., & Aluru, S. (2018, November). Optimizing high performance distributed memory parallel hash tables for DNA k-mer counting. In SC18: International Conference for High Performance Computing, Networking, Storage and Analysis (pp. 135-147). IEEE.

[13] Javaid, Q., Zafar, A., Awais, M., & Shah, M. A. (2017). Cache memory: An analysis on replacement algorithms and optimization techniques. Mehran University Research Journal of Engineering & Technology, 36(4), 831-840.

[14] Suh, G. E., Rudolph, L., & Devadas, S. (2004). Dynamic partitioning of shared cache memory. The Journal of Supercomputing, 28(1), 7-26.

7

Multiple-Valued Quantum-DNA Memory Devices

7.1 Introduction

Both DNA and quantum computers have the potential to exceed the power of conventional digital computers, though substantial technical difficulties first must be overcome. Through coherent superposition of states, quantum computers are more powerful than classical Turing machines. DNA computers are evolvable through biotechnology techniques. By combining DNA and quantum computers, both of these characteristics might be captured. DNA computers could be used to self-assemble quantum logic circuits from gates attached to DNA strands. Moreover, quantum computers might be implemented directly using the physical characteristics of the DNA molecule. In DNA computing biomolecules and biomolecular reactions are designed to implement computational algorithms. In quantum computing, computation is done at a scale where quantum mechanical effects are important. Both of these relatively new computing paradigms have been mentioned as successors to current solid-state computers. They offer significant benefits over traditional computing, as well as significant challenges to overcome in order to achieve effective implementation. In this part, the possibility of quantum and DNA computers working together is explored. The motivation is that each paradigm has unique features that complement the other, and therefore, a system that uses both would have more capability than an implementation of one alone. The essence of the complementary features of DNA and quantum computing has probably best been expressed by Michael Conrad. In any computing system, there is a trade-off between programmability, computational efficiency, and evolutionary adaptability.

If programs can be delivered to a system in a timely and accurate manner, it is programmable. The percentage of a system's available interactions that contribute to a calculation is called computational efficiency. Evolutionary adaptability characterizes the system's ability to function in changing and unknown environments. Conventional electronic, quantum, and biomolecular computers each primarily exhibit one of these properties. Molecular biology techniques and enzymes can be used to evolve biomolecular computers so that they can adapt to changing environments and input. It is the adaptability and robustness of biological systems to external change that has inspired evolutionary programs and artificial neural networks. On the other hand, the complexity of the interactions in a biomolecular computer, and the nature of the

DOI: 10.1201/9781003381921-7

biochemistry makes it difficult to program them to behave in an exact way. Moreover, biomolecular computers are not very efficient in their computations. Many interactions are wasted because of errors, or because they do not directly contribute to the desired computational result. Quantum computers excel at computational efficiency because of the entanglement and superposition of an exponential number of states. They can be effectively programmed, though not as easily as conventional computers. Their adaptability, however, is nonexistent because of their extreme sensitivity to the effects of environmental changes. In fact, quantum computer should be isolated from the external environment. They are the most programmable, but are less adaptable than biological and less efficient than quantum. This chapter will describe the memory devices in quantum-DNA computing with multi-valued logic system.

7.2　Multiple-Valued Quantum-DNA Random Access Memory

Quantum computing and DNA computing are gaining attention as alternatives to classical digital computing systems. In this chapter, combined Quantum and DNA computing are discussed for designing different memory devices considering a binary logic system. The design of various memory devices includes quantum and DNA logic gates and decoder for Quantum-DNA 9-to-1 Random Access memory. Both DNA and Quantum computers have the potential to significantly affect traditional digital computers in terms of processing power and depot. Quantum computers are more powerful than traditional Turing machines computing because of their coherent superposition of states and biotechnology techniques can be used to evolve DNA computers. Both of these properties might be captured by integrating DNA and Quantum computing. An overview of quantum computation is, there exist some quantum algorithms which are significantly faster than their conventional counterparts and these processes establish quantum computing as a superior future technology that involves quantum circuits and quantum gates. Traditional silicon computers consume much more power as compared to the computing systems, based on Deoxyribonucleic Acid (DNA), whereas DNA-based logic gates are stable and reusable. To achieve a balance between these two computer systems, a storage device is required. Since DNA computing provides us with additional storage capacity, the output portions of circuits will perform in DNA computing. Therefore, it is needed to transform Quantum qubit to DNA sequence as a result, it is required to use a transformation mechanism by which an excited magnetic state returns to its equilibrium distribution. Quantum computing generates additional heat in circuits, which may be utilized to conduct DNA logic operations in DNA computing. Consequently, a heat transfer circuit between Quantum and DNA processing is required. Thus the proposed designs will improve the result of any logic device for both binary and ternary systems rather than a conventional computing platform. Finally, the information flow of memory devices in Quantum-DNA will: Quantum-> Qubit Storage -> Transformation -> DNA.

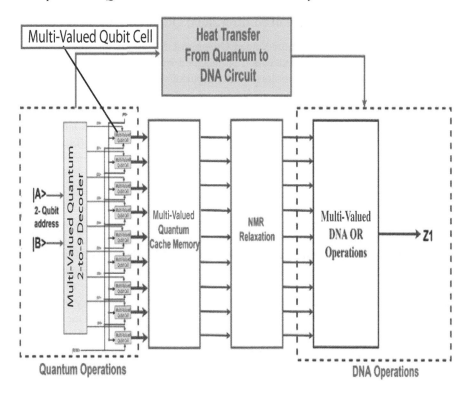

FIGURE 7.1
Block Diagram of Quantum-DNA 9-to-1 RAM

7.2.1 General Organization of Multiple-Valued Quantum-DNA RAM

Figure 7.1 represents the block diagram of Quantum-DNA 9-to-1 RAM, consisting of 9 separate Words of memory and each is single sequence wide.

7.2.2 Architecture of Basic Components of Multiple-Valued Quantum-DNA RAM

RAM consists of three basic components- Decoder, Qubit Cells, and OR operations. To execute Quantum-DNA 9-to-1 RAM following operations are required:

1. A Quantum 2-to-9 Decoder
2. Qubit cells
3. Quantum Cache Memory
4. NMR Relaxation
5. Heat transfer circuit and
6. DNA OR operations for corresponding minterms

7.2.3   Circuit Architecture of Multi-Valued Quantum-DNA RAM

For the design algorithm of Multi-Valued Quantum-DNA 9-to-1 RAM, a quantum 2-to-9 ternary decoder is needed which is designed by using two quantum multi-valued 1-to-3 Quantum decoders to get the output of nine Quantum ternary RAM cells operations.

Finally, the multi-valued 9-to-1 ternary RAM cells perform qubit cell output qubits from |Q0> to |Q8> will go through NMR for further minterms ternary OR operation of DNA sequences to perform desired DNA output sequence (Figure 7.2).

7.2.4   Working Principle

Figure 7.2 represents the implementation of 9-to-1 Quantum-DNA RAM. This quantum RAM consists of 9 separate "Words" of memory and each is 1 qubit wide. The quantum RAM Cell has three inputs and one output. A word consists of nine quantum RAM cells arranged in such a way so that both qubits can be accessed simultaneously. Nine words of memory need two address lines. |A> and |B> are the two-qubit address lines input that goes through a 2-to-9 decoder that selects one of the nine words. The memory-enabled input enables the decoder. If the memory enables is |0>, all output of the decoder will be |0> and in that case, none of the memory addresses will be selected. But when the memory enables is |1> then one of the four words is selected. The word is selected by the value in the two address lines. When a word has been selected, the read/write input determines the operation. During the read operation, the four qubits of the selected word pass to the DNA OR gates to the output Z1 terminals. But during the write operation, the data which is available in the input lines are transferred into the four quantum cells of the selected word. The quantum RAM cells that are not selected have become disabled and their previous qubit never changes. But when the memory enable input that passes into the decoder is equal to |0>, none of the words are selected, and then all quantum cells remain unchanged regardless of the value of the read/ write input.

7.3   Multiple-Valued Quantum-DNA Read-Only Memory

Multi-Valued Quantum computing and DNA computing are gaining attention as alternatives to classical digital computing systems. In this chapter, combined multiple-valued Quantum and DNA computing are discussed novel theoretical methods for designing different memory devices considering a binary logic system. The design of various memory devices includes multi-valued quantum and DNA logic operations and decoder for multi-valued Quantum-DNA 9-to-2 Read-only memory. Both multi-valued DNA and Quantum computers have the potential to significantly affect traditional digital computers in terms of processing power and depot. Quantum computers are more powerful than traditional Turing machines computing because of their

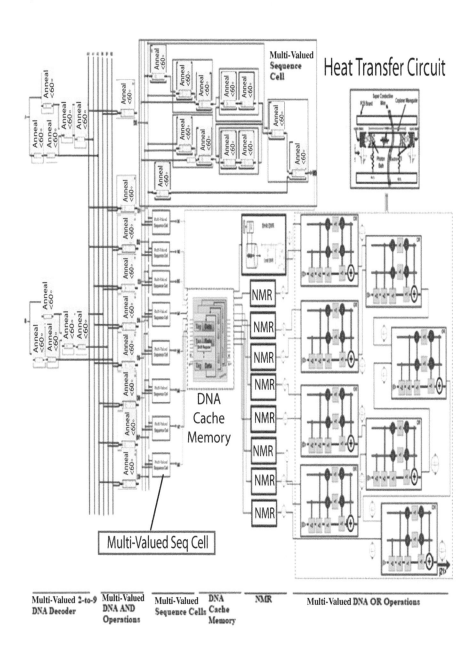

FIGURE 7.2
Circuit Architecture of Quantum-DNA 9-to-1 RAM

coherent superposition of states and biotechnology techniques can be used to evolve DNA computers. Both of these properties might be captured by integrating DNA and Quantum computing. Therefore, an integrated computing system as multiple-valued Quantum-to-DNA and vice versa for multi-valued 9-to-2 ROM memory is the main concern of this section.

7.3.1 Basic Definitions

For information processing and storing, bits are used as traditional computing systems, qubits in quantum computing, and sequences in DNA computing systems. Conventional computing use ternary codes: 0 or 1 or 2 to compute and store information. Quantum computing use multi-valued Qubits ($|0>$, $|1>$ and $|2>$) for ternary system and multi-valued DNA Computing system use combination of A (Adenine), T (Thymine), C (Cytosine), G (Guanine) of 2-to-16 length Molecule sequences (**ACCTAG** = $|0>$ **CAAGCT** = $|1>$ **TGGATC** = $|2>$) to compute and store information. Initially, the qubit will perform as the input to combinational logic circuits in the proposed integrated multi-valued Quantum-DNA memory devices (Figure 7.3). These are used in initial logic circuits for faster calculation than sequences. To achieve a balance between these two computer systems, a storage device is required. Since multi-valued DNA computing provides us with additional storage capacity, the output portions of circuits will perform in multi-valued DNA computing. Therefore, it is needed to transform multi-valued Quantum qubits into multi-valued DNA sequences as a result, it is required to use a transformation mechanism by which an excited magnetic state returns to its equilibrium distribution. Multi-Valued quantum computing generates additional heat in circuits, which may be utilized to conduct multi-valued DNA logic operations in multi-valued DNA computing. Consequently, a heat transfer circuit between multi-valued Quantum and DNA processing is required. Thus, the proposed designs will improve the result of any logic device for a ternary system rather than a conventional computing platform. Finally, the information flow of memory devices in Quantum-DNA will: Multi-Valued Quantum-> Multi-Valued Qubit Storage -> Multi-Valued Transformation -> Multi-Valued DNA. Figure 7.3 shows the prototype of multi-valued integrated quantum-DNA circuit for memory devices.

7.3.2 General Organization of Multiple-Valued Quantum-DNA ROM

Figure 7.4 shows the general block diagram of Quantum-DNA 9-to-2 ROM. Input qubits $|A>$ and $|B>$ in multi-valued quantum computing operations, store the qubits in cache memory and transform those qubits to DNA sequences to form output F1 and F2 in multi-valued DNA computing operations.

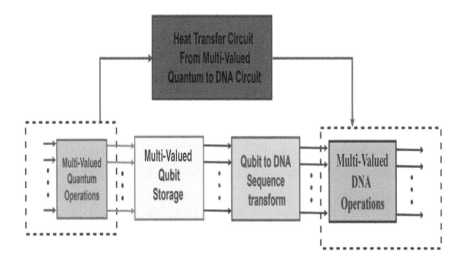

FIGURE 7.3
Prototype of Multi-Valued Integrated Quantum-DNA Circuit for Memory Devices

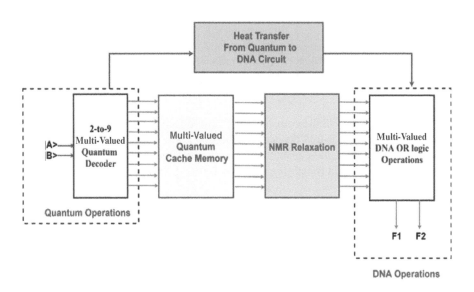

FIGURE 7.4
General Block Diagram of Quantum-DNA 9-to-2 ROM

7.3.3   Architecture of Basic Components of Multiple-Valued Quantum-DNA ROM

Multi-Valued ROM consists of two basic components- multi-valued Decoder and multi-valued OR operations. To execute Multiple-Valued Quantum-DNA 9-to-2 ROM, following operations are required:

1.   A Multi-Valued Quantum 2-to-9 Decoder
2.   Multi-Valued Quantum Cache Memory
3.   NMR Relaxation
4.   Heat transfer circuit and
5.   Multi-Valued DNA OR operations for corresponding minterms.

7.3.4   Circuit Architecture of Multiple-Valued Quantum-DNA ROM

To design a multi-valued Quantum-DNA 9-to-2 ROM circuit (considering NMR relaxation), the input values are in qubit (multi-valued Quantum Computing) and the output will be a multi-valued DNA sequence (multi-valued DNA Computing) at normal room temperature. The circuit architecture of Multiple-Valued Quantum-DNA ROM is shown in Figure 7.5.

Step 1:
First, draw two input qubits |A> and |B>. Two possible qubits are |0>, |1>, or |2>. These 3 will produce 9 combinations of 2 input qubits.

Step 2:
To design a multi-valued 2-to-9 quantum decoder, draw two multi-valued 1-to-3 quantum decoders and then nine multi-valued quantum AND operations.

Step 3:
After getting qubits of all the corresponding values from nine multi-valued quantum AND operations of the 2-to-9 decoder, it is needed to store all the multi-valued quantum bits into a multi-valued quantum cache memory. Multi-Valued Quantum cache memory is actually used for storing logical qubits and error correction.

Step 4:
After storing qubits of all the corresponding values from nine multi-valued quantum AND operations in multi-valued cache memory, it is needed to perform NMR relaxation to transform qubits into DNA sequences. Here doing NMR relaxation at room temperature and in NMR relaxation does not need to emit EMR because it is needed to down the spin.

Step 5:
Finally, after NMR relaxation, the multi-valued DNA sequences will be found from multi-valued qubits and these will go through multi-valued DNA OR gate to produce desired multi-valued Quantum-DNA 9-to-2 ROM output sequences.

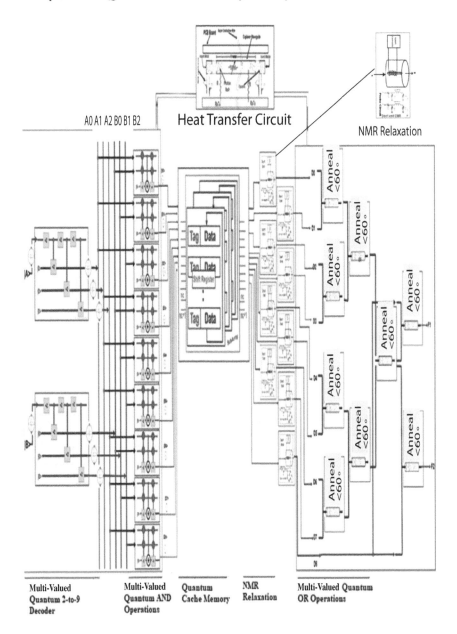

AO A1 A2 B0 B1 B2

Heat Transfer Circuit

NMR Relaxation

Multi-Valued
Quantum 2-to-9
Decoder

Multi-Valued
Quantum AND
Operations

Quantum
Cache Memory

NMR
Relaxation

Multi-Valued Quantum
OR Operations

FIGURE 7.5
Multi-Valued Quantum-DNA 9-to-2 ROM with NMR Relaxation in Normal Room
Temperature

The circuit architecture of multi-valued Quantum – DNA 9-to-2 ROM with NMR
relaxation in normal room temperature is shown in Figure 7.5.

TABLE 7.1
Truth Table of Multi-Valued Quantum-DNA 9-to-2
ROM

|A>	|B>	F1	F2
|0>	|0>	TGGATC	TGGATC
|0>	|1>	TGGATC	TGGATC
|0>	|2>	TGGATC	TGGATC
|1>	|0>	TGGATC	TGGATC
|1>	|1>	TGGATC	TGGATC
|1>	|2>	TGGATC	TGGATC
|2>	|0>	TGGATC	TGGATC
|2>	|1>	TGGATC	TGGATC
|2>	|2>	TGGATC	TGGATC

7.3.5 Working Principle

Table 7.1 shows the truth table of multi-valued Quantum-DNA 9-to-2 ROM. Here inputs are |A> and |B>, outputs are F1 and F2.

According to the truth table of multi-valued Quantum-DNA 9-to-2 ROM (Table 7.1), it is necessary to do the following operations to perform desired output sequences:

[1]
For input sequences |A>, |B>= |0>, |0>

|A>= |A0>, |A1>, |A2> =|2>, |0>, |0> and |B> = |B0>, |B1>, |B2>= |2>, |0>, |0>. With this input combinations, only input |A0> and |B0> is true. Thus the |A0> and |B0> are connected to output line |D0> so only the |D0> will **|2>** and rest of the gates will produce **|0>**.

For Functions F1 and F2, perform multi-valued DNA OR operations among all decoder (D0-D8) outputs. So, max (D0, D1, D2, D3, D4, D5, D6, D7, D8) = max (2,0,0,0,0,0,0,0,0) = 2 "**TGGATC**" will produce as the function's outputs F1 and F2 of multi-valued 9-to-2 Quantum- DNA ROM.

[2]
For input sequences |A>, |B>= |0>, |1>

|A>= |A0>, |A1>, |A2> = |2>, |0>, |0> and |B> = |B0>, |B1>, |B2> = |0>, |2>, |0>. With this input combinations, only input |A0> and |B1> is true. Thus the |A0> and |B1> are connected to output line |D1> so only the |D1> will **|2>** and rest of the gates will produce **|0>**.

For Functions F1 and F2, perform OR operations among all decoder (D0-D8) outputs. So, max (D0, D1, D2, D3, D4, D5, D6, D7, D8) = max (0,2,0,0,0,0,0,0,0) = 2

"**TGGATC**" will produce as the function's outputs F1 and F2 of multi-valued 9-to-2 Quantum-DNA ROM.

[3]
For input sequences |A>, |B>= **|0>, |2>**

|A>= |A0>, |A1>, |A2> = |2>, |0>, |0> and |B0>, |B1>, |B2> = |0>, |0>, |2>. With this input combinations, only input |A0> and |B2> is true. Thus the |A0> and |B2> are connected to output line |D2> so only the D2 will **|2>** and rest of the gates will produce **|0>**.

For Functions F1 and F2, perform OR operations among all decoder (D0-D8) outputs. So, max (D0, D1, D2, D3, D4, D5, D6, D7, D8) = max (0,0,2,0,0,0,0,0,0) = 2 "**TGGATC**" will produce as the function's outputs F1 and F2 of multi-valued 9-to-2 Quantum-DNA ROM.

[4]
For input sequences |A>, |B>= **|1>, |0>**

|A>= |A0>, |A1>, |A2> = |0>, |2>, |0> and |B0>, |B1>, |B2> = |2>, |0>, |0>. With this input combinations, only input |A1> and |B0> is true. Thus the |A1> and |B0> are connected to output line |D3> so only the |D3> will **|2>** and rest of the gates will produce **|0>**.

For Functions F1 and F2, perform OR operations among all decoder (D0-D8) outputs. So, max (D0, D1, D2, D3, D4, D5, D6, D7, D8) = max (0,0,0,2,0,0,0,0,0) = 2 "**TGGATC**" will produce as the function's outputs F1 and F2 of multi-valued 9-to-2 Quantum-DNA ROM.

[5]
For input sequences |A>, |B>= **|1>, |1>**

|A>= |A0>, |A1>, |A2> = |0>, |2>, |0> and |B0>, |B1>, |B2> = |0>, |2>, |0>. With this input combinations, only input |A1> and |B1> is true. Thus the |A1> and |B1> are connected to output line |D4> so only the |D4> will **|2>** and rest of the gates will produce **|0>**.

For Functions F1 and F2, perform OR operations among all decoder (D0-D8) outputs. So, max (D0, D1, D2, D3, D4, D5, D6, D7, D8) = max (0,0,0,0,2,0,0,0,0) = 2 "**TGGATC**" will produce as the functions outputs F1 and F2 of multi-valued 9-to-2 Quantum- DNA ROM.

[6]
For input sequences |A>, |B>= **|1>, |2>**

|A>= |A0>, |A1>, |A2> = |0>, |2>, |0> and |B0>, |B1>, |B2> = |0>, |0>, |2>. With this input combinations, only input |A1> and |B2> is true. Thus the |A1> and |B2> are connected to output line |D5> so only the |D5> will **|2>** and rest of the gates will produce **|0>.**

For Functions F1 and F2, perform OR operations among all decoder (D0-D8) outputs. So, max (D0, D1, D2, D3, D4, D5, D6, D7, D8) = max (0,0,0,0,0,2,0,0,0) = 2 **"TGGATC"** will produce as the functions outputs F1 and F2 of multi-valued 9-to-2 Quantum-DNA ROM.

[7]
For input sequences |A>, |B>= **|2>, |0>**

|A>= |A0>, |A1>, |A2> = |0>, |0>, |2> and |B0>, |B1>, |B2> = |2>, |0>, |0>. With this input combinations, only input |A2> and |B0> is true. Thus the |A2> and |B0> are connected to output line |D6> so only the |D6> will **|2>** and rest of the gates will produce **|0>.**

For Functions F1 and F2, perform OR operations among all decoder (D0-D8) outputs. So, max (D0, D1, D2, D3, D4, D5, D6, D7, D8) = max (0,0,0,0,0,0,2,0,0) = 2 **"TGGATC"** will produce as the function's outputs F1 and F2 of multi-valued 9-to-2 Quantum-DNA ROM.

[8]
For input sequences |A>, |B>= **|2>, |1>**

|A>= |A0>, |A1>, |A2> = |0>, |0>, |2> and |B0>, |B1>, |B2> = |0>, |2>, |0>. With this input combinations, only input |A2> and |B1> is true. Thus the |A2> and |B1> are connected to output line |D7> so only the |D7> will **|2>** and rest of the gates will produce **|0>.**

For Functions F1 and F2, perform OR operations among all decoder (D0-D8) outputs. So, max (D0, D1, D2, D3, D4, D5, D6, D7, D8) = max (0,0,0,0,0,0,0,2,0) = 2 **"TGGATC"** will produce as the function's outputs F1 and F2 of multi-valued 9-to-2 Quantum-DNA ROM.

[9]
For input sequences |A>, |B>= **|2>, |2>,**

|A>= |A0>, |A1>, |A2> = |0>, |0>, |2> and |B0>, |B1>, |B2> = |0>, |0>, |2>. With this input combinations, only input |A2> and |B2> is true. Thus the |A2> and |B2> are connected to output line |D8> so only the |D8> will **|2>** and rest of the gates will produce **|0>.**

For Functions F1 and F2, perform OR operations among all decoder (D0-D8) outputs. So, max (D0, D1, D2, D3, D4, D5, D6, D7, D8) = max (0,0,0,0,0,0,0,0,2) = 2

"**TGGATC**" will produce as the function's outputs F1 and F2 of multi-valued 9-to-2 Quantum- DNA ROM.

7.4 Multiple-Valued Quantum-DNA Programmable Read Only Memory

Multi-Valued Quantum computing and DNA computing are gaining attention as alternatives to classical digital computing systems. In this chapter, combined multiple-valued Quantum and DNA computing are discussed novel theoretical methods for designing different memory devices considering a binary logic system. The design of various memory devices includes multi-valued quantum and DNA logic operations and decoder for multi-valued Quantum-DNA 9-to-2 Read-only memory. Both multi-valued DNA and Quantum computers have the potential to significantly affect traditional digital computers in terms of processing power and depot. Quantum computers are more powerful than traditional Turing machines computing because of their coherent superposition of states and biotechnology techniques can be used to evolve DNA computers. Both of these properties might be captured by integrating DNA and Quantum computing. Therefore, an integrated computing system as multiple-valued Quantum-to-DNA and vice versa for multi-valued 9-to-2 ROM memory is the main concern of this section.

7.4.1 Basic Definitions

For information processing and storing, bits are used as traditional computing systems, qubits in quantum computing, and sequences in DNA computing systems. Conventional computing use ternary codes: 0 or 1 or 2 to compute and store information. Quantum computing use multi-valued Qubits ($|0>$, $|1>$ and $|2>$) for ternary system and multi-valued DNA Computing system use combination of A (Adenine), T (Thymine), C (Cytosine), G (Guanin) of 2-to-16 length Molecule sequences (**ACCTAG** = $|0>$ **CAAGCT** = $|1>$ **TGGATC** = $|2>$) to compute and store information. Initially, the qubit will perform as the input to combinational logic circuits in the proposed integrated multi-valued Quantum-DNA memory devices. These are used in initial logic circuits for faster calculation than sequences. To achieve a balance between these two computer systems, a storage device is required. Since multi-valued DNA computing provides us with additional storage capacity, the output portions of circuits will perform in multi-valued DNA computing. Therefore, it is needed to transform multi-valued Quantum qubits into multi-valued DNA sequences as a result, it is required to use a transformation mechanism by which an excited magnetic state returns to its equilibrium distribution. Multi-Valued quantum computing generates additional heat in circuits, which may be utilized to conduct multi-valued DNA logic operations in multi-valued DNA computing. Consequently, a heat transfer circuit between multi-valued Quantum and DNA processing is required. Thus the proposed designs will improve the result of any logic device for the ternary system rather than a conventional computing platform. Finally, the information flow of memory devices in Quantum-DNA will: Multi-Valued Quantum-> Multi-Valued Qubit Storage -> Multi-Valued Transformation -> Multi-Valued DNA.

7.4.2   General Organization of Multiple-Valued Quantum-DNA PROM

Input qubits |A> and |B> in multi-valued quantum computing operations, store the qubits in cache memory and transform those qubits to DNA sequences to form output F1 and F2 in multi-valued DNA computing operations. Figure 7.6 shows the general block diagram of Quantum-DNA 9-to-2 PROM.

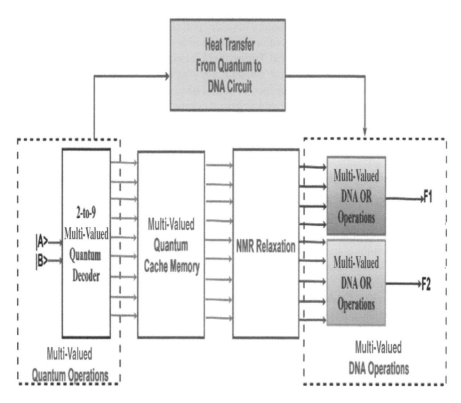

FIGURE 7.6
General Block Diagram of Quantum-DNA 9-to-2 PROM

7.4.3 Basic Components of Multiple-Valued Quantum-DNA PROM Memory

Multi-Valued PROM consists of two basic components- multi-valued Decoder and multi-valued OR operations. To execute Multiple-Valued Quantum-DNA 9-to-2 PROM following operations are required:

1. A Multi-Valued Quantum 2-to-9 Decoder
2. Multi-Valued Quantum Cache Memory
3. NMR Relaxation
4. Heat transfer circuit and
5. Multi-Valued DNA OR operations for corresponding minterms.

7.4.4 Circuit Architecture of Multiple-Valued Quantum-DNA PROM

To design a multi-valued Quantum-DNA 9-to-2 PROM circuit (considering NMR relaxation), the input values are in qubit (multi-valued Quantum Computing) and the output will be a multi-valued DNA sequence (multi-valued DNA Computing) at normal room temperature.

Step 1:
First, draw two input qubits |A> and |B>. Two possible qubits are |0>, |1> or |2>. These 3 will produce 9 combinations of 2 input qubits.

Step 2:
To design a multi-valued 2-to-9 quantum decoder, draw two multi-valued 1-to-3 quantum decoders and then nine multi-valued quantum AND operations.

Step 3:
After getting qubits of all the corresponding values from nine multi-valued quantum AND operations of 2-to-9 decoder, all the multi-valued quantum bits are needed to be stored into a multi-valued quantum cache memory.

Step 4:
After storing qubits of all the corresponding values from nine multi-valued quantum AND operations in multi-valued cache memory, it is needed to perform NMR relaxation to transform qubits into DNA sequences. Here doing NMR relaxation at room temperature and in NMR relaxation, there is no need to emit EMR because to down the spin.

Step 5:
Finally, after NMR relaxation, the multi-valued DNA sequences are found from multi-valued qubits and these will go through multi-valued DNA OR gate to produce desired multi-valued Quantum-DNA 9-to-2 PROM output sequences.

Figure 7.7 depicts the multi-valued Quantum – DNA 9-to-2 PROM with NMR relaxation in normal room temperature.

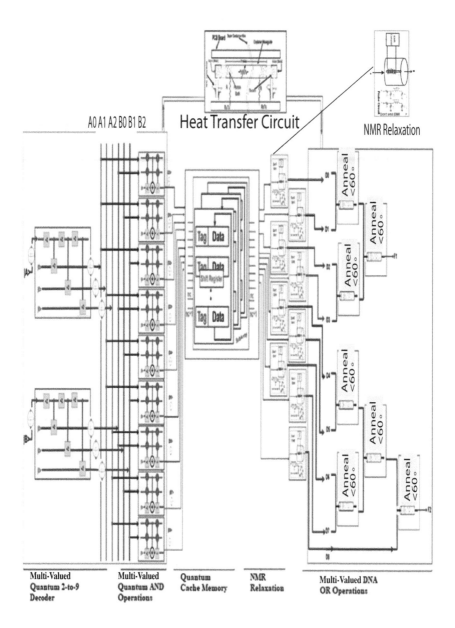

FIGURE 7.7
Multi-Valued Quantum-DNA 9-to-2 PROM with NMR Relaxation in Normal Room
Temperature

7.4.5 Working Principle

Table 7.2 shows the truth table of multi-valued Quantum-DNA 9-to-2 PROM where
inputs are |A> and |B>, outputs are F1 and F2.

TABLE 7.2
Truth Table of Multi-Valued Quantum-
DNA 9-to-2 PROM

$|A>$	$|B>$	F1	F2
$|0>$	$|0>$	TGGATC	ACCTAG
$|0>$	$|1>$	TGGATC	ACCTAG
$|0>$	$|2>$	TGGATC	ACCTAG
$|1>$	$|0>$	TGGATC	ACCTAG
$|1>$	$|1>$	ACCTAG	TGGATC
$|1>$	$|2>$	ACCTAG	TGGATC
$|2>$	$|0>$	ACCTAG	TGGATC
$|2>$	$|1>$	ACCTAG	TGGATC
$|2>$	$|2>$	ACCTAG	TGGATC

According to the truth table of multi-valued Quantum-DNA 9-to-2 PROM (Table 7.2), it is necessary to do the following operations to perform desire output sequences:

[1]
For input sequences $|A>$, $|B>= |0>, |0>$

$|A>= |A0>, |A1>, |A2> = |2>, |0>, |0>$ and $|B> = |B0>, |B1>, |B2>= |2>, |0>, |0>$. With this input combinations, only input $|A0>$ and $|B0>$ is true. Thus the $|A0>$ and $|B0>$ are connected to output line $|D0>$ so only the $|D0>$ will $|2>$ and rest of the gates will produce $|0>$.

For output sequence F1, perform OR operations among (D0-D3) outputs. So, max (D0, D1, D2, D3) = max (2,0,0,0) = 2 "**TGGATC**" and for F2 perform OR operations among (D4, D5, D6, D7, D8) = max (0,0,0,0,0) =0 "**ACCTAG**" of multi-valued Quantum-DNA 9-to-2 PROM.

[2]
For input sequences $|A>$, $|B>= |0>, |1>$

$|A>= |A0>, |A1>, |A2> = |2>, |0>, |0>$ and $|B> = |B0>, |B1>, |B2> = |0>, |2>, |0>$. With this input combinations, only input $|A0>$ and $|B1>$ is true. Thus the $|A0>$ and $|B1>$ are connected to output line $|D1>$ so only the $|D1>$ will $|2>$ and rest of the gates will produce $|0>$.

For output sequence F1, perform OR operations among (D0-D3) outputs. So, max (D0, D1, D2, D3) = max (0,2,0,0) = 2 "**TGGATC**" and for F2 perform OR operations among (D4, D5, D6, D7, D8) = max (0,0,0,0,0) =0 "**ACCTAG**" of multi-valued Quantum-DNA 9-to-2 PROM.

[3]
For input sequences $|A>$, $|B>= |0>, |2>$

|A>= |A0>, |A1>, |A2> = |2>, |0>, |0> and |B0>, |B1>, |B2> = |0>, |0>, |2>. With this input combinations, only input |A0> and |B2> is true. Thus the |A0> and |B2> are connected to output line |D2> so only the D2 will |2> and rest of the gates will produce |0>.

For output sequence F1, perform OR operations among (D0-D3) outputs. So, max (D0, D1, D2, D3) = max (0,0,2,0) = 2 "**TGGATC**" and for F2 perform OR operations among (D4, D5, D6, D7, D8) = max (0,0,0,0,0) =0 "**ACCTAG**" of multi-valued Quantum-DNA 9-to-2 PROM.

[4]
For input sequences |A>, |B>= **|1>, |0>**

|A>= |A0>, |A1>, |A2> = |0>, |2>, |0> and |B0>, |B1>, |B2> = |2>, |0>, |0>. With this input combinations, only input |A1> and |B0> is true. Thus the |A1> and |B0> are connected to output line |D3> so only the |D3> will |2> and rest of the gates will produce |0>.

For output sequence F1, perform OR operations among (D0-D3) outputs. So, max (D0, D1, D2, D3) = max (0,0,0,2) = 2 "**TGGATC**" and for F2 perform OR operations among (D4, D5, D6, D7, D8) = max (0,0,0,0,0)=0 "**ACCTAG**" of multi-valued Quantum-DNA 9-to-2 PROM.

[5]
For input sequences |A>, |B>= **|1>, |1>**

|A>= |A0>, |A1>, |A2> = |0>, |2>, |0> and |B0>, |B1>, |B2> = |0>, |2>, |0>. With this input combinations, only input |A1> and |B1> is true. Thus the |A1> and |B1> are connected to output line |D4> so only the |D4> will |2> and rest of the gates will produce |0>.

For output sequence F1, perform OR operations among (D0-D3) outputs. So, max (D0, D1, D2, D3) = max (0,0,0,0) = 0 "**ACCTAG**" and for F2 perform OR operations among (D4, D5, D6, D7, D8) = max (2,0,0,0,0) = 2 "**TGGATC**" of multi-valued Quantum-DNA 9-to-2 PROM.

[6]
For input sequences |A>, |B>= **|1>, |2>**

|A>= |A0>, |A1>, |A2> = |0>, |2>, |0> and |B0>, |B1>, |B2> = |0>, |0>, |2>. With this input combinations, only input |A1> and |B2> is true. Thus the |A1> and |B2> are connected to output line |D5> so only the |D5> will |2> and rest of the gates will produce |0>.

For output sequence F1, perform OR operations among (D0-D3) outputs. So, max (D0, D1, D2, D3) = max (0,0,0,0) = 0 "**ACCTAG**" and for F2 perform OR operations among (D4, D5, D6, D7, D8) = max (0,2,0,0,0) = 2 "**TGGATC**" of multi-valued Quantum-DNA 9-to-2 PROM.

[7]
For input sequences |A>, |B>= **|2>, |0>**

|A>= |A0>, |A1>, |A2> = |0>, |0>, |2> and |B0>, |B1>, |B2> = |2>, |0>, |0>. With this input combinations, only input |A2> and |B0> is true. Thus the |A2> and |B0> are connected to output line |D6> so only the |D6> will **|2>** and rest of the gates will produce **|0>**.

For output sequence F1, perform OR operations among (D0-D3) outputs. So, max (D0, D1, D2, D3) = max (0,0,0,0) = 0 "**ACCTAG**" and for F2 perform OR operations among (D4, D5, D6, D7, D8) = max (0,0,2,0,0) = 2 "**TGGATC**" of multi-valued Quantum-DNA 9-to-2 PROM.

[8]
For input sequences |A>, |B>= **|2>, |1>**

|A>= |A0>, |A1>, |A2> = |0>, |0>, |2> and |B0>, |B1>, |B2> = |0>, |2>, |0>. With this input combinations, only input |A2> and |B1> is true. Thus the |A2> and |B1> are connected to output line |D7> so only the |D7> will **|2>** and rest of the gates will produce **|0>**.

For output sequence F1, perform OR operations among (D0-D3) outputs. So, max (D0, D1, D2, D3) = max (0,0,0,0) = 0 "**ACCTAG**" and for F2 perform OR operations among (D4, D5, D6, D7, D8) = max (0,0,0,2,0) = 2 "**TGGATC**" of multi-valued Quantum-DNA 9-to-2 PROM.

[9]
For input sequences |A>, |B>= **|2>, |2>,**

|A>= |A0>, |A1>, |A2> = |0>, |0>, |2> and |B0>, |B1>, |B2> = |0>, |0>, |2>. With this input combinations, only input |A2> and |B2> is true. Thus the |A2> and |B2> are connected to output line |D8> so only the |D8> will **|2>** and rest of the gates will produce **|0>**.

For output sequence F1, perform OR operations among (D0-D3) outputs. So, max (D0, D1, D2, D3) = max (0,0,0,0) = 0 "**ACCTAG**" and for F2 perform OR operations among (D4, D5, D6, D7, D8) = max (0,0,0,0,2) = 2 "**TGGATC**" of multi-valued Quantum-DNA 9-to-2 PROM.

7.5   Multiple-Valued Quantum-DNA Cache Memory

The design of various memory devices includes quantum and DNA logic gates and decoder for Quantum-DNA 9-to-1 Cache memory. Both DNA and Quantum computers have the potential to significantly affect traditional digital computers in terms of processing power and depot. Quantum computers are more powerful than traditional Turing machines computing because of their coherent superposition of states and biotechnology techniques can be used to evolve DNA computers. Both of these properties might be captured by integrating DNA and Quantum computing. A novel logic gate design based on chemical reactions is presented in which observance of double-stranded sequences indicates a truth evaluation. Therefore, integrated computing systems as Quantum-to-DNA and vice versa for multi-valued cache memory are the main concern of this section.

7.5.1   General Organization of Multiple-Valued Quantum-DNA Cache Memory

The block diagram of Quantum-DNA 9-to-1 cache memory is shown in Figure 7.8. Cache memory consists of three basic components- Decoder, Qubit RAM Cells and OR operations. To execute Quantum-DNA 9-to-1 Cache memory following operations are required:

1. A Quantum 2-to-9 Decoder
2. Qubit RAM cells
3. Quantum Cache Memory
4. NMR Relaxation
5. Heat transfer circuit and
6. DNA OR operations for corresponding minterms

7.5.2   Circuit Architecture of Multiple-Valued Quantum-DNA Cache Memory

Figure 7.9 depicts the circuit architecture of multiple-valued Quantum-DNA Cache Memory.

7.5.3   Working Principle

Figure 7.9 represents the implementation of 9-to-1 sequence Quantum-DNA Cache memory. This quantum Cache memory consists of 9 separate "Words" of memory and each is 1 sequence wide. The quantum Cache memory Cell has three inputs and one output. The complete circuit of a quantum Cache memory cell is described in

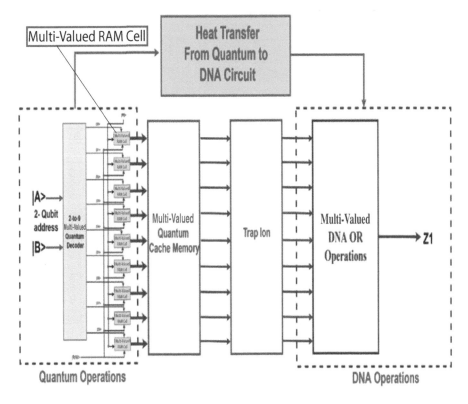

FIGURE 7.8
Block Diagram of Quantum-DNA 9-to-1 Cache Memory

Figure 7.9 with proper explanation. A word consists of two quantum Cache memory cells arranged in such a way so that both sequences can be accessed simultaneously. Nine words of memory need two address lines. A and B are the two-sequence address lines input that goes through a 2-to-9 decoder that selects one of the nine words. The memory-enabled input enables the decoder. If the memory enables is **TGGATC**, all output of the decoder will be **TGGATC** and in that case, none of the memory addresses will be selected. But when the memory enables is **ACCTAG** then one of the four words is selected. The word is selected by the value in the two address lines. When a word has been selected, the read/write input determines the operation. During the read operation, the four sequences of the selected word pass to the quantum OR gates to the output Z1 terminals. But during the write operation, the data which is available in the input lines are transferred into the four quantum cells of the selected word. The quantum Cache memory cells that are not selected are become disabled and their previous sequence never changes. But when the memory enable input that passes into the decoder is equal to **TGGATC**, none of the words are selected, and then all quantum cells remain unchanged regardless of the value of the read/ write input.

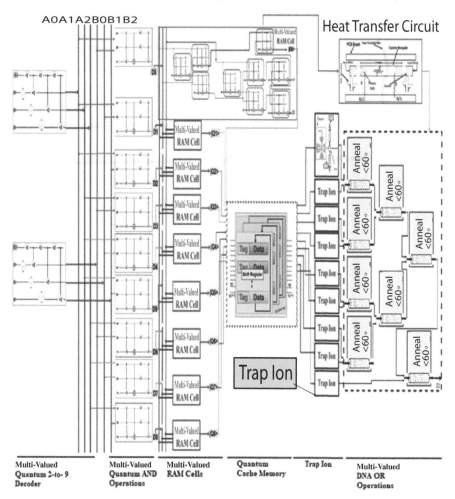

FIGURE 7.9
Multiple-Valued Quantum-DNA Cache Memory

7.6 Summary

The software can "partition" a portion of a computer's Cache memory, allowing it to act as a much faster hard drive that is called a Cache memory disk. A Cache memory disk loses the stored data when the computer is shut down unless memory is arranged to have a standby battery source. Most modern operating systems employ a method of extending Cache memory capacity, known as "virtual memory". A portion of the computer's hard drive is set aside for a paging file or a scratch partition, and the combination of physical Cache memory and the paging file form the system's total memory. (For example, if a computer has 2 GB of Cache memory and a 1 GB page file, the operating system has 3 GB total memory available to it.) When the system runs low on physical memory, it can "swap" portions of Cache memory to the paging file to make room for new data, as well as to read previously swapped information back into Cache memory. Excessive use of this mechanism results in thrashing and generally hampers overall system performance, mainly because hard drives are far slower than Cache memory. This chapter has presented the multi-valued memory devices in quantum-DNA computing.

Bibliography

[1] Goodman, J. R. (1983, June). Using cache memory to reduce processor-memory traffic. In Proceedings of the 10th Annual International Symposium on Computer Architecture (pp. 124-131).

[2] Mano, M. M. (1993). Computer System Architecture. Prentice-Hall, Inc.

[3] Smith, A. J. (1987). Line (block) size choice for CPU cache memories. IEEE transactions on computers, 100(9), 1063-1075.

[4] Iyer, R., Zhao, L., Guo, F., Illikkal, R., Makineni, S., Newell, D., & Reinhardt, S. (2007). QoS policies and architecture for cache/memory in CMP platforms. ACM SIGMETRICS Performance Evaluation Review, 35(1), 25-36.

[5] Resconi, G., & Nikravesh, M. (2008). Morphic Computing: Quantum and Fields. In Forging New Frontiers: Fuzzy Pioneers II (pp. 1-19). Springer, Berlin, Heidelberg.

8

Multiple-Valued DNA-Quantum Memory Devices

8.1 Introduction

DNA computing is a recent topic that is attracting the growing technology. Multi-Valued DNA-Quantum computing is one of the most intriguing new scientific topics to have emerged in recent years: it is a combination of DNA and quantum computing. The construction of Multi-Valued DNA-Quantum computers at present time would be incredibly challenging, even if it were conceivable. This chapter is aimed at the development of Multi-Valued DNA-Quantum memory devices. It is necessary to implement the functionality of this circuit by using the fundamental multi-valued quantum and DNA operational gates (AND, OR, and NOT) of the Multi-Valued DNA-Quantum system. The combination of these two technologies is quantum molecular biology. Quantum molecular biology is of two type; one is quantum-DNA computing which is discussed in previous chapter for multi-valued memory devices, this chapter will discuss about multi-valued DNA-quantum memory devices. This chapter will describe multi-valued DNA-quantum RAM, ROM, PROM and Cache Memory.

8.2 Multiple-Valued DNA-Quantum Random Access Memory

Multi-Valued DNA-quantum RAM is the opposite of quantum-DNA RAM. Here the first portion is DNA operation and last portion is quantum operations which is complementary of multi-valued quantum-DNA RAM.

Figure 8.1 represents the DNA-Quantum 9-to-1 RAM general organization of block diagram, consisting of 9 separate "Words" of memory and each is single sequence wide.

8.2.1 Circuit Architecture of Multiple-Valued DNA-Quantum RAM

For the circuit architecture of Multi-Valued DNA- Quantum 9-to-1 RAM, a DNA 2-to-9 ternary decoder is needed which is designed by using two DNA multi-valued 1-to-3 Quantum decoders to get the output of nine DNA ternary RAM cells oper-

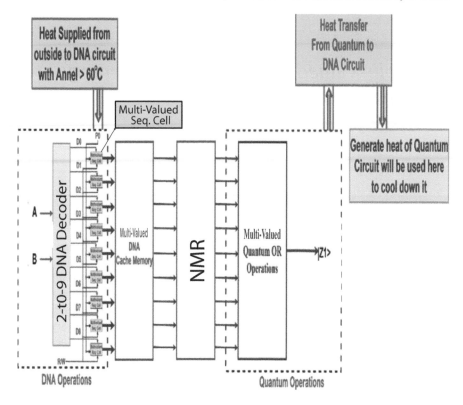

FIGURE 8.1

Block Diagram of DNA- Quantum 9-to-1 RAM

ations. Finally, the multi-valued 9-ternary RAM cells perform sequence cell output sequences from Q0 to Q8 and will go through NMR for further minterms ternary OR operation of Quantum qubits to perform desire qubits as output. Figure 8.2 depicts the circuit architecture of DNA-Quantum 9-to-1 RAM.

8.2.2 Working Principle

Figure 8.2 represents the implementation of 9-to-1 DNA-Quantum RAM. This DNA-Quantum RAM consists of 9 separate "Words" of memory and each is 1 sequence wide. The DNA RAM Cell has three inputs and one output. A word consists of two DNA RAM cells arranged in such a way so that both sequences can be accessed simultaneously. Four words of memory need two address lines. A and B are the two-sequence address lines input that goes through a DNA 2-to-9 decoder that selects one of the four words. The memory-enabled input enables the decoder.

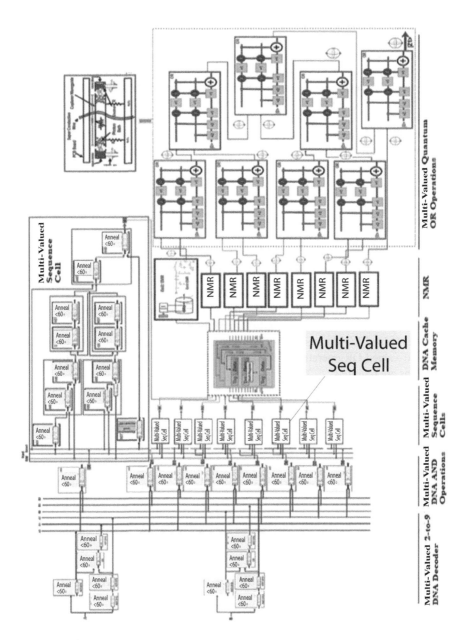

FIGURE 8.2
Circuit Architecture of DNA-Quantum 9-to-1 RAM

If the memory enables is **TGGATC**, all output of the decoder will be **TGGATC** and in that case, none of the memory addresses will be selected. But when the memory enables is **ACCTAG** then one of the four words is selected. The word is selected by the value in the two address lines. When a word has been selected, the read/write input determines the operation. During the read operation, the four sequences of the selected word pass to the Quantum OR operations to the output $|Z1>$ terminals. But during the write operation, the data which is available in the input lines are transferred into the nine DNA cells of the selected word. The DNA RAM cells that are not selected have become disabled and their previous sequence never changes. But when the memory enable input that passes into the decoder is equal to **TGGATC**, none of the words are selected, and then all DNA cells remain unchanged regardless of the value of the read/ write input. This is the working procedure of 9-to-1 DNA-Quantum RAM.

8.3   Multiple-Valued DNA-Quantum Read Only Memory

Input sequences A and B in multi-valued DNA computing operations, store the sequences in multi-valued DNA cache memory and transform those sequences to qubits to form outputs $|F1>$ and $|F2>$ in multi-valued quantum computing operations.

Multi-Valued ROM consists of two basic components- A multi-valued Decoder and multi-valued OR operations. To execute DNA-Quantum 9-to-2 ROM following operations are required:

1.   A multi-valued DNA 2-to-9 Decoder

2.   Multi-Valued DNA Cache Memory

3.   NMR

4.   Heat transfer circuit and

5.   Multi-Valued Quantum OR operations for corresponding minterms.

Figure 8.3 depicts general block diagram of Multi-Valued DNA-Quantum 9-to-2 ROM.

8.3.1   Circuit Architecture of Multiple-Valued DNA-Quantum ROM

Multi-Valued DNA-quantum ROM circuit needs 2-to-9 multi-valued DNA decoder and multi-valued quantum OR operations. Here inputs are DNA sequences and outputs are qubits. Figure 8.4 depicts the multi-valued DNA-Quantum 9-to-2 ROM with NMR in normal room temperature.

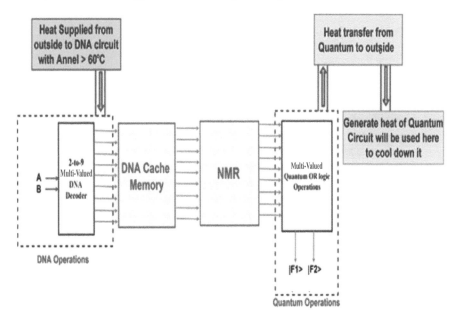

FIGURE 8.3
General Block Diagram of Multi-Valued DNA-Quantum 9-to-2 ROM

8.3.2 Design Procedure

To perform multi-valued ROM at normal room temperature, a 2-to-9 decoder and minterms of decoder output are needed as the OR operations input to produce the desired ROM function. In the multi-valued DNA- Quantum computing (considering NMR), the input values are in DNA sequences (multi-valued DNA Computing) and the output will be qubits (multi-valued Quantum Computing) at normal room temperature.

Step 1: First, draw two input DNA sequences A and B. Two possible sequences are **ACCTAG or CAAGCT or TGGATC**. These 3 will produce 9 combinations of 2 input sequences.

Step 2: To design a multi-valued 2-to-9 DNA decoder, draw two multi-valued 1-to-3 DNA decoders and nine multi-valued DNA AND operations.

Step 3: After getting sequences of all the corresponding values from nine DNA AND operations of the 2-to-9 decoder, it is needed to store all the DNA sequences into a multi-valued DNA cache memory. DNA cache memory is actually used for storing logical qubits and error correction.

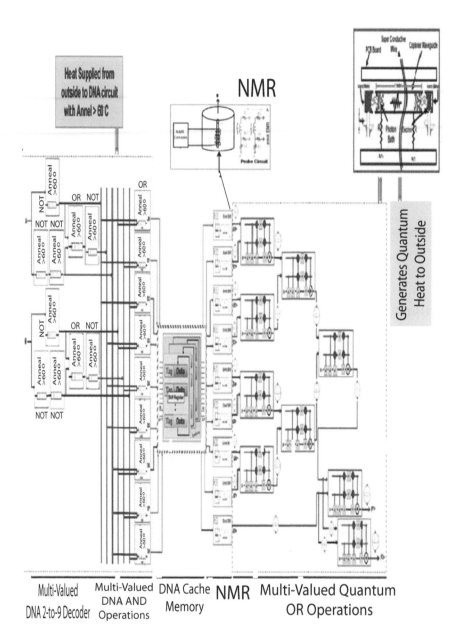

FIGURE 8.4
Multi-Valued DNA-Quantum 9-to-2 ROM with NMR in Normal Room Temperature

TABLE 8.1

Truth Table of DNA-Quantum 9-to-2 ROM

A	B	|F1>	|F2>
ACCTAG	ACCTAG	|2>	|2>
ACCTAG	CAAGCT	|2>	|2>
ACCTAG	TGGATC	|2>	|2>
CAAGCT	ACCTAG	|2>	|2>
CAAGCT	CAAGCT	|2>	|2>
CAAGCT	TGGATC	|2>	|2>
TGGATC	ACCTAG	|2>	|2>
TGGATC	CAAGCT	|2>	|2>
TGGATC	TGGATC	|2>	|2>

Step 4: After storing sequences of all the corresponding values from nine multi-valued DNA AND operations in cache memory, need to perform NMR to transform DNA sequences to qubits.

Step 5: Finally, the qubits are found from DNA sequences and these will go through multi-valued Quantum OR operations to produce desired multi-valued DNA-Quantum 9-to-2 ROM output qubits.

8.3.3 Working Principle

Table 8.1 shows the truth table of DNA-Quantum 9-to-2 ROM. Here A and B inputs, outputs are |F1> and |F2>.

[1]
For input sequences A, B= **ACCTAG, ACCTAG**

1. To perform value of input A0, A will go through the NOT (NTI), and produces TGGATC in the A0.

2. To perform the value of input A2, A will go through the NOT (PTI) and produce TGGATC then go through NOT (NTI) and produces ACCTAG in the A2.

3. To perform the value of input A1, the value of A0 and A2 go through DNA NOR (OR, NOT) Operations and produces ACCTAG in the A1.

4. To perform the value of input B0, B will go through the NOT (NTI), and produces TGGATC in the B0.

5. To perform the value of input B2, B will go through the NOT (PTI) and produce TGGATC then go through NOT (NTI) and produces ACCTAG in the B2.

6. To perform the value of input B1, the value of B0 and B2 go through DNA NOR (OR, NOT) Operations and produces ACCTAG in the B1.

ACCTAG = |0> **CAAGCT** =|1> **TGGATC** = |2

A= A0, A1, A2 = 2, 0, 0 and B= B0, B1, B2= 2, 0, 0. With these input combinations, only input A0 and B0 are true. Thus the A0 and B0 are connected to output line D0 so only the D0 will **TGGATC** and the rest of the gates will produce **ACCTAG**.

To perform 9 to 2 multi-valued DNA ROM functions outputs,

[i] For input sequences |D0>, |D1> = |2>, |0>, the multi-valued OR gates will perform maximum operations of input sequences. So the output sequence of |2>.

[ii] For input sequences |D2>, |D3> = |0>, |0>, the multi-valued OR gates will perform maximum operations of input sequences. So the output sequence of |0>.

[iii] For input sequences |D4>, |D5> = |0>, |0> the multi-valued OR gates will perform maximum operations of input sequences. So the output sequence of |0>.

[iv] For input sequences |D6>, |D7> = |0>, |0> the multi-valued OR gates will perform maximum operations of input sequences. So the output sequence of |0>.

[v] Combine the output sequences |2>, |0> from [i] and [ii] respectively as input sequences, the multi-valued OR gates will perform maximum operations of input sequences. So the output sequence of |2>.

[vi] Combine the output sequences |0>, |0> from [iii] and [iv] respectively as input sequences, the multi-valued OR gates will perform maximum operations of input sequences. So the output sequence of |0>.

[vii] Combine the output sequences |2>, |0> from [v] and [vi] respectively as input sequences, the multi-valued OR gates will perform maximum operations of input sequences. So the output sequence of |2>.

[viii] Finally, Combine the output sequence |2> from [vii] and |D8> (|0>) as input sequences, the multi-valued OR gates will perform maximum operations of input sequences. So the output sequence of |2> for both functions output F1 and F2 multi-valued 9-to-2 DNA-Quantum ROM.

[2]
For input sequences A, B= **ACCTAG, CAAGCT**

A= A0, A1, A2 = 2, 0, 0 and B= B0, B1, B2= 0, 2, 0. With these input combinations, only input A0 and B1 are true. Thus, the A0 and B1 are connected to output line D1 so only the D1 will **TGGATC** and the rest of the gates will produce **ACCTAG**.

For Functions F1 and F2, perform OR operations among all decoder (|D0>-|D8>) outputs. So, max (|D0>, |D1>, |D2>, |D3>, |D4>, |D5>, |D6>, |D7>, |D8>) = max

(|0>, |2>, |0>, |0>, |0>, |0>, |0>, |0>, |0>) = **|2>** will produce as the function's outputs F1 and F2 of multi-valued 9-to-2 DNA-Quantum ROM.

[3]
For input sequences A, B= **ACCTAG, TGGATC**

A= A0, A1, A2 = 2, 0, 0 and B= B0, B1, B2= 0, 0, 2. With these input combinations, only input A0 and B2 are true. Thus the A0 and B2 are connected to output line D2 so only the D2 will **TGGATC** and the rest of the gates will produce **ACCTAG**.

For Functions F1 and F2, perform OR operations among all decoder (|D0>-|D8>) outputs. So, max (|D0>, |D1>, |D2>, |D3>, |D4>, |D5>, |D6>, |D7>, |D8>) = max (|0>, |0>, |2>, |0>, |0>, |0>, |0>, |0>, |0>) = **|2>** will produce as the function's outputs F1 and F2 of multi-valued 9-to-2 DNA-Quantum ROM.

[4]
For input sequences A, B= **CAAGCT, ACCTAG**

A= A0, A1, A2 = 0, 2, 0 and B= B0, B1, B2= 2, 0, 0. With these input combinations, only input A1 and B0 is true. Thus, the A1 and B0 are connected to output line D3 so only the D3 will **TGGATC** and the rest of the gates will produce **ACCTAG**.

For Functions F1 and F2, perform OR operations among all decoder (|D0>-|D8>) outputs. So, max (|D0>, |D1>, |D2>, |D3>, |D4>, |D5>, |D6>, |D7>, |D8>) = max (|0>, |0>, |0>, |2>, |0>, |0>, |0>, |0>, |0>) = **|2>** will produce as the function's outputs F1 and F2 of multi-valued 9-to-2 DNA-Quantum ROM.

[5]
For input sequences A, B= **CAAGCT, CAAGCT**

A= A0, A1, A2 = 0, 2, 0 and B= B0, B1, B2= 0, 2, 0. With these input combinations, only input A1 and B1 is true. Thus, the A1 and B1 are connected to output line D4 so only the D4 will **TGGATC** and the rest of the gates will produce **ACCTAG**.

For Functions F1 and F2, perform OR operations among all decoder (|D0>-|D8>) outputs. So, max (|D0>, |D1>, |D2>, |D3>, |D4>, |D5>, |D6>, |D7>, |D8>) = max (|0>, |0>, |0>, |0>, |2>, |0>, |0>, |0>, |0>) = **|2>** will produce as the function's outputs F1 and F2 of multi-valued 9-to-2 DNA-Quantum ROM.

[6]
For input sequences A, B= **CAAGCT, TGGATC**

A= A0, A1, A2 = 0, 2, 0 and B= B0, B1, B2= 0, 0, 2. With these input combinations, only input A1 and B2 is true. Thus, the A1 and B2 are connected to output line D5 so only the D5 will **TGGATC** and the rest of the gates will produce **ACCTAG**.

For Functions F1 and F2, perform OR operations among all decoder ($|D0>$-$|D8>$) outputs. So, max ($|D0>$, $|D1>$, $|D2>$, $|D3>$, $|D4>$, $|D5>$, $|D6>$, $|D7>$, $|D8>$) = max ($|0>$, $|0>$, $|0>$, $|0>$, $|0>$, $|2>$, $|0>$, $|0>$, $|0>$) = **$|2>$** will produce as the function's outputs F1 and F2 of multi-valued 9-to-2 DNA-Quantum ROM.

[7]
For input sequences A, B= **TGGATC, ACCTAG**

A= A0, A1, A2 = 0, 0, 2 and B= B0, B1, B2= 2, 0, 0. With these input combinations, only input A2 and B0 are true. Thus, the A2 and B0 are connected to output line D6 so only the D6 will **TGGATC** and the rest of the gates will produce **ACCTAG**.

For Functions F1 and F2, perform OR operations among all decoder ($|D0>$-$|D8>$) outputs. So, max ($|D0>$, $|D1>$, $|D2>$, $|D3>$, $|D4>$, $|D5>$, $|D6>$, $|D7>$, $|D8>$) = max ($|0>$, $|0>$, $|0>$, $|0>$, $|0>$, $|0>$, $|2>$, $|0>$, $|0>$) = **$|2>$** will produce as the function's outputs F1 and F2 of multi-valued 9-to-2 DNA-Quantum ROM.

[8]
For input sequences A, B= **TGGATC, CAAGCT**

A= A0, A1, A2 = 0, 0, 2 and B= B0, B1, B2= 0, 2, 0. With these input combinations, only input A2 and B1 is true. Thus, the A2 and B1 are connected to output line D7 so only the D7 will **TGGATC** and the rest of the gates will produce **ACCTAG**.

For Functions F1 and F2, perform OR operations among all decoder ($|D0>$-$|D8>$) outputs. So, max ($|D0>$, $|D1>$, $|D2>$, $|D3>$, $|D4>$, $|D5>$, $|D6>$, $|D7>$, $|D8>$) = max ($|0>$, $|0>$, $|0>$, $|0>$, $|0>$, $|0>$, $|0>$, $|2>$, $|0>$) = **$|2>$** will produce as the function's outputs F1 and F2 of multi-valued 9-to-2 DNA-Quantum ROM.

[9]
For input sequences A, B= **TGGATC, TGGATC**

A= A0, A1, A2 = 0, 0, 2 and B= B0, B1, B2= 0, 0, 2. With these input combinations, only input A2 and B2 is true. Thus, the A2 and B2 are connected to output line D8 so only the D8 will **TGGATC** and the rest of the gates will produce **ACCTAG**.

For Functions F1 and F2, perform OR operations among all decoder ($|D0>$-$|D8>$) outputs. So, max ($|D0>$, $|D1>$, $|D2>$, $|D3>$, $|D4>$, $|D5>$, $|D6>$, $|D7>$, $|D8>$) = max ($|0>$, $|0>$, $|0>$, $|0>$, $|0>$, $|0>$, $|0>$, $|0>$, $|2>$) = **$|2>$** will produce as the function's outputs F1 and F2 of multi-valued 9-to-2 DNA-Quantum ROM.

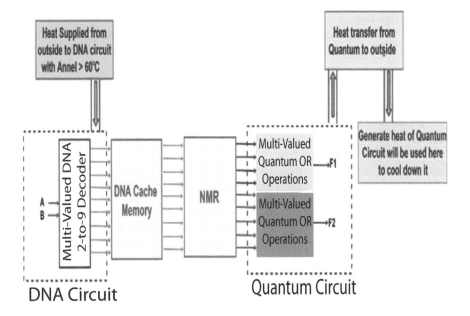

FIGURE 8.5
General Block Diagram of Multi-Valued DNA-Quantum 9-to-2 PROM

8.4 Multiple-Valued DNA-Quantum Programmable Read Only Memory

The vice versa of multi-valued Quantum-DNA, the proposed integrated multi-valued DNA-Quantum memory devices, will perform according to where the info flow of circuit design is multi-valued DNA-> multi-valued Sequence Storage -> multi-valued Transformation -> multi-valued Quantum.

The general block diagram of multi-valued DNA-Quantum 9-to-2 PROM is shown in Figure 8.5.

Multi-Valued PROM consists of two basic components- A multi-valued Decoder and multi-valued OR operations. To execute DNA-Quantum 9-to-2 PROM following operations are required:

1. A multi-valued DNA 2-to-9 Decoder

2. Multi-Valued DNA Cache Memory

3. NMR

4. Heat transfer circuit and

Multi-Valued Quantum OR operations for corresponding minterms.

Input sequences A and B in multi-valued DNA computing operations, store the sequences in multi-valued DNA cache memory and transform those sequences to qubits to form outputs |F1> and |F2> in multi-valued quantum computing operations

8.4.1 Circuit Architecture of Multiple-Valued DNA-Quantum PROM

For the design algorithm of Multi-Valued DNA-Quantum 9-to-2 PROM, a DNA 2-to-9 ternary decoder is needed which is designed by using two DNA multi-valued 1-to-3 Quantum decoders to get the output sequences from Q0 to Q8 will go through NMR process for further minterms ternary OR operation of Quantum qubits to perform desire qubit as output. Following the design algorithms of multiple-valued DNA-Quantum 9-to-2 PROM Memory. Figure 8.6 depicts the multi-valued DNA-quantum 9-to-2 ROM with NMR in Normal Room Temperature.

8.4.2 Design Procedure

To perform multi-valued ROM at normal room temperature, need a 2-to-9 decoder and minterms of decoder output as the OR operations input to produce the desired ROM function. In the multi-valued DNA- Quantum computing (considering NMR), the input values are in DNA sequences (multi-valued DNA Computing) and the output will be qubits (multi-valued Quantum Computing) at normal room temperature.

Step 1:
First, draw two input DNA sequences A and B. Two possible sequences are **ACC-TAG or CAAGCT or TGGATC**. These 3 will produce 9 combinations of 2 input sequences.

Step 2:
To design a multi-valued 2-to-9 DNA decoder, draw two multi-valued 1-to-3 DNA decoders and nine multi-valued DNA AND operations.

Step 3:
After getting sequences of all the corresponding values from nine DNA AND operations of the 2-to-9 decoder, it is needed to store all the DNA sequences into a multi-valued DNA cache memory. DNA cache memory is actually used for storing logical qubits and error correction.

Step 4:
After storing sequences of all the corresponding values from nine multi-valued DNA AND operations in cache memory, it is needed to perform NMR to transform DNA sequences to qubits.

Step 5:
Finally, the qubits will be found from DNA sequences and these will go through multi-valued Quantum OR operations to produce desired multi-valued DNA-Quantum 9-to-2 ROM output qubits.

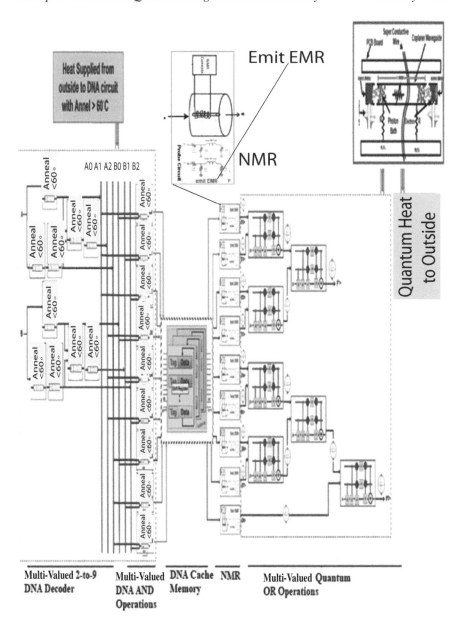

FIGURE 8.6
Multi-Valued DNA-Quantum 9-to-2 ROM with NMR in Normal Room Temperature

TABLE 8.2
Truth Table of DNA-Quantum 9-to-2 PROM

A	B	|F1>	|F2>
ACCTAG	ACCTAG	|2>	|0>
ACCTAG	CAAGCT	|2>	|0>
ACCTAG	TGGATC	|2>	|0>
CAAGCT	ACCTAG	|2>	|0>
CAAGCT	CAAGCT	|0>	|2>
CAAGCT	TGGATC	|0>	|2>
TGGATC	ACCTAG	|0>	|2>
TGGATC	CAAGCT	|0>	|2>
TGGATC	TGGATC	|0>	|2>

8.4.3 Working Principle

Table 8.2 shows the truth table of DNA-Quantum 9-to-2 PROM. Here A and B inputs, outputs are |F1> and |F2>.

[1]
For input sequences A, B= **ACCTAG, ACCTAG**

1. To perform the value of input A0, A will go through the NOT (NTI), and produces TGGATC in the A0.
2. To perform the value of input A2, A will go through the NOT (PTI) and produce TGGATC then go through NOT (NTI) and produces ACCTAG in the A2.
3. To perform the value of input A1, the value of A0 and A2 go through DNA NOR (OR, NOT) Operations and produces ACCTAG in the A1.
4. To perform the value of input B0, B will go through the NOT (NTI), and produces TGGATC in the B0.
5. To perform the value of input B2, B will go through the NOT (PTI) and produce TGGATC then go through NOT (NTI) and produces ACCTAG in the B2.
6. To perform the value of input B1, the value of B0 and B2 go through DNA NOR (OR, NOT) Operations and produces ACCTAG in the B1.

ACCTAG = |0> **CAAGCT** =|1> **TGGATC** = |2

A= A0, A1, A2 = 2, 0, 0 and B= B0, B1, B2= 2, 0, 0. With these input combinations, only input A0 and B0 are true. Thus the A0 and B0 are connected to output line D0 so only the D0 will **TGGATC** and the rest of the gates will produce **ACCTAG**.

To perform 9 to 2 multi-valued DNA ROM functions outputs,

[i] For input sequences |D0>, |D1> = **|2>, |0>**, the multi-valued OR gates will perform maximum operations of input sequences. So the output sequence of **|2>.**

[ii] For input sequences |D2>, |D3> = **|0>, |0>**, the multi-valued OR gates will perform maximum operations of input sequences. So the output sequence of **|0>**.

[iii] For input sequences |D4>, |D5> = **|0>, |0>** the multi-valued OR gates will perform maximum operations of input sequences. So the output sequence of **|0>**.

[iv] For input sequences |D6>, |D7> = **|0>, |0>** the multi-valued OR gates will perform maximum operations of input sequences. So the output sequence of **|0>**.

[v] Combine the output sequences **|2>, |0>** from [i] and [ii] respectively as input sequences, the multi-valued OR gates will perform maximum operations of input sequences. So the output sequence of **|2>**.

[vi] Combine the output sequences **|0>, |0>** from [iii] and [iv] respectively as input sequences, the multi-valued OR gates will perform maximum operations of input sequences. So the output sequence of **|0>**.

[vii] Combine the output sequences **|2>, |0>** from [v] and [vi] respectively as input sequences, the multi-valued OR gates will perform maximum operations of input sequences. So the output sequence of **|2>**.

[viii] Finally, Combine the output sequence **|2>** from [vii] and |D8> (**|0>**) as input sequences, the multi-valued OR gates will perform maximum operations of input sequences. So the output sequence of **|2> (v)** for the function output |F1> and **|0> (vii)** for |F2> multi-valued 9-to-2 Quantum PROM.

[2]
For input sequences A, B= **ACCTAG, CAAGCT**

A= A0, A1, A2 = 2, 0, 0 and B= B0, B1, B2= 0, 2, 0. With these input combinations, only input A0 and B1 are true. Thus the A0 and B1 are connected to output line D1 so only the D1 will **TGGATC** and the rest of the gates will produce **ACCTAG**.

For output sequence |F1>, perform OR operations among (|D0> - |D3>) outputs. So, max (|D0>, |D1>, |D2>, |D3>) = max (|0>, |2>, |0>, |0) = **|2>** and for |F2> perform OR operations among (|D4>, |D5>, |D6>, |D7>, |D8>) = max (|0>, |0>, |0>, |0>, |0>) = **|0>** of multi-valued 9-to-2 DNA-Quantum PROM.

[3]
For input sequences A, B= **ACCTAG, TGGATC**

A= A0, A1, A2 = 2, 0, 0 and B= B0, B1, B2= 0, 0, 2. With these input combinations, only input A0 and B2 are true. Thus the A0 and B2 are connected to output line D2 so only the D2 will **TGGATC** and the rest of the gates will produce **ACCTAG**.

For output sequence |F1>, perform OR operations among (|D0> - |D3>) outputs. So, max (|D0>, |D1>, |D2>, |D3>) = max (|0>, |0>, |2>, |0) = **|2>** and for |F2> perform OR operations among (|D4>, |D5>, |D6>, |D7>, |D8>) = max (|0>, |0>, |0>, |0>, |0>) = **|0>** of multi-valued 9-to-2 DNA-Quantum PROM.

[4]
For input sequences A, B= **CAAGCT, ACCTAG**

A= A0, A1, A2 = 0, 2, 0 and B= B0, B1, B2= 2, 0, 0. With these input combinations, only input A1 and B0 is true. Thus the A1 and B0 are connected to output line D3 so only the D3 will **TGGATC** and the rest of the gates will produce **ACCTAG**.

For output sequence |F1>, perform OR operations among (|D0> - |D3>) outputs. So, max (|D0>, |D1>, |D2>, |D3>) = max (|0>, |0>, |0>, |2) = **|2>** and for |F2> perform OR operations among (|D4>, |D5>, |D6>, |D7>, |D8>) = max (|0>, |0>, |0>, |0>, |0>) = **|0>** of multi-valued 9-to-2 DNA-Quantum PROM.

[5]
For input sequences A, B= **CAAGCT, CAAGCT**

A= A0, A1, A2 = 0, 2, 0 and B= B0, B1, B2= 0, 2, 0. With these input combinations, only input A1 and B1 is true. Thus the A1 and B1 are connected to output line D4 so only the D4 will **TGGATC** and the rest of the gates will produce **ACCTAG**.

For output sequence |F1>, perform OR operations among (|D0> - |D3>) outputs. So, max (|D0>, |D1>, |D2>, |D3>) = max (|0>, |0>, |0>, |0) = **|0>** and for |F2> perform OR operations among (|D4>, |D5>, |D6>, |D7>, |D8>) = max (|2>, |0>, |0>, |0>, |0>) = **|2>**of multi-valued 9-to-2 DNA-Quantum PROM.

[6]
For input sequences A, B= **CAAGCT, TGGATC**

A= A0, A1, A2 = 0, 2, 0 and B= B0, B1, B2= 0, 0, 2. With these input combinations, only input A1 and B2 is true. Thus the A1 and B2 are connected to output line D5 so only the D5 will **TGGATC** and the rest of the gates will produce **ACCTAG**.

For output sequence |F1>, perform OR operations among (|D0> - |D3>) outputs. So, max (|D0>, |D1>, |D2>, |D3>) = max (|0>, |0>, |0>, |0) = **|0>** and for |F2> perform OR operations among (|D4>, |D5>, |D6>, |D7>, |D8>) = max (|0>, |2>, |0>, |0>, |0>) = **|2>** of multi-valued 9-to-2 DNA-Quantum PROM.

[7]
For input sequences A, B= **TGGATC, ACCTAG**

A= A0, A1, A2 = 0, 0, 2 and B= B0, B1, B2= 2, 0, 0. With these input combinations, only input A2 and B0 are true. Thus the A2 and B0 are connected to output line D6 so only the D6 will **TGGATC** and the rest of the gates will produce **ACCTAG**.

For output sequence $|F1>$, perform OR operations among ($|D0>$ - $|D3>$) outputs. So, max ($|D0>$, $|D1>$, $|D2>$, $|D3>$) = max ($|0>$, $|0>$, $|0>$, $|0$) = **$|0>$** and for $|F2>$ perform OR operations among ($|D4>$, $|D5>$, $|D6>$, $|D7>$, $|D8>$) = max ($|0>$, $|0>$, $|2>$, $|0>$, $|0>$) = **$|2>$** of multi-valued 9-to-2 DNA-Quantum PROM.

[8]
For input sequences A, B= **TGGATC, CAAGCT**

A= A0, A1, A2 = 0, 0, 2 and B= B0, B1, B2= 0, 2, 0. With this input combinations, only input A2 and B1 is true. Thus the A2 and B1 are connected to output line D7 so only the D7 will **TGGATC** and rest of the gates will produce **ACCTAG**.

For output sequence $|F1>$, perform OR operations among ($|D0>$ - $|D3>$) outputs. So, max ($|D0>$, $|D1>$, $|D2>$, $|D3>$) = max ($|0>$, $|0>$, $|0>$, $|0$) = **$|0>$** and for $|F2>$ perform OR operations among ($|D4>$, $|D5>$, $|D6>$, $|D7>$, $|D8>$) = max ($|0>$, $|0>$, $|0>$, $|2>$, $|0>$) = **$|2>$** of multi-valued 9-to-2 DNA-Quantum PROM.

[9]
For input sequences A, B= **TGGATC, TGGATC**

A= A0, A1, A2 = 0, 0, 2 and B= B0, B1, B2= 0, 0, 2. With these input combinations, only input A2 and B2 is true. Thus the A2 and B2 are connected to output line D8 so only the D8 will **TGGATC** and the rest of the gates will produce **ACCTAG**.

For output sequence $|F1>$, perform OR operations among ($|D0>$ - $|D3>$) outputs. So, max ($|D0>$, $|D1>$, $|D2>$, $|D3>$) = max ($|0>$, $|0>$, $|0>$, $|0$) = **$|0>$** and for $|F2>$ perform OR operations among ($|D4>$, $|D5>$, $|D6>$, $|D7>$, $|D8>$) = max ($|0>$, $|0>$, $|0>$, $|0>$, $|2>$) = **$|2>$** of multi-valued 9-to-2 DNA-Quantum PROM.

8.5   Multiple-Valued DNA-Quantum Cache Memory

Cache memory is a small-sized type of volatile computer memory that provides high-speed data access to a processor and stores frequently used computer programs, applications and data. The temporary storage of memory, the cache makes data

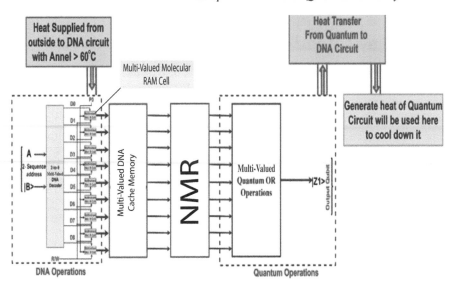

FIGURE 8.7
Block Diagram of Multiple-Valued DNA-Quantum Cache Memory

retrieving easier and more efficient. Multi-Valued DNA-quantum cache memory is the combination circuits of DNA operations and quantum operations. The input sequences are DNA molecules and the output will be quantum bits. A multi-valued 2-to-9 DNA decoder is needed in DNA operation and in quantum operation, a multi-valued quantum OR operations are needed. The block diagram of multi-valued DNA-quantum cache memory is given below (Figure 8.7).

8.5.1 Working Principle of Multiple-Valued DNA-Quantum Cache Memory

Figure 8.8 represents the implementation of 9-to-1 sequence ternary DNA-Quantum Cache memory. This quantum Cache memory consists of 9 separate "Words" of memory and each is 1 sequence wide. The quantum Cache memory Cell has three inputs and one output. The complete circuit of a quantum Cache memory cell is described in figure 8 with proper explanation. A word consists of two quantum Cache memory cells arranged in such a way so that both sequences can be accessed simultaneously. Nine words of memory need two address lines. A and B are the two-sequence address lines input that goes through a 2-to-9 decoder that selects one of the nine words. The memory-enabled input enables the decoder. If the memory enables is **TGGATC**, all output of the decoder will be **TGGATC** and in that case, none of the memory addresses will be selected. But when the memory enables is **ACCTAG** then one of the four words is selected. The word is selected by the value in the two address lines. When a word has been selected, the read/write input determines the

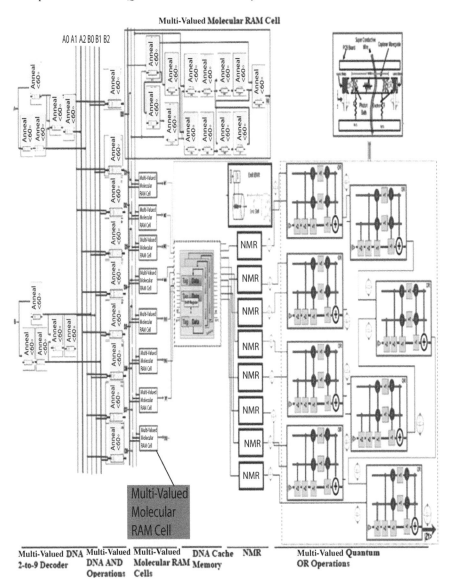

FIGURE 8.8
Multiple-Valued DNA-Quantum Cache Memory

operation. During the read operation, the four sequences of the selected word pass to the quantum OR gates to the output Z1 terminals. But during the write operation, the data which is available in the input lines are transferred into the four quantum cells of the selected word. The quantum Cache memory cells that are not selected have become disabled and their previous sequence never changes. But when the memory enable input that passes into the decoder is equal to **TGGATC**, none of the words are selected, and then all quantum cells remain unchanged regardless of the value of the read/ write input.

8.6 Summary

A ROM consists of arrays of AND and OR gates. A system designer implements a logic design with a device programmer that blows fuses on the ROM to control gate operation. A programmable read-only memory (PROM) has a fixed AND gate array, which links to a programmable OR gate array, which can then be conditionally complemented to produce an output. Quantum computing provides fast computation in logic devices than DNA computing. So, it is needed to use a storage device (CACHE MEMORY) for making a balance between these two computing systems. DNA computing gives more storage capacity, as a result, the output operation of the logic devices will perform in DNA computing. Quantum Computing produces more heat in circuits which can be used in DNA computing to perform DNA logic operations. This chapter has provided information and knowledge about multi-valued memory devices in DNA-quantum computing.

Bibliography

[1] Pan, T. C., Misra, S., & Aluru, S. (2018, November). Optimizing high performance distributed memory parallel hash tables for DNA k-mer counting. In SC18: International Conference for High Performance Computing, Networking, Storage and Analysis (pp. 135-147). IEEE.

[2] Wang, Y., & Kang, Q. (2014). Palladium-catalyzed allylic esterification via C–C bond cleavage of a secondary homoallyl alcohol. Organic Letters, 16(16), 4190-4193.

Part III

Multiple-Valued Programmable Logic Devices in Quantum Molecular Biology

Overview

Multi-valued Logic has been the topic of interest of many researchers due to its merits over common binary logic. Much of the previous work on ternary logic is purely theoretical. Due to the problems with the binary logic to reduce interconnect complexity and reduce chip area, it is giving motivation for the investigation of many hardware implementations of ternary logic. Work on the hardware implementation of three value devices has been more recent. Ternary logic is an effective approach to the default binary logic design technique because it allows for defining one or more voltage levels which are 0, Vdd / 2, and Vdd. It allows a circuit to be simple in design and energy-efficient due to its property of reduction in circuit overhead such as interconnects and chip area. The main advantage of ternary logic is its computing power and lower demand for memory. However, electronic implementation of ternary logic gates is not as straightforward as in the case of binary logic gates three types of ternary logic inverters i.e. STI, PTI and NTI are implemented. With the increased number of logic states, the bit handling capability of ternary logic circuits will increase. PLD digital logic devices have a wide range of applications in digital systems. In this part, various PLDs (PLA, PAL, FPGA, CPLD) performed by multi-valued logic system using quantum ternary AND, OR, XOR, NAND, inverter, etc. Multiple-Valued PLDs are the programmable logic device that has both Programmable ternary AND array & Programmable Quantum array. Hence, PLDs are more flexible. All these multi-valued PLDs will be implemented for quantum molecular biology. That means quantum computing, DNA computing, quantum-DNA computing and DNA-quantum computing.

9

Multiple-Valued Programmable Logic Devices in Quantum Computing

9.1 Introduction

Compared to fixed logic devices, programmable logic devices simplify the design of complex logic and may offer superior performance. Unlike for microprocessors, programming a PLD changes the connections made between the gates in the device. Multi-Valued programmable logic devices are electronic component used to construct electronic circuits. In quantum computing, it is known that the devices that perform quantum computations are known as quantum computers. Multi-Valued logic means more than two values of information. Here programmable logic devices are programmable logic array, programmable array logic, field programmable gate array, complex programmable logic devices. This chapter will describe the multiple-valued programmable logic devices in quantum computing.

9.2 Multiple-Valued Quantum Programmable Logic Array

Multi-Valued quantum Programmable Logic Array (PLA) has both programmable Quantum AND array and programmable Quantum OR array. In a combinational circuit, because of don't care conditions, not all the minterms are used. Programmable Logic Array (PLA) is a fixed architecture logic device with programmable Quantum AND gates followed by programmable Quantum OR gates. PLA is a type of programmable logic device used to build a reconfigurable digital circuit. PLDs have an undefined function at the time of manufacturing, but they are programmed before being made into use. PLA is a combination of memory and logic.

9.2.1 General Organization of Multi-Valued Quantum PLA

Multiple-Valued Quantum PLA is a programmable logic device that has both Programmable Quantum AND array & Programmable Quantum OR array . Hence, it

DOI: 10.1201/9781003381921-9

343

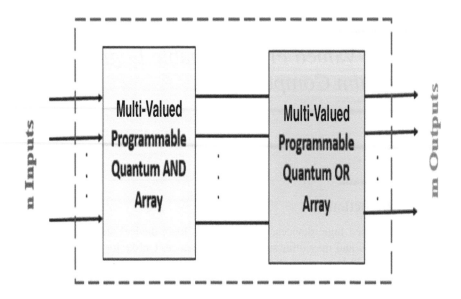

FIGURE 9.1
Block Diagram of Multi-Valued Quantum PLA

is the most flexible PLD. The block diagram of Multiple-Valued Quantum PLA is shown in Figure 9.1.

Here, the inputs of Quantum ternary AND operations are programmable. That means each Quantum ternary AND operation has both normal and complemented inputs of variables. So, based on the requirement, it is possible to program any of those inputs. So, required product terms can be generated by using these Quantum ternary AND operations.

Here, the inputs of Quantum ternary OR operations are also programmable. So, any number of required product terms can be programmed, since all the outputs of Quantum ternary AND operations are applied as inputs to each Quantum ternary OR operation. Therefore, the outputs of Quantum ternary PAL will be in the form of the sum of products form.

9.2.2   Circuit Architecture of Multi-Valued Quantum PLA

Consider the design of ternary quantum PLA for the following functions:

$|F1>=|A0>.|B1>+|A1>.|B1>+|1>. (|A2>.|B0>)$

$|F2>=|A0>.|B2>+|1>. (|A1>.|B2> + |A2>.|B1>)$

TABLE 9.1
Truth Table of Ternary
Quantum PLA

|A>	|B>	|F1>	|F2>
|0>	|0>	|0>	|0>
|0>	|0>	|2>	|0>
|0>	|2>	|0>	|2>
|1>	|0>	|0>	|0>
|1>	|1>	|2>	|0>
|1>	|2>	|0>	|1>
|2>	|0>	|1>	|0>
|2>	|1>	|0>	|1>
|2>	|2>	|0>	|0>

Table 9.1 shows the truth table of Ternary Quantum PLA.
Let us implement the following **Boolean functions** using ternary Quantum PLA.

|F1>=|A0>.|B1>+|A1>.|B1>+|1>. (|A2>.|B0>)

|F2>=|A0>.|B2>+|1>. (|A1>.|B2> + |A2>.|B1>)

The given two functions are sum of products form. The number of product terms present in the given Boolean functions |F1>, |F2> are three.

Eight programmable ternary quantum AND gates & four programmable ternary quantum OR gates are required for producing those two functions. The corresponding Quantum **PLA** is shown in the following figure (Figure 9.2).

For the given problem, there are two inputs (|A>,| B>) and two outputs (|F1>, |F2>). Quantum ternary decoders are used for these two inputs. Thus the realization has six input lines.

9.2.3 Working Principle

According to the truth table (Table 9.1) of ternary Quantum PLA, it is necessary to do the following operations to perform desire output qubits:

1. For input Qubits |A>, |B> =|0>, |0> function |F1> and |F2> produce |0>

2. For input Qubits |A>, |B> =|0>, |1> function |F1> produce |2> and |F2> produce |0>

3. For input Qubits |A>, |B> =|0>, |2> unction |F1> produce |0> and |F2> produce |2>

4. For input Qubits |A>, |B> =|1>, |0> function |F1> and |F2> produce |0>

5. For input Qubits |A>, |B> =|1>, |1> unction |F1> produce |2> and |F2> produce |0>

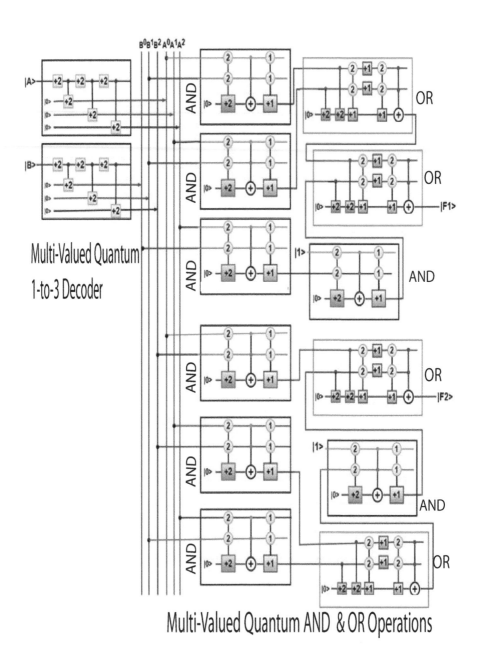

FIGURE 9.2
Ternary Quantum PLA

6. For input Qubits |A>, |B> =|1>, |2> unction |F1> produce |0> and |F2> produce |1>

7. For input Qubits |A>, |B> =|2>, |0> unction |F1> produce |1> and |F2> produce |0>

8. For input Qubits |A>, |B> =|2>, |1> unction |F1> produce |0> and |F2> produce |1>

9. For input Qubits |A>, |B> =|2>, |2> function |F1> and |F2> produce |0>

9.3   Multiple-Valued Quantum Programmable Array Logic

Programmable Array Logic (PAL) is a logic device, which has a programmable AND array and fixed OR array. It is used to realize a logic function. In this PLD, only AND gates are programmable and hence it is easier to work with PAL.

Figure 9.3 shows the internal structure of Programmable Array Logic. The product terms can be programmed through the fuse link. It means the user can decide the connection between the inputs and the AND gates. If a particular input line is to be connected to the AND gate, then the fuse link must be placed at the interconnection. The AND gate outputs are then fed as an input to the fixed OR gate. Depending upon the required function, the output line of the AND gate is connected to the corresponding input of the OR gate.

9.3.1   General Organization of Multi-Valued Quantum PAL

Multiple-valued Quantum Programmable Array Logic (PAL) is a type of Programmable Logic Device (PLD) used to realize a particular logical function. Quantum PALs comprise a quantum AND gate array followed by a quantum OR gate array. However, it is to be noted that here only the multiple-valued quantum AND gate array is programmable, unlike the Multiple-valued quantum OR gate array which has a fixed logic. This is because here the inputs are fed to the multiple-valued quantum AND gates through fuses which act as programmable links. Programmable-AND and fixed-OR structure of multiple-valued quantum PALs make them less flexible from the programming point of view when compared with Programmable Logic Arrays (PLAs). However, due to the same reason PALs are less expensive than PLAs. Hence, it is the most flexible multiple-valued quantum PLD. The block diagram of multiple-valued Quantum PAL is shown in the following figure (Figure 9.4).

Here, the inputs of multiple-valued Quantum AND operations are programmable. That means each multiple-valued Quantum AND operation has both normal and complemented inputs of variables. So, based on the requirement, it is possible to program any of those inputs and can generate only the required product terms by using these multiple-valued quantum AND operations. But, the inputs of multiple-valued Quantum OR operations are fixed.

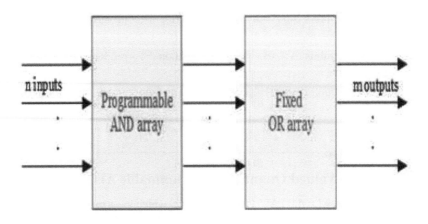

FIGURE 9.3
Block Diagram of PAL

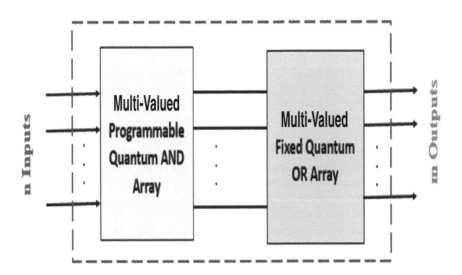

FIGURE 9.4
Block Diagram of Multi-Valued Quantum PAL

TABLE 9.2
Truth Table of Ternary Quantum PLA

|A>	|B>	|F1>	|F2>
|0>	|0>	|0>	|0>
|0>	|0>	|2>	|0>
|0>	|2>	|0>	|2>
|1>	|0>	|0>	|0>
|1>	|1>	|2>	|2>
|1>	|2>	|0>	|0>
|2>	|0>	|1>	|0>
|2>	|1>	|0>	|0>
|2>	|2>	|0>	|0>

9.3.2 Circuit Architecture of Multi-Valued Quantum PAL

Consider the design of ternary quantum PLA for the following functions:

|F1>=|A0>.|B1>+|A1>.|B1>+|1>. (|A2>.|B0>)

|F2>=|A0>.|B2>+|A1>.|B1>

Table 9.2 shows the truth table of Ternary Quantum PLA.
Let us implement the following Boolean functions using ternary Quantum PLA.

|F1>=|A0>.|B1>+|A1>.|B1>+|1>. (|A2>.|B0>)

|F2>=|A0>.|B2>+|A1>.|B1>

The given two functions are the sum of products form. The number of product terms present in the given Boolean functions |F1>, |F2> are three and 2 respectively.

Seven programmable ternary quantum AND gates & four fixed ternary quantum OR gates are required for producing those two functions. The corresponding Quantum PAL is shown in the following figure (Figure 9.5).

For the given problem, there are two inputs (|A>,| B>) and two outputs (|F1>, |F2>). Quantum ternary decoders are used for these two inputs. For these two inputs, two 3-to-1 quantum ternary decoders are used. The given first expression has three product terms and second expression has two product terms. But as the OR operation are fixed so the fuses are placed in the corresponding literals to obtain the product terms.

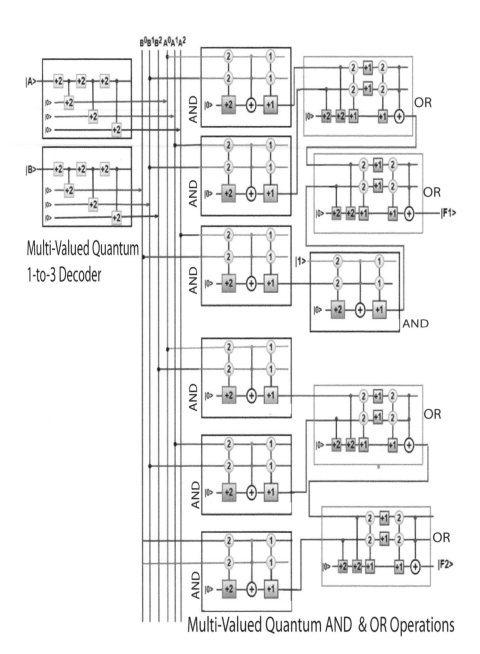

FIGURE 9.5
Multi-Valued Quantum PAL

9.3.3 Working Principle

According to truth table (Table 9.2) of ternary Quantum PLA , it is necessary to do the following operations to perform desire output qubits:

1. For input Qubits |A>, |B> =|0>, |0> function |F1> and F2> produce |0>
2. For input Qubits |A>, |B> =|0>, |1> function |F1> produce |2> and |F2> produce |0>
3. For input Qubits |A>, |B> =|0>, |2> function |F1> produce |0> and |F2> produce |2>
4. For input Qubits |A>, |B> =|1>, |0> function |F1> and |F2> produce |0>
5. For input Qubits |A>, |B> =|1>, |1> function |F1> and |F2> produce |2>
6. For input Qubits |A>, |B> =|1>, |2> function |F1> and |F2> produce |0>
7. For input Qubits |A>, |B> =|2>, |0> function |F1> produce |1> and |F2> produce |0>
8. For input Qubits |A>, |B> =|2>, |1> function |F1> and |F2> produce |0>
9. For input Qubits |A>, |B> =|2>, |2> function |F1> and |F2> produce |0>

9.4 Multi-Valued Quantum Field Programmable Gate Array

Customary silicon-based Field Programmable Gate Array (FPGA) is defenseless to security attacks as an outcome of its static design. When a static circuit is acquired by an attacker, it is a matter of time before one can figure out its configuration. To circumvent such altering, circuits should be dynamic naturally. Considering this vision, biological methodologies have been created to mirror existing silicon-based innovations in data manipulation. Inside the advanced world, information control includes information generation, storage, recovery, and preparation. A DNA-based plan empowers hardware to be founded on biochemical and natural stimuli. DNA computing introduces an approach by which one could create information that could reliably be stored and recovered inside a DNA sequence.

Modern DNA sequencing technologies have incredibly decreased their actual measurements, while yet having the option to handle huge quantities of molecules. The high data thickness of DNA molecules and gigantic parallelism engaged with the DNA reactions make DNA computing an integral asset. DNA search algorithm is applied on the Boolean equation for solving routing alternatives using the properties of DNA computation. The simulated results are good and give the sign of applicability of DNA computing for solving the Routing problem.

There are numerous old-style algorithms for finding routing in FPGA. DNA computing can solve the routes proficiently and quickly. The run time complexity of DNA algorithms is considerably less than other traditional algorithms which are utilized for

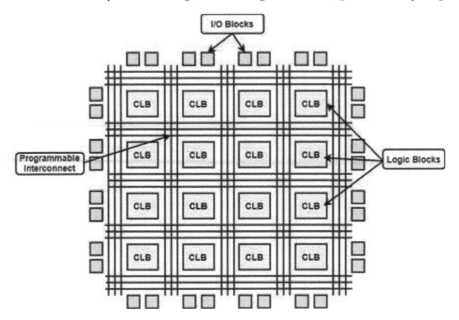

FIGURE 9.6
Circuit Diagram of FPGA

solving routing in FPGA. The high data thickness of DNA molecules and monstrous parallelism associated with the DNA reactions make DNA computing a powerful tool.

A Field Programmable Gate Array (FPGA) is a Programmable Logic Device (PLD). It is a digital integrated circuit that contains configurable logic blocks along with programmable interconnection between these blocks. It can be configured by end-users to implement specific applications. Programmable interconnections are to be had for customers or designers to carry out given functions without difficulty. A typical model FPGA chip is shown in the Figure 9.6. There are I/O blocks, which are designed and numbered consistent with function. There are CLB's (Configurable Logic Blocks) for implementing different functions.

9.4.1   General Organization of Multi-Valued Quantum FPGA

A basic multi-valued Quantum FPGA architecture consists of thousands of fundamental elements called configurable logic blocks (CLBs) surrounded by a system of programmable interconnects, called a fabric that routes signals between multi-valued Quantum CLBs. A multi-valued Quantum configurable logic block (CLB) (Figure 9.7) is the basic repeating logic block on a multi-valued Quantum FPGA.

There are hundreds of similar logic blocks available onto the multi-valued Quantum FPGA connected via routing resources. The purpose of these logic blocks is to implement combinational and sequential logic.

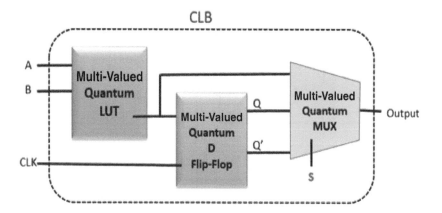

FIGURE 9.7
Configurable Logic Block of FPGA

There are three essential multi-valued Quantum CLBs components:

1. Multi-Valued Quantum Flip-Flops
2. Multi-Valued Quantum Look-up Tables (LUTs)
3. Multi-Valued Quantum Multiplexers

9.4.2 Circuit Architecture of Basic Components

Flip-Flop: The D Flip Flop is by far the most important of all the clocked flip-flops. By adding an inverter (NOT gate) between the Set and Reset inputs, the S and R inputs become complements of each other ensuring that the two inputs S and R are never equal (0 or 1) to each other at the same time allowing us to control the toggle action of the flip-flop using one single D (Data) input. Then this Data input labeled "D" is used in place of the "Set" signal, and the inverter is used to generate the complementary "Reset" input thereby making a level-sensitive D-type flip-flop from a level-sensitive SR-latch. The Quantum D Flip-Flop is shown in Figure 9.8.

Look-up Table (LUT): A LUT is the heart of the FPGA. It contains all the logically possible output of the design. The LUT is programmed by the Digital Designer to perform a Boolean algebra equation. All possible combinations of Boolean expressions need to be able to be programmed into the Look-Up Table. Figure 9.9 shows 2 Inputs Look-Up Table.

Multiplexer: The multiplexer is one of the most important block designs of a binary digital system. It is such a device that allows only one input from several input signals and the input which is selected by Multiplexer is transmitted into a single medium.

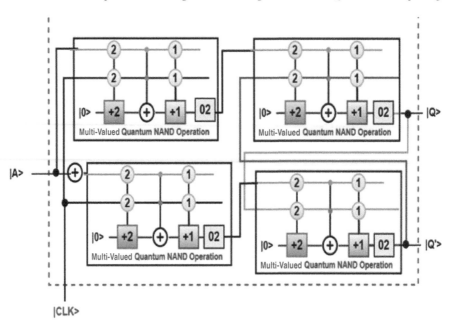

FIGURE 9.8
Quantum D Flip-Flop

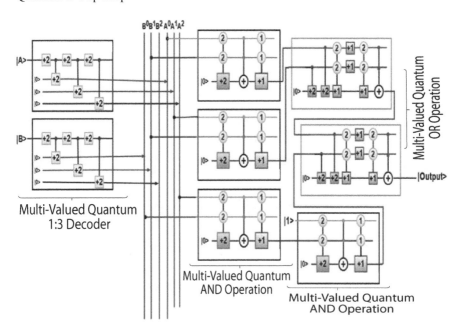

FIGURE 9.9
2 Inputs Look-Up Table

Multiplexers aid to improve the efficiency of the communication system. It allows the transmission of data such as audio, video, etc. from different channels via cables. And here, this will be implemented using MVL which helps us to simulate the circuit so easily than the digital binary multiplexer. Then the design will be proposed of DNA 1-to-3 ternary multiplexer circuit based on a combinational digital 1-to-3 ternary circuit

For performing quantum 3-to-1 multiplexer operation, firstly it is needed to operate a 1-to-3 decoder. The purpose of using this decoder is to take three inputs. As it is a ternary multiplexer, it is needed to access three inputs such as 0, 1, and 2. If 1-to-3 decoders is not used, it will not be possible to access all combinations of these three inputs. After performing the decoder operation, a single output will enter into the Mux which performs TAND (Ternary AND) operation. Figure 9.10 shows the Quantum 3-to-1 MUX .

Ternary Decoder: A ternary decoder is a combinational circuit that has 'n' input lines and a maximum of 3^n output lines. One of these outputs will be active High based on the combination of inputs present when the decoder is enabled. That means the decoder detects a particular code. The outputs of the decoder are nothing but the midterms of 'n' input variables lines when it is enabled. The following figure (Figure 9.11) shows the design of a ternary 1-to-3 decoder.

The Quantum Ternary 1-to-3 decoder operation requires the shifting operations that include (+2) shifting only. Where three (+2) operations are not controlled, and three (+2) operation is controlled by respective one input.

Cache Memory: Cache memory is a supplementary memory system that temporarily stores frequently used instructions and data for quicker processing. It is an extremely fast memory type that holds frequently requested data and instructions so that they are immediately available for further processing. In most microprocessors, Static Random-Access Memory (SRAM) is used as cache memory as SRAM has a very high speed. Due to this high speed, the cache memory can be used to store data temporarily that will be designed for the quantum computer.

Cache memory can be designed using the D flip-flops which will create a one-bit SRAM along with Read/Write and select bit as input. A Delay flip flop of D flip flop is an electronic circuit used to delay the change of state of its output signal (Q) until the next rising edge of a clock timing input signal occurs. Block Diagram of Quantum Cache Memory is shown in Figure 9.12.

9.4.3   Circuit Architecture of Multi-Valued Quantum FPGA

FPGA logic block is designed by connecting Flip-Flops, Look-up Tables (LUT) and Multiplexers. Here a simple Quantum FPGA logic block is designed by using a D Flip-Flop, a Look-up Table (LUT) and a multiplexer. A simple 2 input multi-valued quantum LUT consists of four Quantum AND, and two Quantum OR operations.

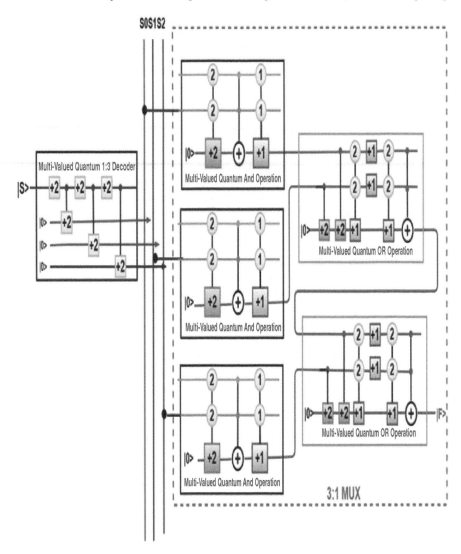

FIGURE 9.10
Quantum 3-to-1 MUX

The output of the LUT will go through the multi-valued quantum D Flip-Flop and multi-valued quantum multiplexer as input. The multi-valued quantum D Flip-Flop is a sequential circuit that consists of four multi-valued quantum NAND operations and one multi-valued quantum NOT operation. The output of the multi-valued quantum D Flip-Flop will go through the multi-valued quantum multiplexer as input. A multi-valued quantum 3-to-1 MUX is designed using three multi-valued quantum AND operations, and two multi-valued quantum OR operations. The multiplexer generates

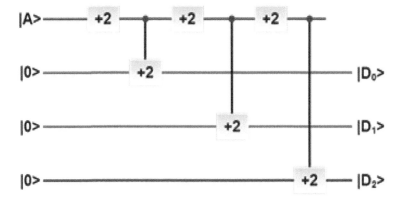

FIGURE 9.11
Quantum Ternary 1-to-3 Decoder

the desired output for the FPGA logic block. Figure 9.13 shows the Multi-Valued Quantum FPGA.

9.4.4 Working Principle

Three multi-valued Quantum Ternary 1-to-3 decoder designs for input |A>,|B>, and |S>. Multi-Valued Quantum Ternary 1-to-3 decoder operation requires the shifting operations that include (+2) shifting only. Where three (+2) operations are not controlled, and three (+2) operation is controlled by respective one input.

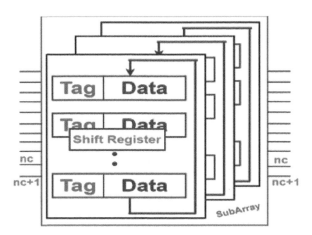

FIGURE 9.12
Block Diagram of Quantum Cache Memory

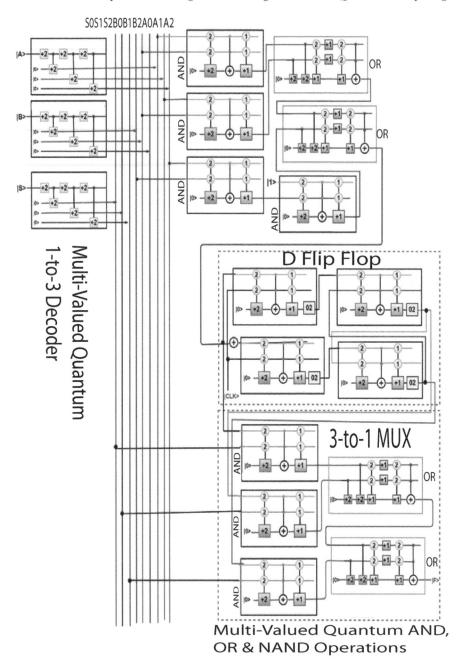

FIGURE 9.13
Multi-Valued Quantum FPGA

TABLE 9.3

Truth Table of Multi-Valued LUT

A	B	LUT output
0	0	0
0	1	2
0	2	0
1	0	0
1	1	2
1	2	0
2	0	1
2	1	0
2	2	0

The output of the decoder for input |A> and |B> first go through the Lookup Tables (LUT). The LUT use four multi-valued quantum TAND operation and two multi-valued Quantum TOR operation. The LUT implements the function (A0B1 + A1B1) + 1 (A2B0) to generate the LUT output. The outputs of the LUT are given in the Table 9.3.

2 inputs, one is the output of LUT and another is clock input will go thought the D flip-flop. The D flip-flop use one Quantum TNOT operation and four multi-valued Quantum TNAND operations. The D flip-flop transfer the LUT output same as |D>, if the CLK input sequences is |1> or |2>. If the CLK input is |0>, one of the inputs to each of the last two Quantum TNAND operations will be |2>, thus output of the D flip-flop remains unchanged regardless of the values of the LUT output.

The ternary selection input S is decoded by the ternary decoder. Then it will be transferred to the multiplexer as a selection input (S0, S1, and S2). A 3-to-1 multiplexer contains three input signals (outputs of D flip-flop (Q and Q') and LUT)) and three select lines. Each selected input line and one input signal perform ternary AND operation. For ternary AND operation, Output will be the minimum value. After this, the obtained output From TAND is connected with the ternary OR operation as an input and another two inputs of TOR will come to another Two ternary AND operations. And, finally, these three inputs give an output. For ternary OR operation, the output will always be maximum. Therefore, among the three inputs, the maximum will be the final output.

9.5 Multi-Valued Quantum Complex Programmable Logic Devices

The acronym of the CPLD is "Complex programmable logic devices", it is one kind of integrated circuit that application designers design to implement digital hardware like mobile phones. These can handle knowingly higher designs than SPLDs (simple

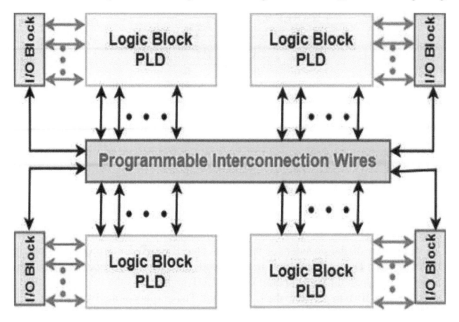

FIGURE 9.14
Block Diagram of CPLD

programmable logic devices) but offer less logic than FPGAs (field-programmable gate arrays). CPLDs include numerous logic blocks; each of the blocks includes 8-16 macrocells. Because every logic block executes a specific function, all of the macrocells in a logic block are fully connected. Depending upon the use, these blocks may or may not be connected to one another. Figure 9.14 depicts the block diagram of CPLD.

Most CPLDs (complex programmable logic devices) have macrocells with a sum of logic functions, a D FF (D flip-flop), and a MUX. Depending on the chip, the combinatorial logic function supports from 4 to 16 product terms with inclusive fan-in. CPLDs also differ in terms of shift registers and logic gates. Due to this reason, CPLDs with a huge number of logic gates may be used instead of FPGAs. Another CPLD specification signifies the number of product terms that a macrocell can accomplish. Product terms are the product of digital signals that execute a specific logic function. CPLDs are available in several IC package forms and logic families. CPLDs also differ in terms of supply voltage, operating current, standby current and power dissipation.

In terms of complexity, CPLD (complex programmable logic device) lies in between SPLD (simple programmable logic device) and FPGA and thus, inherits features from both these devices. CPLDs are more complex than SPLDs but less complex than FPGAs.

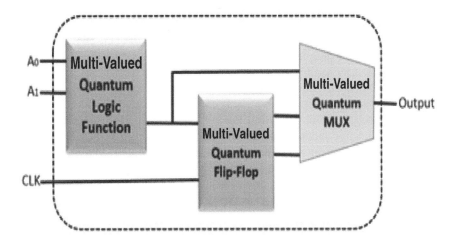

FIGURE 9.15
CPLD Functional Block

9.5.1   General Organization of Multi-Valued Quantum CPLD

A complex programmable logic device comprises a group of programmable multi-valued Quantum FBs (functional blocks). Most multi-valued Quantum CPLDs (complex programmable logic devices) have macrocells with a sum of multi-valued Quantum logic functions, a multi-valued Quantum D FF (D flip-flop), and a multi-valued Quantum MUX. Figure 9.15 shows the CPLD Functional Block.

A multi-valued Quantum functional block (FB) is the basic repeating logic block on a multi-valued Quantum CPLD. The function blocks have programmable interconnections. The programmable multi-valued Quantum FB looks like the array of logic gates, where an array of AND gates can be programmed and OR gates are stable. A switch matrix is used for function blocks to function blocks interconnections. Further, the switch matrix in a multi-valued Quantum CPLD may or may not be fully connected. The complexity of a typical PAL device is around a few hundred logic gates whereas the complexity of multi-valued Quantum CPLD is around tens of thousands of logic gates. The multi-valued Quantum CPLDs have predictable timing characteristics hence are suitable for critical control applications and other applications where a high-performance level is required. Further, due to low power consumption and low cost, multi-valued Quantum CPLDs are mostly used for battery-operated portable applications such as mobile phones, digital assistants, etc. There are three essential multi-valued Quantum FBs components:

1. Multi-Valued Quantum Flip-Flops
2. Multi-Valued Quantum Logic Function
3. Multi-Valued Quantum Multiplexers

9.5.2   Circuit Architecture of Multi-Valued Quantum CPLD

Multi-Valued quantum CPLD functional block is designed by connecting Flip-Flops, Logic function and Multiplexers. Here a simple multi-valued Quantum CPLD functional block is designed by using a multi-valued Quantum D Flip-Flop, a multi-valued Quantum logic function and a multi-valued Quantum multiplexer (Figure 9.16).

Multi-Valued quantum CPLD functional block is designed by connecting Flip-Flops, Logic function and Multiplexers. Here a simple multi-valued Quantum CPLD functional block is designed by using a multi-valued Quantum D Flip-Flop, a multi-valued Quantum logic function and a multi-valued Quantum multiplexer. A simple 2-input multi-valued quantum logic function consists of four Quantum AND, and two Quantum OR operations. The output of the logic function will go through XOR with 0 value. All the input first goes through a ternary decoder. The output of the XOR will go through the multi-valued quantum D Flip-Flop and multi-valued quantum multiplexer as input. The multi-valued quantum D Flip-Flop is a sequential circuit that consists of four multi-valued quantum NAND operations and one multi-valued quantum NOT operation. The output of the multi-valued quantum D Flip-Flop will go through the multi-valued quantum multiplexer as input. A multi-valued quantum 3-to-1 MUX is designed using three multi-valued quantum AND operations, and two multi-valued quantum OR operations. The multiplexer generates the desired output for the CPLD functional block.

9.5.3   Working Principle

Three multi-valued Quantum Ternary 1-to-3 decoder designs for input $|A>$, $|B>$ and $|S>$. Multi-Valued Quantum Ternary 1-to-3 decoder operation requires the shifting operations that include (+2) shifting only. Where three (+2) operations are not controlled, and three (+2) operation is controlled by respective one input.

The output of the decoder for input $|A>$ and $|B>$ first go through the logic function. The logic function use four multi-valued quantum TAND operation and two multi-valued Quantum TOR operation. The logic function implements the function $(A0.B1 + A1.B1) + 1 (A2.B0)$ to generate the function output. The outputs of the function are given in the table. The output of the logic function will XOR with $|0>$. Table 9.4 shows the Multi-Valued Logic Function.

2 inputs, one is the output of XOR and another is clock input will go through the D flip-flop. The D flip-flop uses one Quantum TNOT operation and four multi-valued Quantum TNAND operations. The D flip-flop transfer the XOR output the same as $|D>$, if the CLK input sequences are $|1>$ or $|2>$. If the CLK input is $|0>$, one of the inputs to each of the last two Quantum TNAND operations will be $|2>$, thus the output of the D flip-flop remains unchanged regardless of the values of the logic function output.

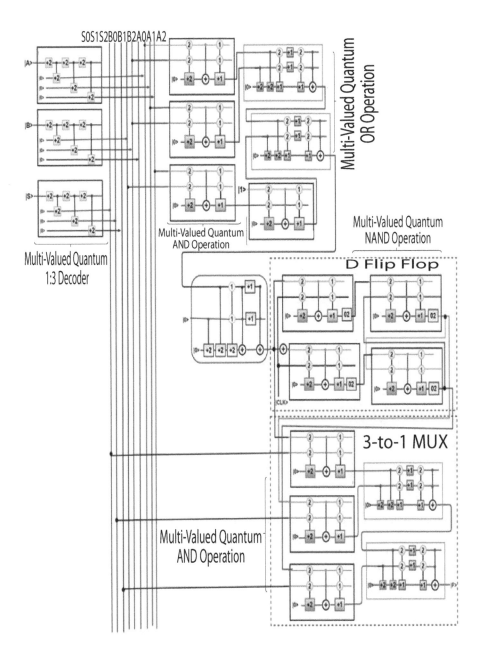

FIGURE 9.16
Multi-Valued Quantum CPLD

TABLE 9.4

Multi-Valued Logic Function

A	B	LUT output
0	0	0
0	1	2
0	2	0
1	0	0
1	1	2
1	2	0
2	0	1
2	1	0
2	2	0

The ternary selection input S is firstly decoded by the ternary decoder. Then it will be transferred to the multiplexer as a selection input (S0, S1, and S2). A 3-to-1 multiplexer contains three input signals (outputs of D flip-flop (Q and Q') and logic function) and three select lines. Each selected input line and one input signal perform ternary AND operation. For ternary AND operation, Output will be the minimum value. After this, the obtained output From TAND is connected with the ternary OR operation as an input and another two inputs of TOR will come to another Two ternary AND operations. And, finally, these three inputs give an output. For ternary OR operation, the output will always be maximum. Therefore, among the three inputs which are maximum will be the final output.

9.6 Summary

PLA is used for the implementation of various combinational circuits using Quantum AND operation and Quantum OR operation. In Quantum PLA, all the minterms are not realized but only required minterms are implemented. As Quantum PLA has a programmable Quantum AND operation array and programmable Quantum OR operation array, it provides more flexibility but the disadvantage is, it is not easy to use. Quantum PLD is used to provide control over data path and used as a counter and decoder. This chapter has presented all PLDs with their detail explanation and appropriate figures.

Bibliography

[1] Price, M. D., Havel, T. F., & Cory, D. G. (2000). Multiqubit logic gates in NMR quantum computing. New Journal of Physics, 2(1), 10.

[2] Gershenfeld, N., & Chuang, I. L. (1998). Quantum computing with molecules. Scientific American, 278(6), 66-71.

[3] Wang, X., Sørensen, A., & Mølmer, K. (2001). Multibit gates for quantum computing. Physical Review Letters, 86(17), 3907.

10

Multiple-Valued Programmable Logic Devices in DNA Computing

10.1 Introduction

DNA computing (also sometimes referred to as biomolecular computing or molecular computing) is a new computational paradigm that employs (bio) molecule manipulation to solve computational problems, at the same time exploring natural processes as computational models. Research in this area began with an experiment by Leonard Adleman, who surprised the scientific community in 1994 by using the tools of molecular biology to solve a difficult computational problem. Adleman's experiment solved an instance of the Directed Hamiltonian Path Problem solely by manipulating DNA strands. This marked the first solution of a mathematical problem by use of molecular biology. A ternary or three-valued DNA logic function is one that has two inputs that can assume three states (say 0, 1, and 2) and generates one output signal that can have one of these three states. In ternary DNA computing, two DNA sequences are used as inputs and one DNA sequence is used as output. During ternary DNA computing, sequence ACCTAG is considered as "0", sequence CAAGCT strands as "1" and TGGATC stands as "2".

In this part ternary DNA PLDs (PLA, PAL, FPGA, CPLD) are designed using Ternary AND, NAND, OR, NOR, XOR, and XNOR logic gates. For AND logic gate its output value depends on the minimum value of its inputs. Similarly, in the case of the OR logic gate, its output value depends on the maximum value of its inputs. For the XOR gate, its output value is the difference of the value of its inputs. Finally, the outputs of NAND, NOR, and XNOR logic gates become the inverted of AND, OR and XNOR logic gates. All these operations are described in earlier chapters, this chapter will discuss the multi-valued PLDs in DNA computing.

10.2 Multiple-Valued DNA Programmable Logic Array

Programmable Logic Array (PLA) is a type of PLD, which has both a programmable Quantum AND array and a programmable Quantum OR array. In

DOI: 10.1201/9781003381921-10

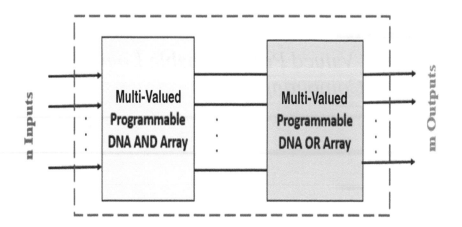

FIGURE 10.1
Block Diagram of Multi-Valued DNA PLA

a combinational circuit, because of don't care conditions, not all the minterms are used. Programmable Logic Array (PLA) is a fixed architecture logic device with programmable Quantum AND gates followed by programmable Quantum OR gates. PLA is a type of programmable logic device used to build a reconfigurable digital circuit. PLDs have an undefined function at the time of manufacturing, but they are programmed before being made into use. PLA is a combination of memory and logic.

10.2.1   General Organization of Multiple-Valued DNA PLA

Multiple-Valued DNA PLA is a programmable logic device that has both Programmable DNA AND array & Programmable DNA OR array. Hence, it is the most flexible PLD. The block diagram of Multiple-Valued DNA PLA is shown in the following figure (Figure 10.1).

Here, the inputs of DNA ternary AND operations are programmable. That means each DNA ternary AND operation has both normal and complemented inputs of variables. So, based on the requirement, any of those inputs can be programmed and can generate only the required product terms by using these DNA ternary AND operations.

Here, the inputs of DNA ternary OR operations are also programmable. So, any number of required product terms can be programmed, since all the outputs of DNA ternary AND operations are applied as inputs to each Quantum ternary OR operation. Therefore, the outputs of DNA ternary PAL will be in the form of the sum of products form.

TABLE 10.1

Truth Table of Ternary DNA PLA

A	B	F1	F2
ACCTAG	ACCTAG	ACCTAG	ACCTAG
ACCTAG	ACCTAG	TGGATC	ACCTAG
ACCTAG	TGGATC	ACCTAG	TGGATC
CAAGCT	ACCTAG	ACCTAG	ACCTAG
CAAGCT	CAAGCT	TGGATC	ACCTAG
CAAGCT	TGGATC	ACCTAG	CAAGCT
TGGATC	ACCTAG	CAAGCT	ACCTAG
TGGATC	CAAGCT	ACCTAG	CAAGCT
TGGATC	TGGATC	ACCTAG	ACCTAG

10.2.2   Circuit Architecture of Multi-Valued DNA PLA

Consider the design of ternary quantum PLA for the following functions:

$$F1 = A0.B1+A1.B1+ 1. (A2.B0)$$
$$F2 = A0.B2+ 1. (A1.B2+A2.B1)$$

Table 10.1 shows the truth table of ternary DNA PLA.
Express 0 & 1 using DNA molecule:

ACCTAG = 0

CAAGCT = 1

TGGATC = 2

The given two functions are in the sum of products form. The number of product terms present in the given Boolean functions |F1>, |F2> are three.

Eight programmable ternary Quantum AND gates & four programmable ternary Quantum OR gates are required for producing those two functions. The corresponding Quantum **PLA** is shown in the following figure (Figure 10.2).

For the given problem, there are two inputs (|A>,| B>) and two outputs (|F1>, |F2>). Quantum ternary decoders are used for these two inputs. Thus the realization has six input lines.

10.2.3   Working Principle

According to the truth table (Table 10.1) of ternary Quantum PLA, it is necessary to do the following operations to perform desired output qubits:

1.   For input Qubits A, B = ACCTAG , ACCTAG function F1 and F2 produce ACCTAG

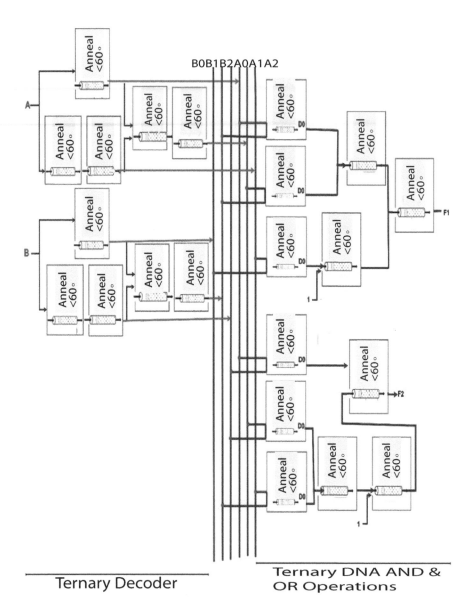

B0B1B2A0A1A2

Ternary Decoder

Ternary DNA AND &
OR Operations

FIGURE 10.2
Ternary DNA PLA

2. For input Qubits A, B = ACCTAG, CAAGCT function F1 produce TG-GATC and |F2> produce ACCTAG

3. For input Qubits A, B = ACCTAG, | TGGATC unction F1 produce ACC-TAG and F2 produces TGGATC

4. For input Qubits A, B = CAAGCT ACCTAG function F1 and F2 produce ACCTAG

5. For input Qubits A, B = CAAGCT , CAAGCT unction F1 produce TG-GATC and F2 produce ACCTAG

6. For input Qubits A, B = CAAGCT, TGGATC function F1 produce ACC-TAG and F2 produces CAAGCT

7. For input Qubits A, B = TGGATC, ACCTAG unction F1 produce CAAGCT and F2 produce ACCTAG

8. For input Qubits A, B = TGGATC, CAAGCT unction F1 produce ACC-TAG and F2 produce CAAGCT

9. For input Qubits A, B = TGGATC , TGGATC function F1 and F2 produce ACCTAG

10.3 Multiple-Valued DNA Programmable Array Logic

Programmable Array Logic (PAL) is a logic device, which has a programmable AND array and fixed OR array. It is used to realize a logic function. In this PLD, only AND gates are programmable and hence it is easier to work with PAL. The block diagram of PAL is shown in Figure 10.3.

The product terms can be programmed through the fuse link. It means the user can decide the connection between the inputs and the AND gates. If a particular

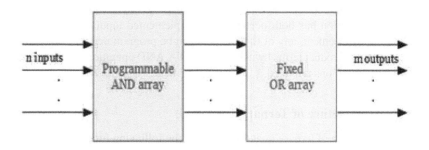

FIGURE 10.3
Block Diagram of PAL

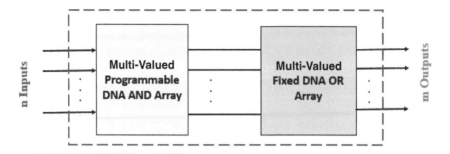

FIGURE 10.4
Block Diagram of Multi-Valued DNA PAL

input line is to be connected to the AND gate, then the fuse link must be placed at the interconnection. The AND gate outputs are then fed as an input to the fixed OR gate. Depending upon the required function, the output line of the AND gate is connected to the corresponding input of the OR gate.

10.3.1 General Organization Multiple-Valued DNA PAL

DNA Programmable Array Logic (PAL) is a type of Programmable Logic Device (PLD) used to realize a particular logical function. DNA PALs comprise a DNA AND gate array followed by a DNA OR gate array. However, it is to be noted that here only the quantum AND gate array is programmable, unlike the quantum OR gate array which has a fixed logic. This is because here the inputs are fed to the quantum AND gates through fuses which act as programmable links. Programmable-AND and fixed-OR structures of quantum PALs make them less flexible from a programming point of view when compared with Programmable Logic Arrays (PLAs). However, due to the same reason PALs are less expensive than PLAs. Hence, it is the most flexible quantum PLD. The block diagram of multi-valued DNA PAL is shown in the following figure (Figure 10.4).

Here, the inputs of DNA AND operations are programmable. That means each DNA AND operation has both normal and complemented inputs of variables. So, based on the requirement, any of those inputs can be programmed and can generate only the required product terms by using these DNA AND operations. But, the inputs of DNA OR operations are fixed.

10.3.2 Architecture of Ternary DNA PAL

Consider the design of ternary quantum PAL for the following functions:

$$F1 = A0.B1 + A1.B1 + 1.(A2.B0)$$
$$F2 = A0.B2 + A1.B1$$

TABLE 10.2
Truth Table of Ternary DNA PAL

A	B	F1	F2
ACCTAG	ACCTAG	ACCTAG	ACCTAG
ACCTAG	ACCTAG	TGGATC	ACCTAG
ACCTAG	TGGATC	ACCTAG	TGGATC
CAAGCT	ACCTAG	ACCTAG	ACCTAG
CAAGCT	CAAGCT	TGGATC	TGGATC
CAAGCT	TGGATC	ACCTAG	ACCTAG
TGGATC	ACCTAG	CAAGCT	ACCTAG
TGGATC	CAAGCT	ACCTAG	ACCTAG
TGGATC	TGGATC	ACCTAG	ACCTAG

Table 10.2 shows the truth table of ternary DNA PAL.
Express 0 & 1 using DNA molecule:

ACCTAG = 0
CAAGCT = 1
TGGATC = 2

The given two functions are in sum of products form. The number of product terms present in the given Boolean functions |F1>, |F2> are three and 2 respectively.

Seven programmable ternary DNA AND gates & four fixed ternary DNA OR gates for producing those two functions are needed. The corresponding DNA **PAL** is shown in the following figure (Figure 10.5).

For the given problem, there are two inputs (A, B) and two outputs (F1, F2). DNA ternary decoders are used for these two inputs. For these two inputs, two 3-to-1 DNA ternary decoders are used. The given first expression has three product terms and the second expression has two product terms. But as the OR operation are fixed the fuses are placed in the corresponding literal to obtain the product terms.

10.3.3 Working Principle

According to the truth table (Table 10.2) of ternary Quantum PLA , it is necessary to do the following operations to perform desired output qubits:

1. For input Qubits A, B = ACCTAG , ACCTAG function F1 and F2 produce ACCTAG

2. For input Qubits A, B = ACCTAG, CAAGCT function F1 produce TG-GATC and |F2> produce ACCTAG

3. For input Qubits A, B = ACCTAG, | TGGATC unction F1 produce ACC-TAG and F2 produces TGGATC

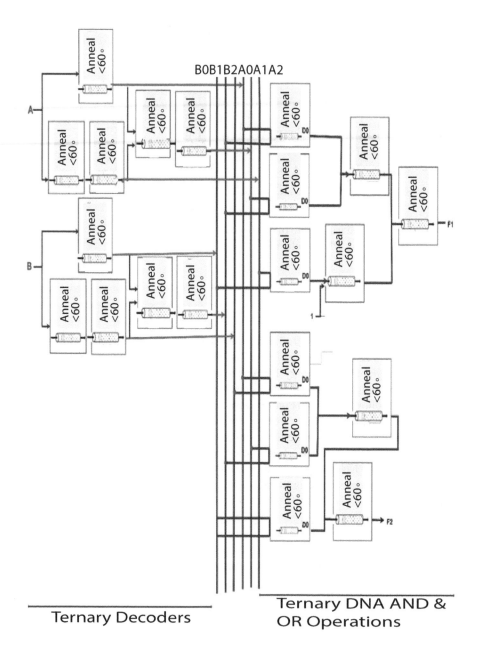

B0B1B2A0A1A2

Ternary Decoders

Ternary DNA AND &
OR Operations

FIGURE 10.5
Ternary DNA PAL

4. For input Qubits A, B = CAAGCT ACCTAG function F1 and F2 produce ACCTAG

5. For input Qubits A, B = CAAGCT , CAAGCT function F1 and F2 produce TGGATC

6. For input Qubits A, B = CAAGCT, TGGATC function F1 produce ACCTAG and F2 produce ACCTAG

7. For input Qubits A, B = TGGATC, ACCTAG unction F1 produce CAAGCT and F2 produce ACCTAG

8. For input Qubits A, B = TGGATC, CAAGCT function F1 and F2 produce ACCTAG

9. For input Qubits A, B = TGGATC , TGGATC function F1 and F2 produce ACCTAG

10.4 Multi-Valued DNA Field Programmable Gate Array

A ternary logic function has inputs that can assume three states (say 0, 1, and 2) and generates one output signal that can have one of these three states. In ternary DNA computing, two DNA sequences are used as inputs and one DNA sequence is used as output. During ternary DNA computing, sequence **ACCTAG** is considered as "0", sequence **CAAGCT** strands as "1" and **TGGATC** stands as "2". In ternary DNA computing, the fluorescent level is used to detect the DNA sequence. Fluorescence is the temporary absorption of electromagnetic wavelengths from the visible light spectrum by fluorescent molecules, and the subsequent emission of light at a lower energy level.

10.4.1 Architecture of Basic Components

Flip-Flop: The D Flip Flop is by far the most important of all the clocked flip-flops. By adding an inverter (NOT gate) between the Set and Reset inputs, the S and R inputs become complements of each other ensuring that the two inputs S and R are never equal (0 or 1) to each other at the same time allowing us to control the toggle action of the flip-flop using one single D (Data) input. Then this Data input is labeled "D", which is used in place of the "Set" signal, and the inverter is used to generate the complementary "Reset" input thereby making a level-sensitive D-type flip-flop from a level-sensitive SR-latch. Figure 10.6 shows the DNA D Flip-Flop.

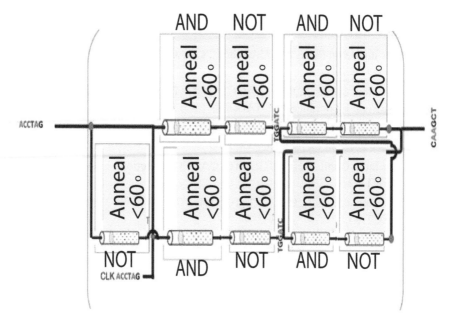

FIGURE 10.6
DNA D Flip-Flop

Look-up Table (LUT): A LUT is the heart of the FPGA. It contains all the logically possible output of the design. The LUT is programmed by the Digital Designer to perform a Boolean algebra equation. All possible combinations of Boolean expressions need to be able to be programmed into the Look-Up Table. The circuit diagram of 2 Input Look-Up Table is shown in Figure 10.7.

Multiplexer: A multiplexer (MUX) is a device that can receive multiple input signals and synthesize a single output signal in a recoverable manner for each input signal. It is also an integrated system that usually contains a certain number of data inputs and a single output. Figure 10.8 shows the DNA 3-to-1 MUX.

For performing DNA 3-to-1 multiplexer operation, firstly it is needed to operate a 1-to-3 decoder. The purpose of using this decoder is to take three inputs. As it is a ternary multiplexer, then access three inputs such as 0, 1, and 2. If 1-to-3 decoders are not used, it will not be possible to access all combinations of these three inputs. After performing decoder operation, a single output will enter into the Mux which performs TAND (Ternary AND) operation.

Ternary Decoder: Decoder in the circuit generates unary functions for input variable x as x0, x1, and x2 which is used for ternary function implementation. Table 10.3 shows the truth table of Ternary Decoder.

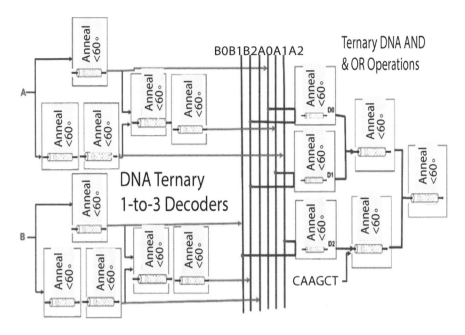

FIGURE 10.7
2 Input Look-Up Table

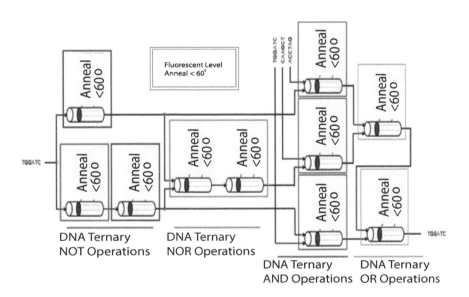

FIGURE 10.8
DNA 3-to-1 MUX

TABLE 10.3

Truth Table of Ternary Decoder

A	A2	A1	A0
0	0	0	2
1	0	2	0
2	2	0	0

Figure 10.9 depicts the circuit architecture of the DNA Ternary Decoder.

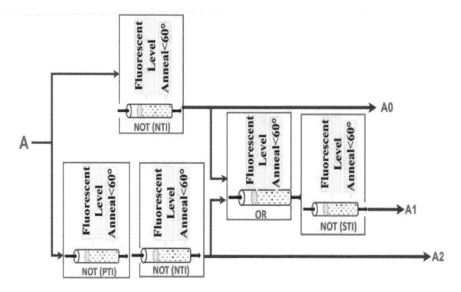

FIGURE 10.9

DNA Ternary Decoder

Any input A,

1. First, go through NOT (NTI) to produce A0.

2. Then again same input goes through NOT (PTI) and NOT (NTI) to produce A2.

3. Finally, the result of steps 1 and 2 will go through NOR operation to produce A1.

10.4.2   General Organization of Multi-Valued DNA FPGA

A ternary or three-valued logic function is one that has two inputs that can assume three states (say 0, 1, and 2) and generates one output signal that can have one of these three states. In ternary DNA computing two DNA sequence are used as inputs and one DNA sequence is used as output. During ternary DNA computing, sequence

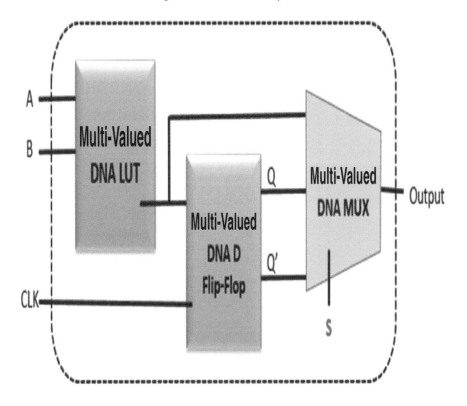

FIGURE 10.10
Block Diagram of DNA FPGA

ACCTAG is considered as "0", sequence **CAAGCT** strands as "1" and **TGGATC** stands as "2". In this Ternary DNA computing, the fluorescent level is used to detect the DNA sequence. Fluorescence is the temporary absorption of electromagnetic wavelengths from the visible light spectrum by fluorescent molecules, and the subsequent emission of light at a lower energy level. The block diagram of DNA FPGA is shown in Figure 10.10.

10.4.3   Circuit Architecture of Multi-Valued DNA FPGA

Ternary FPGA logic block is designed by connecting Flip-Flops, Look-up Tables (LUT) and Multiplexers. Here a simple multi-valued FPGA logic block is designed by using a ternary D Flip-Flop, a ternary Look-up Table (LUT), and a 3-to-1 multiplexer. Figure 10.11 depicts the Ternary DNA FPGA.

The ternary input A and B signals are firstly decoded by the ternary decoder. Then they will be transferred to the LUT. A simple 2 input LUT is designed using four AND operations and two OR operations. The output equations is = (A0.B1 +

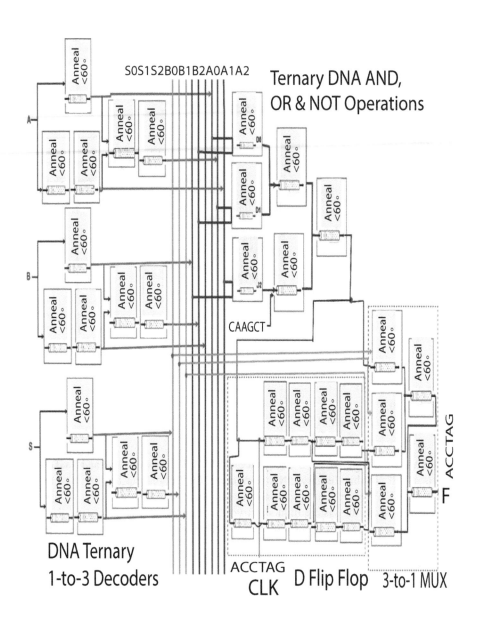

FIGURE 10.11
Ternary DNA FPGA

A1.B1) + 1. (A2.B0). Note that the output of the decoder has only two logic values, i.e., "2" and "0", corresponding to logic "1" and "0" in binary logic. The output of the LUT is used as input in the D Flip-Flop and multiplexer. Ternary D Flip-Flop is a sequential circuit that consists of four ternary NAND operations and one NOT operation. Two Output (Q and Q') of the D Flip-Flop are used as input in the multiplexer. The ternary selection input S is firstly decoded by the ternary decoder. Then it will be transferred to the multiplexer as a selection input. A 3-to-1 MUX is designed using three ternary AND operations and two ternary OR operations. The multiplexer generates the desired output for the FPGA logic block. In ternary AND operation, output value depends on the minimum value of its inputs. Similarly, in the case of ternary OR operation, output value depends on the maximum value of its inputs. Finally, the inverted ternary AND operation generate the output of the ternary NAND operation. The fluorescent level is used here to detect the sequences and logical value of the sequences.

10.4.4 Working Procedure

A two-input ternary Look-up Tables (LUT) is used to design the FPGA control logic block. Input A and B firstly decoded by the ternary decoder and generated the output lines A0, A1, A2 for input A, and B0, B1, B2 for input B. These outputs will go through the Lookup Tables (LUT). The LUT use four ternary AND operation and two OR operation to implement the function (A0.B1 + A1.B1) + 1. (A2.B0). During ternary DNA computing, sequence ACCTAG is considered as "0", sequence CAAGCT strands as "1" and TGGATC stands as "2". In ternary AND operation, output value depends on the minimum value of its inputs and in ternary OR operation, output value depends on the maximum value of its inputs.). The output of the decoder has only two sequences, "TGGATC (2) and "ACCTAG (0)".

1. When the input sequences A, B = ACCTAG (0), ternary decoder generate A0, A1, A2 = (2, 0, 0) and B0, B1, B2 = (2, 0, 0). None of the ternary AND operations (D0-D2) are connected to A0 and B0 lines. As a result, none of the ternary AND operation values will be true. So the output of the LUT will be ACCTAG (0).

2. When the input sequences A = ACCTAG (0) and B = CAAGCT (1), ternary decoder generate A0, A1, A2 = (2, 0, 0) and B0, B1, B2 = (0, 2, 0). D0 ternary AND operation is connected to A0 and B1 lines. As a result, the D0 ternary AND operation value will be true (TGGATC). So the output of the LUT will be TGGATC.

3. When the input sequences A = ACCTAG (0) and B = TGGATC (2), ternary decoder generate A0, A1, A2 = (2, 0, 0) and B0, B1, B2 = (0, 0, 2). None of the ternary AND operations (D0-D2) are connected to A0 and B2 lines. As a result, none of the ternary AND operation values will be true. So the output of the LUT will be ACCTAG (0).

4. When the input sequences A = CAAGCT (1) and B = ACCTAG (0), ternary decoder generate A0, A1, A2 = (0, 2, 0) and B0, B1, B2 = (2,

0, 0). None of the ternary AND operations (D0-D2) are connected to A1 and B0 lines. As a result, none of the ternary AND operation values will be true. So the output of the LUT will be ACCTAG (0).

5. When the input sequences A, B = CAAGCT (1), ternary decoder generate A0, A1, A2 = (0, 2, 0) and B0, B1, B2 = (0, 2, 0). D1 ternary AND operation is connected to A1 and B1 lines. As a result D1 ternary AND operation value will be true (TGGATC). So the output of the LUT will be TGGATC.

6. When the input sequences A = CAAGCT (1), and B = TGGATC (2), ternary decoder generate A0, A1, A2 = (0, 2, 0) and B0, B1, B2 = (0, 0, 2).). None of the ternary AND operations (D0-D2) are connected to A1 and B2 lines. As a result, none of the ternary AND operation value will be true. So the output of the LUT will be ACCTAG (0).

7. When the input sequences A = TGGATC (2), and B = ACCTAG (0), ternary decoder generate A0, A1, A2 = (0, 0, 2) and B0, B1, B2 = (2, 0, 0). D2 ternary AND operation is connected to A2 and B0 lines. As a result D ternary AND operation value will be true (TGGATC). Now it will proceed with ternary AND operation with CAAGCT (1) and produce minimum as CAAGCT as LUT output.

8. When the input sequences A = TGGATC (2), and B = CAAGCT (1), ternary decoder generate A0, A1, A2 = (0, 0, 2) and B0, B1, B2 = (0, 2, 0).). None of the ternary AND operations (D0-D2) are connected to A2 and B1 lines. As a result, none of the ternary AND operation values will be true. So the output of the LUT will be ACCTAG (0).

9. When the input sequences A, B = TGGATC (2), ternary decoder generate A0, A1, A2 = (0, 0, 2) and B0, B1, B2 = (0, 0, 2).). None of the ternary AND operations (D0-D2) are connected to A2 and B2 lines. As a result, none of the ternary AND operation values will be true. So the output of the LUT will be ACCTAG (0).

2 input, one is the output of LUT and another is clock input will go through the ternary D flip-flop. The ternary D flip-flop use one NOT gate four NAND gates.

1. If the LUT output and CLK input both are ACCTAG (0), then the output of the ternary D flip-flop will remain unchanged.

2. If the LUT output is ACCTAG (0) and CLK input is CAAGCT (1), then the output of the D flip-flop will be ACCTAG (0).

3. If the LUT output is ACCTAG (0) and CLK input is TGGATC (2), then the output of the D flip-flop will be ACCTAG (0).

4. If the LUT output is CAAGCT (1) and CLK input is ACCTAG (0), then the output of the D flip-flop will remain unchanged.

5. If the LUT output and CLK input both are CAAGCT (1), then the output of the D flip-flop will be CAAGCT (1).

6. If the LUT output is CAAGCT (1) and CLK input is TGGATC (2), then the output of the D flip-flop will be CAAGCT (1).

7. If the LUT output is TGGATC (2) and CLK input is ACCTAG (0), then the output of the D flip-flop will remain unchanged.

8. If the LUT output is TGGATC (2) and CLK input is CAAGCT (1), then the output of the D flip-flop will be CAAGCT (1).

9. If the LUT output and CLK input both are TGGATC (2), then the output of the D flip-flop will be TGGATC (2).

The ternary selection input S is firstly decoded by the ternary decoder. Then it will be transferred to the multiplexer as a selection input (S0, S1, and S2). A 3-to-1 multiplexer contains three input signals (outputs of D flip-flop (Q and Q') and LUT)) and three select lines. Each selected input line and one input signal perform ternary AND operation. For ternary AND operation, Output will be the minimum value. After this, the obtained output From TAND is connected with the ternary OR operation as an input and another two inputs of TOR will come to another Two ternary AND operations. And, finally, these three inputs give an output. For ternary OR operation, the output will always be maximum. Therefore, among the three inputs which are maximum will be the final output.

Among three ternary AND operations, S0 and LUT output are connected to the first ternary AND operation. It produces a true sequence when both the input sequence are true TGGATC (2). S1 and output of D flip-flop (Q) are connected to the second ternary AND operation. It produces a true sequence when both the input sequence (S1 and Q) are true TGGATC (2). S2 and output of D flip-flop (Q') are connected to the third ternary AND operation. It produces a true sequence when both the input sequence (S1 and Q,) are true TGGATC (2). The output of these three ternary AND operations will go through the ternary OR operation to generate the FPGA logical block output.

1. If all three input sequences (output of three ternary AND operations) are ACCTAG (0), then the output of the ternary OR operation is ACCTAG (0).

2. If one of the three input sequences is TGGATC (2), then the output of the ternary OR operation is TGGATC (2).

3. If input sequences are in ACCTAG (0), and CAAGCT (2) then the output of the ternary OR operation is CAAGCT (1).

4. If input sequences are in ACCTAG (0), and TGGATC (2) then the output of the ternary OR operation is TGGATC (2).

5. If input sequences are in CAAGCT (1), and TGGATC (2) then the output of the ternary OR operation is TGGATC (2).

10.5   Multi-Valued DNA Complex Programmable Logic Devices

A ternary logic function has inputs that can assume three states (say 0, 1, and 2) and generates one output signal that can have one of these three states. In ternary DNA computing, two DNA sequences are used as inputs and one DNA sequence is used as output. During ternary DNA computing, sequence **ACCTAG** is considered as "0", sequence **CAAGCT** strands as "1" and **TGGATC** stands as "2". In ternary DNA computing, the fluorescent level is used to detect the DNA sequence. Fluorescence is the temporary absorption of electromagnetic wavelengths from the visible light spectrum by fluorescent molecules, and the subsequent emission of light at a lower energy level.

The acronym of the CPLD is "Complex programmable logic devices", it is one kind of integrated circuit that application designers design to implement digital hardware like mobile phones. These can handle knowingly higher designs than SPLDs (simple programmable logic devices), but offer less logic than FPGAs (field-programmable gate arrays). CPLDs include numerous logic blocks; each of the blocks includes 8-16 macrocells. Because every logic block executes a specific function, all of the macrocells in a logic block are fully connected. Depending upon the use, these blocks may or may not be connected to one another.

10.5.1   General Organization of Multi-Valued DNA CPLD

A multi-valued DNA complex programmable logic device comprises a group of programmable multi-valued DNA FBs (functional blocks). Most multi-valued DNA CPLDs (complex programmable logic devices) have macrocells with a sum of multi-valued DNA logic functions, a multi-valued DNA D FF (D flip-flop) and a multi-valued DNA MUX as shown in Figure 10.12.

A multi-valued DNA functional block (FB) is the basic repeating logic block on a multi-valued DNA CPLD. The function blocks have programmable interconnections. The programmable FB looks like the array of logic gates, where an array of AND gates can be programmed and OR gates are stable. A switch matrix is used for function blocks to function blocks interconnections. Further, the switch matrix in a multi-valued DNA CPLD may or may not be fully connected. The complexity of a typical multi-valued DNA PAL device is around a few hundred logic gates whereas the complexity of multi-valued DNA CPLD is around tens of thousands of logic gates. The multi-valued DNA CPLDs have predictable timing characteristics hence are suitable for critical control applications and other applications where a high-performance level is required. Further, due to low power consumption and low cost, multi-valued DNA CPLDs are mostly used for battery-operated portable applications such as mobile phones, digital assistants, etc.

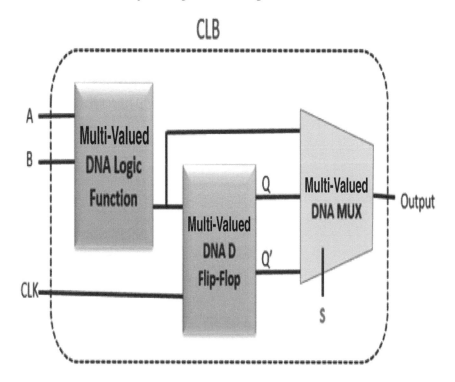

FIGURE 10.12
Functional Block of Multi-Valued DNA CPLD

10.5.2 Circuit Architecture of Multi-Valued DNA CPLD

Ternary CPLD functional block is designed by connecting Flip-Flops, Logic function and Multiplexers. Here a simple multi-valued CPLD functional block is designed by using a ternary D Flip-Flop, a ternary logic function and a 3-to-1 multiplexer. The ternary input A and B signals are firstly decoded by the ternary decoder. Then they will be transferred to the logic function. A simple 2 input logic function is designed using four AND operations and two OR operations. Figure 10.13 depicts the Ternary DNA CPLD.

The output equations is = (A0.B1 + A1.B1) + 1. (A2.B0). Note that the output of the decoder has only two logic values, i.e., "2" and "0", corresponding to logic "1" and "0" in binary logic. The output of the logic function will go through an XOR operation with the value ACCTAG (0). XOR output is used as input in the D Flip-Flop and multiplexer. Ternary D Flip-Flop is a sequential circuit that consists of four ternary NAND operations and one NOT operation. Two Output (Q and Q') of the D Flip-Flop are used as input in the multiplexer. The ternary selection input S is firstly decoded by the ternary decoder. Then it will be transferred to the multiplexer

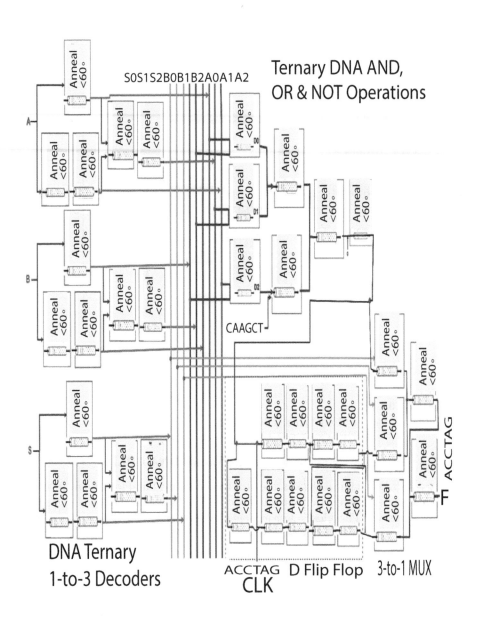

FIGURE 10.13
Ternary DNA CPLD

as a selection input. A 3-to-1 MUX is designed using three ternary AND operations and two ternary OR operations. The multiplexer generates the desired output for the CPLD functional block. In ternary AND operation, output value depends on the minimum value of its inputs. Similarly, in the case of ternary OR operation, output value depends on the maximum value of its inputs. Finally, the inverted ternary AND operation generate the output of the ternary NAND operation. The fluorescent level is used here to detect the sequences and logical value of the sequences.

10.5.3 Working Procedure

A two-input ternary, the output of the logic function is used to design the CPLD functional block. Input A and B firstly decoded by the ternary decoder and generated the output lines A0, A1, A2 for input A, and B0, B1, B2 for input B. These outputs will go through the output of the logic function. The output of the logic function use four ternary AND operation and two OR operation to implement the function (A0.B1 + A1.B1) + 1. (A2.B0). During ternary DNA computing, sequence ACCTAG is considered as "0", sequence CAAGCT strands as "1" and TGGATC stands as "2". In ternary AND operation, output value depends on the minimum value of its inputs and in ternary OR operation, output value depends on the maximum value of its inputs. The output of the decoder has only two sequences, "TGGATC (2)" and "ACCTAG (0)".

1. When the input sequences A, B = ACCTAG (0), ternary decoder generate A0, A1, A2 = (2, 0, 0) and B0, B1, B2 = (2, 0, 0). None of the ternary AND operations (D0-D2) are connected to A0 and B0 lines. As a result, none of the ternary AND operation values will be true. So the output of the logic function will be ACCTAG (0).

2. When the input sequences A = ACCTAG (0) and B = CAAGCT (1), ternary decoder generate A0, A1, A2 = (2, 0, 0) and B0, B1, B2 = (0, 2, 0). D0 ternary AND operation is connected to A0 and B1 line. As a result D0 ternary AND operation value will be true (TGGATC). So the output of the logic function will be TGGATC.

3. When the input sequences A = ACCTAG (0) and B = TGGATC (2), ternary decoder generate A0, A1, A2 = (2, 0, 0) and B0, B1, B2 = (0, 0, 2). None of the ternary AND operations (D0-D2) are connected to A0 and B2 lines. As a result, none of the ternary AND operation values will be true. So the output of the logic function will be ACCTAG (0).

4. When the input sequences A = CAAGCT (1) and B = ACCTAG (0), ternary decoder generate A0, A1, A2 = (0, 2, 0) and B0, B1, B2 = (2, 0, 0). None of the ternary AND operations (D0-D2) are connected to A1 and B0 lines. As a result, none of the ternary AND operation values will be true. So the output of the logic function will be ACCTAG (0).

5. When the input sequences A, B = CAAGCT (1), ternary decoder generate A0, A1, A2 = (0, 2, 0) and B0, B1, B2 = (0, 2, 0). D1 ternary AND operation are connected to A1 and B1 lines. As a result D1 ternary AND

operation value will be true (TGGATC). So the output of the logic function will be TGGATC.

6. When the input sequences A = CAAGCT (1), and B = TGGATC (2), ternary decoder generate A0, A1, A2 = (0, 2, 0) and B0, B1, B2 = (0, 0, 2).). None of the ternary AND operations (D0-D2) are connected to A1 and B2 lines. As a result, none of the ternary AND operation values will be true. So the output of the logic function will be ACCTAG (0).

7. When the input sequences A = TGGATC (2), and B = ACCTAG (0), ternary decoder generate A0, A1, A2 = (0, 0, 2) and B0, B1, B2 = (2, 0, 0). D2 ternary AND operation are connected to A2 and B0 lines. As a result D ternary AND operation value will be true (TGGATC). Now it will proceed with ternary AND operation with CAAGCT (1) and produce minimum as CAAGCT as logic function output.

8. When the input sequences A = TGGATC (2), and B = CAAGCT (1), ternary decoder generate A0, A1, A2 = (0, 0, 2) and B0, B1, B2 = (0, 2, 0).). None of the ternary AND operations (D0-D2) are connected to A2 and B1 lines. As a result, none of the ternary AND operation values will be true. So the output of the logic function will be ACCTAG (0).

9. When the input sequences A, B = TGGATC (2), ternary decoder generate A0, A1, A2 = (0, 0, 2) and B0, B1, B2 = (0, 0, 2).). None of the ternary AND operations (D0-D2) are connected to A2 and B2 lines. As a result, none of the ternary AND operation values will be true. So the output of the logic function will be ACCTAG (0).

The output of the logic function will XOR with ACCTAG (0). 2 inputs, one is the output of XOR and another is clock input will go through the ternary D flip-flop. The ternary D flip-flop use one NOT gate four NAND gates.

1. If the XOR output and CLK input both are ACCTAG (0), then the output of the ternary D flip-flop will remain unchanged.

2. If the XOR output is ACCTAG (0) and CLK input is CAAGCT (1), then the output of the D flip-flop will be ACCTAG (0).

3. If the XOR output is ACCTAG (0) and CLK input is TGGATC (2), then the output of the D flip-flop will be ACCTAG (0).

4. If the XOR output is CAAGCT (1) and CLK input is ACCTAG (0), then the output of the D flip-flop will remain unchanged.

5. If the XOR output and CLK input both are CAAGCT (1), then the output of the D flip-flop will be CAAGCT (1).

6. If the XOR output is CAAGCT (1) and CLK input is TGGATC (2), then the output of the D flip-flop will be CAAGCT (1).

7. If the XOR output is TGGATC (2) and CLK input is ACCTAG (0), then the output of the D flip-flop will remain unchanged.

8. If the XOR output is TGGATC (2) and CLK input is CAAGCT (1), then the output of the D flip-flop will be CAAGCT (1).

9. If the XOR output and CLK input both are TGGATC (2), then the output of the D flip-flop will be TGGATC (2).

The ternary selection input S is firstly decoded by the ternary decoder. Then it will be transferred to the multiplexer as a selection input (S0, S1, and S2). A 3-to-1 multiplexer contains three input signals (outputs of D flip-flop (Q and Q') and logic function)) and three select lines. Each selected input line and one input signal perform ternary AND operation. For ternary AND operation, Output will be the minimum value. After this, the obtained output From TAND is connected with the ternary OR operation as an input and another two inputs of TOR will come to another Two ternary AND operations. And, finally, these three inputs give an output. For ternary OR operation, the output will always be maximum. Therefore, among the three inputs which are maximum will be the final output.

Among three ternary AND operations, S0 and logic function output are connected to the first ternary AND operation. It produces a true sequence when both the input sequence are true TGGATC (2). S1 and output of D flip-flop (Q) are connected to the second ternary AND operation. It produces a true sequence when both the input sequence (S1 and Q) are true TGGATC (2). S2 and output of D flip-flop (Q') are connected to the third ternary AND operation. It produces a true sequence when both the input sequence (S1 and Q,) are true TGGATC (2). The output of these three ternary AND operations will go through the ternary OR operation to generate the CPLD functional block output.

1. If all three input sequences (output of three ternary AND operations) are ACCTAG (0), then the output of the ternary OR operation is ACCTAG (0).

2. If one of the three input sequences is TGGATC (2), then the output of the ternary OR operation is TGGATC (2).

3. If input sequences are in ACCTAG (0), and CAAGCT (1) then the output of the ternary OR operation is CAAGCT (2).

4. If input sequences are in ACCTAG (0), and TGGATC (2) then the output of the ternary OR operation is TGGATC (2).

5. If input sequences are in CAAGCT (1), and TGGATC (2) then the output of the ternary OR operation is TGGATC (2).

10.6 Summary

In this Ternary DNA computing, the fluorescent level is used to detect the DNA sequence. Fluorescence is the temporary absorption of electromagnetic wavelengths from the visible light spectrum by fluorescent molecules, and the subsequent emission of light at a lower energy level. When it occurs in a living organism, it is sometimes called bio-fluorescence. This causes the light that is emitted to be a different color than the light that is absorbed. Stimulating light excites an electron, raising energy to an unstable level. This chapter has presented the multi-valued programmable logic devices in DNA computing. All necessary figures are also shown.

Bibliography

[1] Ishdorj, T. O., & Ionescu, M. (2004, September). Replicative–distribution rules in P systems with active membranes. In International Colloquium on Theoretical Aspects of Computing (pp. 68-83). Springer, Berlin, Heidelberg.

[2] Daley, M., Kari, L., Gloor, G., & Siromoney, R. (1999, September). Circular contextual insertions/deletions with applications to biomolecular computation. In 6th International Symposium on String Processing and Information Retrieval. 5th International Workshop on Groupware (Cat. No. PR00268) (pp. 47-54). IEEE.

[3] Lin, S., Kim, Y. B., & Lombardi, F. (2009). CNTFET-based design of ternary logic gates and arithmetic circuits. IEEE Transactions on Nanotechnology, 10(2), 217-225.

[4] Heung, A., & Mouftah, H. T. (1985). Depletion/enhancement CMOS for a lower power family of three-valued logic circuits. IEEE Journal of Solid-State Circuits, 20(2), 609-616.

[5] Jia, J., Wang, Y., Liu, J., & Jin, G. (2014, July). Effective CGH calculation algorithm with low memory usage using compressed look-up table based on separation of light modulation variable. In Laser Applications to Chemical, Security and Environmental Analysis (pp. JTu4A-25). Optica Publishing Group.

11

Multiple-Valued Programmable Logic Devices in Quantum-DNA Computing

11.1 Introduction

Ternary Quantum-DNA PLDs are designed by combining ternary Quantum and DNA computing. Multiple-valued Quantum-DNA PLDs are programmable logic devices that have both Programmable Quantum AND array & Programmable DNA OR array. Hence, it is the most flexible PLD. Here, the inputs of Multiple-valued Quantum AND operations are programmable. That means each Multiple-valued Quantum AND operation has both normal and complemented inputs of variables. So, based on the requirement, any of those inputs can be programmed and can generate only the required product terms by using these Multiple-valued Quantum-AND operations.

Quantum computing provides fast computation in logic devices than DNA computing. So, it needs to use a storage device (CACHE MEMORY) for making a balance between these two computing systems. DNA computing gives more storage capacity, as a result, the output operation of the logic devices will perform in DNA computing. It is needed to transform qubit into DNA molecule and for this, a transformation process (NMR relaxation) is used by which an excited magnetic state returns to its equilibrium distribution. Here, the outputs operations of Multiple-valued DNA are also programmable. So, it is possible to program any number of required product terms, since all the outputs of Multiple-valued quantum operations are applied as inputs to each DNA OR operation. Therefore, the outputs of Multiple-valued DNA PLDs will be in the form of the sum of products form.

11.2 Multiple-Valued Quantum-DNA Programmable Logic Array

Programmable Logic Array (PLA) is a type of PLD, which has both programmable Quantum AND array and programmable Quantum OR array. In a combinational circuit, because of don't care conditions, not all the minterms are used. Programmable Logic Array (PLA) is a fixed architecture logic device with programmable Quantum

DOI: 10.1201/9781003381921-11

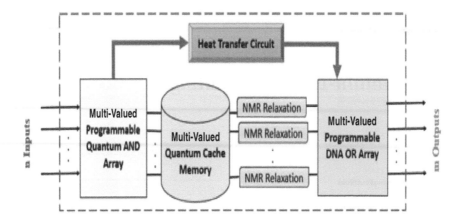

FIGURE 11.1
Block Diagram of Multiple-Valued Quantum-DNA PLA

AND gates followed by programmable Quantum OR gates. PLA is a type of programmable logic device used to build a reconfigurable digital circuit. PLDs have an undefined function at the time of manufacturing, but they are programmed before being made into use. PLA is a combination of memory and logic.

11.2.1   General Organization of Multi-Valued Quantum-DNA PLA

Multiple-valued Quantum-DNA PLA is a programmable logic device that has both Programmable Quantum AND array & Programmable DNA OR array. Hence, it is the most flexible PLD. The block diagram of Multiple-valued Quantum-DNA PLA is shown in the following figure (Figure 11.1).

Here, the inputs of Multiple-valued Quantum AND operations are programmable. That means each Multiple-valued Quantum AND operation has both normal and complemented inputs of variables. So, based on the requirement, any of those inputs can be programmed and can generate only the required product terms by using these Multiple-valued Quantum AND operations.

Quantum computing provides fast computation in logic devices than DNA computing. So, it is needed to use a storage device (CACHE MEMORY) for making a balance between these two computing systems. DNA computing gives more storage capacity, as a result, the output operation of the logic devices will perform in DNA computing. It is needed to transform qubit into DNA molecule by using a transformation process by which an excited magnetic state returns to its equilibrium distribution. Here, the inputs of Multiple-valued DNA OR operations are also programmable. So, it is possible to program any number of required product terms, since all the outputs of Multiple-valued quantum AND operations are applied as inputs to each DNA OR operation. Therefore, the outputs of Multiple-valued DNA PAL will be in the form of

TABLE 11.1
Truth Table of Ternary Quantum-DNA PLA

|A>	|B>	|F1>	|F2>
|0>	|0>	ACCTAG	ACCTAG
|0>	|0>	TGGATC	ACCTAG
|0>	|2>	ACCTAG	TGGATC
|1>	|0>	ACCTAG	ACCTAG
|1>	|1>	TGGATC	ACCTAG
|1>	|2>	ACCTAG	CAAGCT
|2>	|0>	CAAGCT	ACCTAG
|2>	|1>	ACCTAG	CAAGCT
|2>	|2>	ACCTAG	ACCTAG

a sum of products form. Quantum Computing produces more heat in circuits which can be used in DNA computing to perform DNA logic operations. For this, a heat transfer circuit is used between Quantum and DNA operations. The Multiple-valued Quantum-DNA PLA implementation methods will improve the result of any logic device.

11.2.2 Circuit Architecture of Multiple-valued Quantum-DNA PLA

Multiple-valued Quantum-DNA PLA is a programmable logic device that has both Programmable multiple-valued Quantum AND array & Programmable multiple-valued DNA OR array.
The following **Boolean functions** use multiple-valued Quantum-DNA PLA.

$$F1 = A0.B1 + A1.B1 + 1. (A2.B0)$$
$$F2 = A0.B2 + 1. (A1.B2 + A2.B1)$$

Table 11.1 shows the truth table of Ternary Quantum-DNA PLA.

Eight programmable ternary Quantum AND gates & four programmable ternary DNA OR gates are needed for producing those two functions. The corresponding Quantum-DNA **PLA** is shown in Figure 11.2.

For the given problem, there are two inputs (A, B) and two outputs (F1, F2). Eight multiple-valued quantum AND operations are designed using quantum computing. Quantum computing provides fast computation in logic devices than DNA computing. So, a cache memory is used to store the quantum qubits. It is needed to transform qubit into DNA molecule by using NMR-Relaxation by which an excited magnetic state returns to its equilibrium distribution. Here, the inputs of multiple-valued DNA OR operations are also programmable. So, it is possible to program any number of required product terms, since all the outputs of multiple-valued Quantum AND operations are applied as inputs to each multiple-valued DNA OR operation.

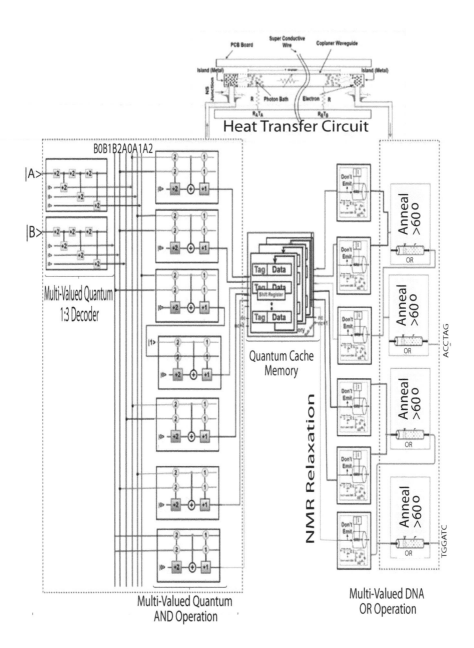

FIGURE 11.2
Multiple-Valued Quantum-DNA PLA

Therefore, the outputs of multiple-valued Quantum-DNA PAL will be in the form of the sum of products form. For the given problem, there are two inputs (|A>,| B>) and two outputs (|F1>, |F2>). Quantum ternary decoders are used for these two inputs. Thus the realization has six input lines.

11.2.3 Working Principle

According to truth table (Table 11.1) of ternary Quantum-DNA PLA, it is necessary to do the following operations to perform desire output qubits:

1. For input Qubits |A>, |B> =|0>, |0> function F1 and F2 produce ACC-TAG

2. For input Qubits |A>, |B> =|0>, |1> function F1 produce TGGATC and F2 produce ACCTAG

3. For input Qubits |A>, |B> =|0>, |2> function F1 produce ACCTAG and F2 produce TGGATC

4. For input Qubits |A>, |B> =|1>, |0> function F1 and F2 produce ACC-TAG

5. For input Qubits |A>, |B> =|1>, |1> function F1 produce TGGATC and F2 produce ACCTAG

6. For input Qubits |A>, |B> =|1>, |2> function F1 produce ACCTAG and F2 produce CAAGCT

7. For input Qubits |A>, |B> =|2>, |0> function F1 produce CAAGCT and F2 produce ACCTAG

8. For input Qubits |A>, |B> =|2>, |1> function F1 produce ACCTAG and F2 produce CAAGCT

9. For input Qubits |A>, |B> =|2>, |2> function F1 and F2 produce ACC-TAG

11.3 Multiple-Valued Quantum-DNA Programmable Array Logic

Programmable Array Logic (PAL) is a logic device, which has a programmable AND array and a fixed OR array. It is used to realize a logic function. In this PLD, only AND gates are programmable and hence it is easier to work with PAL. The Figure 11.3 shows the internal structure of Programmable Array Logic.

The product terms can be programmed through the fuse link. It means the user can decide the connection between the inputs and the AND gates. If a particular input line is to be connected to the AND gate, then the fuse link must be placed at the interconnection. The AND gate outputs are then fed as an input to the fixed OR gate.

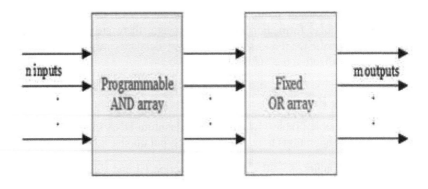

FIGURE 11.3
Block Diagram of PAL

Depending upon the required function, the output line of the AND gate is connected to the corresponding input of the OR gate.

11.3.1   General Organization of Multi-Valued Quantum-DNA PAL

Quantum-DNA PAL is a programmable logic device that has Programmable Quantum AND array & fixed DNA OR array. Hence, it is the most flexible PLD. The block diagram of Quantum-DNA PAL is shown in the following figure (Figure 11.4).

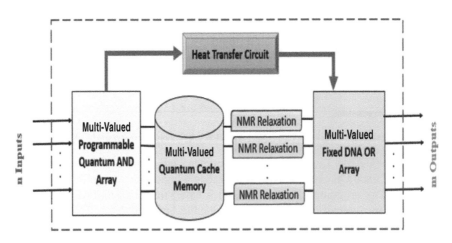

FIGURE 11.4
Block Diagram of Multiple-Valued Quantum-DNA PAL

Here, the inputs of Quantum AND operations are programmable. That means each Quantum AND operation has both normal and complemented inputs of variables. So, based on the requirement, any of those inputs can be programmed and can generate only the required product terms by using these Quantum AND operations.

Quantum computing provides fast computation in logic devices than DNA computing. So, it needs to use a storage device (CACHE MEMORY) for making a balance between these two computing systems. DNA computing gives more storage capacity, as a result, the output operation of the logic devices will perform in DNA computing. It is needed to transform qubit into DNA molecule by using a transformation process by which an excited magnetic state returns to its equilibrium distribution. Here, the inputs of DNA OR operations are fixed. Quantum Computing produces more heat in circuits which can be used in DNA computing to perform DNA logic operations. For this, a heat transfer circuit is used between Quantum and DNA operations. The Quantum-DNA PAL implementation methods will improve the result of any logic device.

11.3.2 Circuit Architecture of Multi-Valued Quantum-DNA PAL

Multiple-valued Quantum-DNA PAL is a programmable logic device that has a Programmable multiple-valued Quantum AND array & fixed multiple-valued DNA OR array.

The following **Boolean functions** use multiple-valued Quantum-DNA PAL.

$F1 = A0.B1 + A1.B1 + 1. (A2.B0)$
$F2 = A0.B2 + A1.B2 + A2B1$

Table 11.2 shows the truth table of ternary Quantum PAL.

The given two functions are in sum of products form. The number of product terms present in the given Boolean functions F1, F2 are three and two respectively.

TABLE 11.2
Truth Table of Ternary Quantum-DNA PAL

|A>	|B>	F1	F2
|0>	|0>	ACCTAG	ACCTAG
|0>	|0>	TGGATC	ACCTAG
|0>	|2>	ACCTAG	TGGATC
|1>	|0>	ACCTAG	ACCTAG
|1>	|1>	TGGATC	TGGATC
|1>	|2>	ACCTAG	ACCTAG
|2>	|0>	CAAGCT	ACCTAG
|2>	|1>	ACCTAG	ACCTAG
|2>	|2>	ACCTAG	ACCTAG

Seven programmable Quantum AND operations & two fixed DNA OR operation are needed for producing those two functions. But, extra two DNA OR operations are needed as it is needed to OR three product terms for each function. The corresponding Quantum **PAL** is shown in the following figure (Figure 11.5).

For the given problem, there are two inputs ($|A>,|B>$) and two outputs (F1, F2). Quantum ternary decoders are used for these two inputs. Thus the realization has six input lines.

The given first expression has three product terms and the second expression has two product terms. But as the DNA OR operation is fixed, the fuses are placed in the corresponding literal to obtain the product terms. Six quantum AND operations are designed using quantum computing. Quantum computing provides fast computation in logic devices than DNA computing. So, cache memory is used to store the quantum qubits. It is needed to transform qubit into DNA molecule by using NMR-Relaxation by which an excited magnetic state returns to its equilibrium distribution. Here, the inputs of DNA OR operations are fixed. All the outputs of Quantum AND operations are applied as inputs to each DNA OR operation. Therefore, the outputs of Quantum-DNA PAL will be in the form of the sum of products form.

11.3.3   Working Principle

According to truth table (Table 11.2) of ternary Quantum-DNA PAL, it is necessary to do the following operations to perform desire output qubits:

1. For input Qubits $|A>$, $|B>$ =$|0>$, $|0>$ function F1 and F2 produce ACC-TAG

2. For input Qubits $|A>$, $|B>$ =$|0>$, $|1>$ function F1 produce TGGATC and F2 produce ACCTAG

3. For input Qubits $|A>$, $|B>$ =$|0>$, $|2>$ function F1 produce ACCTAG and F2 produce TGGATC

4. For input Qubits $|A>$, $|B>$ =$|1>$, $|0>$ function F1 and F2 produce ACC-TAG

5. For input Qubits $|A>$, $|B>$ =$|1>$, $|1>$ function F1 and F2 produce TG-GATC

6. For input Qubits $|A>$, $|B>$ =$|1>$, $|2>$ function F1 and F2 produce ACC-TAG

7. For input Qubits $|A>$, $|B>$ =$|2>$, $|0>$ function F1 produce CAAGCT and F2 produce ACCTAG

8. For input Qubits $|A>$, $|B>$ =$|2>$, $|1>$ function F1 and F2 produce ACC-TAG

9. For input Qubits $|A>$, $|B>$ =$|2>$, $|2>$ function F1 and F2 produce ACC-TAG

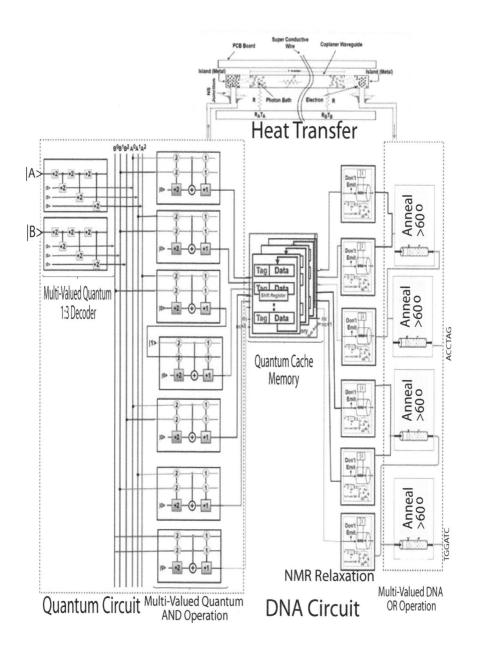

FIGURE 11.5
Ternary Quantum-DNA PAL

11.4   Multi-Valued Quantum-DNA Field Programmable Gate Array

Multi-Valued Quantum computing use Qubits (|0> and |1>) and multi-valued DNA Computing system use Molecule to represent information. In multi-valued Quantum-DNA computing FPGA design, the qubit will work as input for the programmable logic devices. Multi-Valued Quantum computing provides fast computation in logic devices than multi-valued DNA computing. So, need to use a storage device for making a balance between these two computing systems. Multi-Valued DNA computing gives more storage capacity, as a result, the output operation of the logic devices will perform in DNA computing. It is needed to transform qubit into DNA molecule by a transformation process by which an excited magnetic state returns to its equilibrium distribution. Multi-Valued Quantum Computing produces more heat in circuits which can be used in multi-valued DNA computing to perform DNA logic operations. For this, a heat transfer circuit is needed between multi-valued Quantum and DNA operations. The multi-valued Quantum-DNA FPGA implementation methods will improve the result of any logic device. The block diagram of Quantum-DNA FPGA is shown in Figure 11.6.

11.4.1   Circuit Architecture of Multi-Valued Quantum-DNA FPGA

Multi-Valued Quantum-DNA FPGA logic block is designed by connecting Flip-Flops, Look-up Tables (LUT) and Multiplexers. Here a simple Quantum-DNA FPGA logic block is designed by using a D Flip-Flop, a Look-up Table (LUT) and a multi-

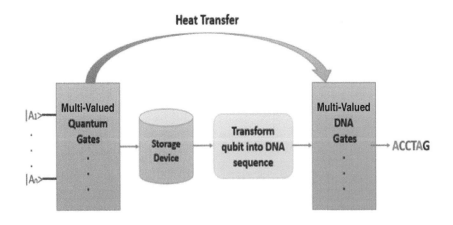

FIGURE 11.6
Block Diagram of Quantum-DNA FPGA

plexer. A simple 2 input multi-valued Quantum LUT consist of four Quantum AND, and two Quantum OR operation. The output of the LUT will go through the multi-valued quantum D Flip-Flop and multi-valued quantum multiplexer as input. The multi-valued quantum D Flip-Flop is a sequential circuit that consists of four multi-valued quantum NAND operations and one multi-valued quantum NOT operation. The output of the multi-valued quantum D Flip-Flop will go through the multi-valued multiplexer as input. A multi-valued 3-to-1 MUX is designed using three multi-valued quantum AND operations, and two multi-valued DNA OR operation. Quantum computing provides fast computation in logic devices than DNA computing. So, need to use a storage device for making a balance between these two computing systems. DNA computing gives more storage capacity, as a result, the output operation of the logic devices will perform in DNA computing. It is needed to transform qubit into DNA molecule by a transformation process by which an excited magnetic state returns to its equilibrium distribution. Quantum Computing produces more heat in circuits which can be used in DNA computing to perform DNA logic operations. For this, a heat transfer circuit is needed to use between Quantum and DNA operations. The Quantum-DNA FPGA implementation methods will improve the result of any logic device. The circuit architecture of Ternary Quantum-DNA FPGA is shown in Figure 11.7.

11.4.2 Working Principle

Three multi-valued Quantum Ternary 1-to-3 decoder designs for input $|A>,|B>$, and $|S>$. Multi-Valued Quantum Ternary 1-to-3 decoder operation requires the shifting operations that include (+2) shifting only. Where three (+2) operations are not controlled, and three (+2) operation is controlled by respective one input.

The output of the decoder for input $|A>$ and $|B>$ first go through the Look-up Tables (LUT). The LUT use four multi-valued quantum TAND operation and two multi-valued Quantum TOR operation. The LUT implements the function (A0.B1 + A1.B1) + 1. (A2.B0) to generate the LUT output. The outputs of the LUT are given in the Table 11.3.

TABLE 11.3

Table of Multi-Valued LUT

A	B	LUT output
0	0	0
0	1	2
0	2	0
1	0	0
1	1	2
1	2	0
2	0	1
2	1	0
2	2	0

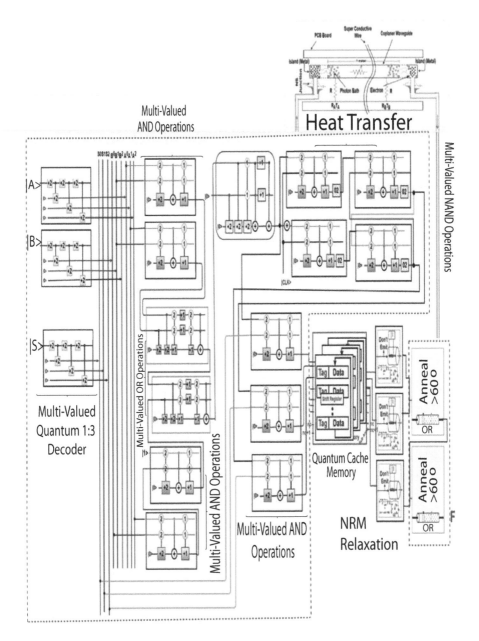

FIGURE 11.7
Ternary Quantum-DNA FPGA

2 input, one is the output of LUT and another is clock input will go thought the D flip-flop. The D flip-flop use one Quantum TNOT operation and four multi-valued Quantum TNAND operations. The D flip-flop transfer the LUT output same as |D>, if the CLK input sequences is |1> or |2>. If the CLK input is |0>, one of the inputs to each of the last two Quantum TNAND operations will be |2>, thus output of the D flip-flop remain unchanged regardless of the values of the LUT output.

The ternary selection input S is firstly decoded by the ternary decoder. Then it will be transferred to the multiplexer as a selection input (S0, S1, and S2). A 3-to-1 multiplexer contains three input signals (outputs of D flip-flop (Q and Q') and LUT)) and three select lines. Each selected input line and one input signal perform ternary AND operation. For ternary AND operation, Output will be the minimum value. The output of these three Quantum AND operations will store in a Quantum cache memory. Cache memory stores these qubits as quantum computing is faster than DNA computing. NMR Relaxation is used to transfer the qubit into the DNA sequence. Two DNA OR operation performs OR of these three DNA sequences. Quantum circuits produce more heat than DNA circuits. For this, a heat transfer circuit is used to transfer the extra heat produced by the Quantum circuit to the DNA circuit. DNA OR operations used this heat to produce the FPGA output.

11.5   Multi-Valued Quantum-DNA Complex Programmable Logic Devices

Multi-Valued Quantum computing uses Qubits (|0> and |1>) and multi-valued DNA Computing system uses molecule to represent information. In multi-valued Quantum-DNA computing CPLD design, the qubit will work as input for the programmable logic devices. Multi-Valued Quantum computing provides fast computation in logic devices than multi-valued DNA computing. So, it needs to use a storage device for making a balance between these two computing systems. Multi-Valued DNA computing gives more storage capacity, as a result, the output operation of the logic devices will perform in DNA computing. It is needed to transform qubit into DNA molecule and for this, a transformation process is used by which an excited magnetic state returns to its equilibrium distribution. Multi-Valued Quantum Computing produces more heat in circuits which can be used in multi-valued DNA computing to perform DNA logic operations. For this, it needs to use a heat transfer circuit between multi-valued Quantum and DNA operations. The multi-valued Quantum-DNA CPLD implementation methods will improve the result of any logic device. Figure 11.8 shows the block diagram of Quantum-DNA CPLD .

11.5.1   Circuit Architecture of Multi-Valued Quantum-DNA CPLD

Multi-Valued Quantum-DNA CPLD functional block is designed by connecting Flip-Flops, logic function and Multiplexers. Here a simple Quantum-DNA CPLD

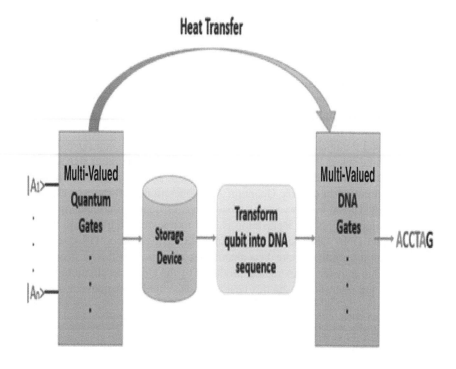

FIGURE 11.8
Block Diagram of Quantum-DNA CPLD

functional block is also designed by using a D Flip-Flop, logic function and a multiplexer. Figure 11.9 depicts the Quantum-DNA CPLD circuit diagram.

A simple 2 input multi-valued Quantum logic function consists of four Quantum AND, and two Quantum OR operation. Output of logic function will XOR with 0. The output of the XOR will go through the multi-valued quantum D Flip-Flop and multi-valued quantum multiplexer as input. The multi-valued quantum D Flip-Flop is a sequential circuit that consists of four multi-valued quantum NAND operations and one multi-valued quantum NOT operation. The output of the multi-valued quantum D Flip-Flop will go through the multi-valued multiplexer as input. A multi-valued 3-to-1 MUX is designed using three multi-valued quantum AND operations, and two multi-valued DNA OR operation. Quantum computing provides fast computation in logic devices than DNA computing. So, a storage device is needed for making a balance between these two computing systems. DNA computing gives more storage capacity, as a result, the output operation of the logic devices will perform in DNA computing. It is needed to transform qubit into DNA molecule by a transformation process by which an excited magnetic state returns to its equilibrium distribution. Quantum Computing produces more heat in circuits which can be used in DNA

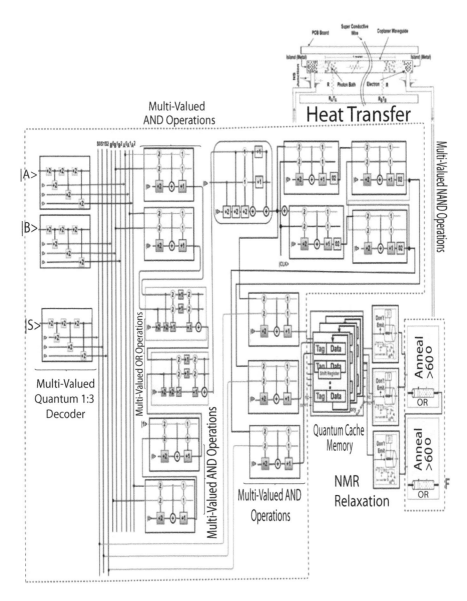

FIGURE 11.9
Quantum-DNA CPLD

computing to perform DNA logic operations. For this, a heat transfer circuit is needed between Quantum and DNA operations. The Quantum-DNA CPLD implementation methods will improve the result of any logic device.

11.5.2   Working Principle

Three multi-valued Quantum Ternary 1-to-3 decoder designs for input |A>, |B>, and |S>. Multi-Valued Quantum Ternary 1-to-3 decoder operation requires the shifting operations that include (+2) shifting only. Where three (+2) operations are not controlled, and three (+2) operation is controlled by respective one input.

The output of the decoder for input |A> and |B> first go through the logic function. The logic function use four multi-valued quantum TAND operation and two multi-valued Quantum TOR operation. The logic function implements the function (A0.B1 + A1.B1) + 1. (A2.B0) to generate the logic function output. The outputs of the logic function are given in the table. The output of logic function will XOR with |0>. Table 11.4 shows the Multi-Valued Logic Function.

2 input, one is the output of XOR and another is clock input will go through the D flip-flop. The D flip-flop uses one Quantum TNOT operation and four multi-valued Quantum TNAND operations. The D flip-flop transfer the XOR output the same as |D>, if the CLK input sequences are |1> or |2>. If the CLK input is |0>, one of the inputs to each of the last two Quantum TNAND operations will be |2>, thus the output of the D flip-flop remains unchanged regardless of the values of the XOR output.

The ternary selection input S is firstly decoded by the ternary decoder. Then it will be transferred to the multiplexer as a selection input (S0, S1, and S2). A 3-to-1 multiplexer contains three input signals (outputs of D flip-flop (Q and Q') and XOR)) and three select lines. Each selected input line and one input signal perform ternary AND operation. For ternary AND operation, Output will be the minimum value. The output of these three Quantum AND operations will store in a Quantum cache memory. Cache memory stores these qubits as quantum computing is faster

TABLE 11.4

Multi-Valued Logic Function

A	B	LUT output
0	0	0
0	1	2
0	2	0
1	0	0
1	1	2
1	2	0
2	0	1
2	1	0
2	2	0

than DNA computing. NMR Relaxation is used to transfer the qubit into the DNA sequence. Two DNA OR operation performs OR of these three DNA sequences. Quantum circuits produce more heat than DNA circuits. For this, a heat transfer circuit is used to transfer the extra heat produced by the Quantum circuit to the DNA circuit. DNA OR operations used this heat to produce the CPLD output.

11.6 Summary

Quantum Computing produces more heat in circuits which can be used in DNA computing to perform DNA logic operations. For this, heat transfer circuit is used between Quantum and DNA operations. The Multiple-valued Quantum-DNA PLDs implementation methods will improve the result of any logic device. Two different technologies are combined to form a new computing technology called quantum molecular biology. This chapter has presented the multi-valued PLDs in quantum-DNA computing which is a form of quantum molecular biology. Another form is called DNA-quantum computing which will be discussed in next chapter.

Bibliography

[1] Jia, J., Wang, Y., Liu, J., & Jin, G. (2014, July). Effective CGH calculation algorithm with low memory usage using compressed look-up table based on separation of light modulation variable. In Laser Applications to Chemical, Security and Environmental Analysis (pp. JTu4A-25). Optica Publishing Group.

[2] Anusudha, K., & Naguboina, G. C. (2017, March). Design and implementation of PAL and PLA using reversible logic on FPGA SPARTAN 3E. In 2017 Fourth International Conference on Signal Processing, Communication and Networking (ICSCN) (pp. 1-6). IEEE.

[3] Kricka, L. J., & Fortina, P. (2009). Analytical ancestry:"Firsts" in fluorescent labeling of nucleosides, nucleotides, and nucleic acids. Clinical Chemistry, 55(4), 670-683.

[4] Sakakibara, Y. (1998). DNA computers: A new computing paradigm. Journal of Photopolymer Science and Technology, 11(4), 681-686.

12

Multiple-Valued Programmable Logic Devices in DNA-Quantum Computing

12.1 Introduction

Multiple-Valued DNA-Quantum PLDs are programmable logic devices that have both Programmable Multiple-Valued DNA AND array and Programmable Multiple-Valued Quantum OR array. Here, the inputs of multiple-valued DNA AND operations are programmable. That means each multiple-valued DNA AND operation has both normal and complemented inputs of variables. So, based on the requirement, any of those inputs can be programmed and can generate only the required product terms by using these multiple-valued Quantum AND operations. A storage device needs to store DNA sequences between these two computing systems. . Quantum computing generates more heat than DNA computing. As a result, extra heat from the outside is required in a multiple-valued DNA circuit. It is needed to transform multiple-valued DNA molecules into qubits by a storage device and a transformation process (Trap-Ion) by which an excited magnetic state returns to its equilibrium distribution. Quantum Computing produces more heat in circuits which can be used in further DNA computing or can transfer the heat in a cooler circuit to cool down.

Here, multiple-valued Quantum OR operations are also programmable. So, any number of required product terms are programmed, since all the outputs of multiple-valued DNA AND operations are applied as inputs to each multiple-valued Quantum OR operation. Therefore, the outputs of multiple-valued DNA-Quantum PLDs will be in the form of the sum of products form.

12.2 Multiple-Valued DNA-Quantum Programmable Logic Array

Programmable Logic Array is a type of PLD, which has both programmable Quantum AND array and programmable Quantum OR array. In a combinational circuit, because of don't care conditions, not all the minterms are used. Programmable Logic Array (PLA) is a fixed architecture logic device with programmable Quantum AND

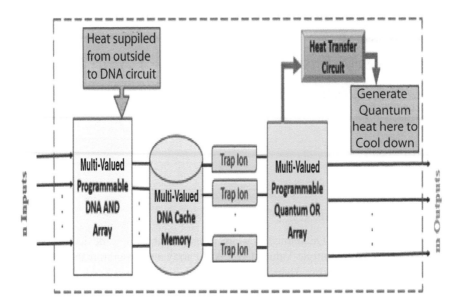

FIGURE 12.1
Block Diagram of Multiple-Valued DNA-Quantum PLA

gates followed by programmable Quantum OR gates. PLA is a type of programmable logic device used to build a reconfigurable digital circuit. PLDs have an undefined function at the time of manufacturing, but they are programmed before being made into use. PLA is a combination of memory and logic.

12.2.1   General Organization of Multiple-Valued DNA-Quantum PLA

Multiple-Valued DNA-Quantum PLA is a programmable logic device that has both Programmable Multiple-Valued DNA AND array and Programmable Multiple-Valued Quantum OR array. Hence, it is the most flexible PLD. The block diagram of Multiple-Valued DNA- Quantum PLA is shown in the Figure 12.1.

Here, the inputs of multiple-valued DNA AND operations are programmable. That means each multiple-valued DNA AND operation has both normal and complemented inputs of variables. So, based on the requirement, any of those inputs can be programmed and can generate only the required product terms by using these multiple-valued Quantum AND operations. A storage device needs to store DNA sequences between these two computing systems. DNA computing produces less heat than quantum computing. As a result, extra heat is supplied from outside in a multiple-valued DNA circuit. It is needed to transform multiple-valued DNA molecules into qubit by using a storage device and a transformation process (Trap-Ion) by which an excited magnetic state returns to its equilibrium distribution.

TABLE 12.1

Truth Table of Ternary DNA-Quantum PLA

$|A>$	$|B>$	$|F1>$	$|F2>$
ACCTAG	ACCTAG	$|0>$	$|0>$
ACCTAG	ACCTAG	$|2>$	$|0>$
ACCTAG	TGGATC	$|0>$	$|2>$
CAAGCT	ACCTAG	$|0>$	$|0>$
CAAGCT	CAAGCT	$|2>$	$|0>$
CAAGCT	TGGATC	$|0>$	$|1>$
TGGATC	ACCTAG	$|1>$	$|0>$
TGGATC	CAAGCT	$|0>$	$|1>$
TGGATC	TGGATC	$|0>$	$|0>$

Quantum Computing produces more heat in circuits which can be used in further DNA computing or can transfer the heat in a cooler circuit to cool down.

Here, multiple-valued Quantum OR operations are also programmable. So, any number of required product terms can be programmed, since all the outputs of multiple-valued DNA AND operations are applied as inputs to each multiple-valued Quantum OR operation. Therefore, the outputs of multiple-valued DNA-Quantum PAL will be in the form of the sum of products form.

12.2.2 Circuit Architecture of Multi-Valued DNA-Quantum PLA

Consider the design of ternary quantum PLA for the following functions:

$F1 = A0.B1 + A1.B1 + 1. (A2.B0)$
$F2 = A0.B2 + 1. (A1.B2 + A2.B1)$

The given two functions are in the sum of product form. The number of product terms present in the given Boolean functions $|F1>$, $|F2>$ are three. Table 12.1 shows the truth table of ternary DNA-Quantum PLA.

Eight programmable ternary DNA AND operation & four programmable ternary Quantum OR gates are needed for producing those two functions. The corresponding DNA-Quantum **PLA** is shown in Figure 12.2.

For the given functions, there are two inputs (A, B) and two outputs (F1, F2). Two ternary DNA 3-to-1 decoders are used for input A and B. Thus the realization has six input lines. Eight ternary DNA AND operations are designed using DNA computing. It is needed to use cache memory to store the DNA sequences. It is needed to transform DNA molecules into qubits by Trap-Ion by which an excited magnetic state returns to its equilibrium distribution. Here, ternary Quantum OR operations are also programmable. So, any number of required product terms can be programmed, since all the outputs of ternary DNA AND operations are applied as inputs to each ternary Quantum OR operation. Therefore, the outputs of ternary DNA- Quantum PAL will be in the form of the sum of products form.

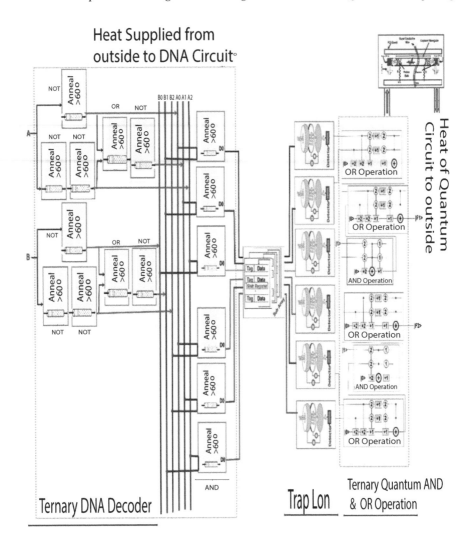

FIGURE 12.2
DNA-Quantum PLA

12.2.3 Working Principle

According to truth table (Table 12.1) of ternary DNA-Quantum PLA, it is necessary to do the following operations to perform desire output qubits:

1. For input Qubits A, B = ACCTAG , ACCTAG function |F1> and |F2> produce |0>

2. For input Qubits A, B = ACCTAG, CAAGCT function |F1> produce |2> and |F2> produce |0>

3. For input Qubits A, B = ACCTAG, | TGGATC function |F1> produce |0> and |F2> produce |2>

4. For input Qubits A, B = CAAGCT ACCTAG function |F1> and |F2> produce |0>

5. For input Qubits A, B = CAAGCT , CAAGCT function |F1> produce |2> and |F2> produce |0>For input Qubits A, B = CAAGCT, TGGATC function |F1> produce |0> and |F2> produce |1>

6. For input Qubits A, B = TGGATC, ACCTAG function |F1> produce |1> and |F2> produce |0>

7. For input Qubits A, B = TGGATC, CAAGCT function |F1> produce |0> and |F2> produce |1>

8. For input Qubits A, B = TGGATC , TGGATC function |F1> and |F2> produce |0>

12.3   Multiple-Valued DNA-Quantum Programmable Array Logic

Programmable Array Logic (PAL) is a logic device, which has a programmable AND array and fixed OR array. It is used to realize a logic function. In this PLD, only AND gates are programmable and hence it is easier to work with PAL. Figure 12.3 shows the internal structure of Programmable Array Logic.

The product terms can be programmed through the fuse link. It means the user can decide the connection between the inputs and the AND gates. If a particular input line is to be connected to the AND gate, then the fuse link must be placed at the interconnection. The AND gate outputs are then fed as an input to the fixed OR gate. Depending upon the required function, the output line of the AND gate is connected to the corresponding input of the OR gate.

12.3.1   General Organization of Multi-Valued DNA-Quantum PAL

DNA- Quantum PLA is a programmable logic device that has Programmable DNA AND array & fixed Quantum OR array. Hence, it is the most flexible PLD. The block diagram of DNA- Quantum PAL is shown in Figure 12.4.

Here, the inputs of DNA AND operations are programmable. That means each DNA AND operation has both normal and complemented inputs of variables. So based on the requirement, any of those inputs can be programmed and can generate only the required product terms by using these Quantum AND operations. A storage device needs to store DNA sequences between these two computing systems. DNA computing produces less heat than quantum computing. As a result, the extra heat from outside in the DNA circuit is supplied. It is needed to transform DNA molecules

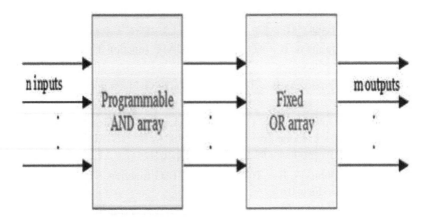

FIGURE 12.3
Block Diagram of PAL

into qubits by a storage device and a transformation process (Trap-Ion) by which an excited magnetic state returns to its equilibrium distribution. Quantum Computing produces more heat in circuits which can be used in further DNA computing or can transfer the heat in a cooler circuit to cool down. Here, Quantum OR operations are fixed. All the outputs of DNA AND operations are applied as inputs to each Quantum

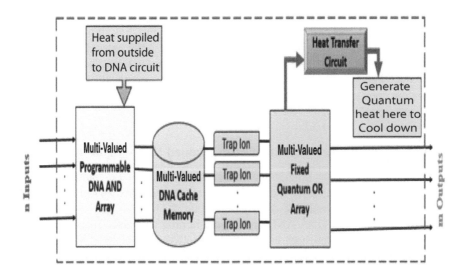

FIGURE 12.4
Block Diagram of Multiple-Valued DNA-Quantum PAL

OR operation. Therefore, the outputs of DNA-Quantum PAL will be in the form of the sum of products form.

12.3.2   Circuit Architecture of Multi-Valued DNA-Quantum PAL

Consider the design of ternary quantum PAL for the following functions:

F1 = A0.B1+A1.B1+ 1. (A2.B0)
F2 = A0.B2+ A1.B2+ A2B1

The given two functions are in the sum of products form. The number of product terms present in the given Boolean functions |F1>, |F2> are three. Table 12.2 shows the truth table of ternary DNA-Quantum PAL

Seven programmable ternary DNA AND operation & four programmable ternary Quantum OR gates are needed for producing those two functions. The corresponding DNA-Quantum **PAL** is shown in Figure 12.5.

For the given functions, there are two inputs (A, B) and two outputs (F1, F2). Two ternary DNA 3-to-1 decoders are used for input A and B. Thus the realization has six input lines. Eight ternary DNA AND operations are designed using DNA computing. Cache memory is used to store the DNA sequences. It is needed to transform DNA molecules into qubits by Trap-Ion by which an excited magnetic state returns to its equilibrium distribution. Here, ternary Quantum OR operations are also programmable.

So, any number of required product terms can be programmed, since all the outputs of ternary DNA AND operations are applied as inputs to each ternary Quantum OR operation. Therefore, the outputs of ternary DNA- Quantum PAL will be in the form of the sum of products form.

TABLE 12.2
Truth Table of Ternary DNA-Quantum PAL

|A>	|B>	|F1>	|F2>
ACCTAG	ACCTAG	|0>	|0>
ACCTAG	ACCTAG	|2>	|0>
ACCTAG	TGGATC	|0>	|2>
CAAGCT	ACCTAG	|0>	|0>
CAAGCT	CAAGCT	|2>	|2>
CAAGCT	TGGATC	|0>	|0>
TGGATC	ACCTAG	|1>	|0>
TGGATC	CAAGCT	|0>	|0>
TGGATC	TGGATC	|0>	|0>

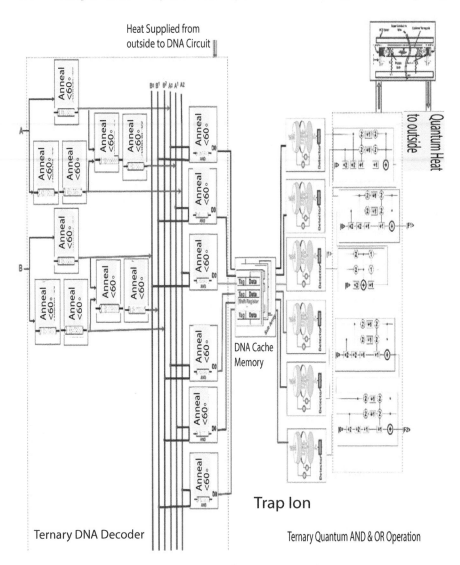

FIGURE 12.5
Ternary DNA-Quantum PAL

12.3.3 Working Principle

According to the truth table (Table 12.2) of ternary DNA-Quantum PLA, it is necessary to do the following operations to perform desire output qubits:

1. For input Qubits A, B = ACCTAG , ACCTAG function |F1> and |F2> produce |0>

2. For input Qubits A, B = ACCTAG, CAAGCT function |F1> produce |2> and |F2> produce |0>

3. For input Qubits A, B = ACCTAG, | TGGATC function |F1> produce |0> and |F2> produce |2>

4. For input Qubits A, B = CAAGCT ACCTAG function |F1> and |F2> produce |0>

5. For input Qubits A, B = CAAGCT , CAAGCT function |F1> produce |2> and |F2> produce |0>For input Qubits A, B = CAAGCT, TGGATC function |F1> produce |0> and |F2> produce |1>

6. For input Qubits A, B = TGGATC, ACCTAG function |F1> produce |1> and |F2> produce |0>

7. For input Qubits A, B = TGGATC, CAAGCT function |F1> produce |0> and |F2> produce |1>

8. For input Qubits A, B = TGGATC , TGGATC function |F1> and |F2> produce |0>

12.4   Multi-Valued DNA-Quantum Field Programmable Gate Array

Field Programmable Gate Arrays (FPGAs) are semiconductor devices that are based around a matrix of configurable logic blocks (CLBs) connected via programmable interconnects. Multiple-valued Quantum computing use Qubits (|0>, |1> and |2>) and Multiple-valued DNA Computing system use Molecule to represent information. In multiple-valued DNA-Quantum computing FPGA design, DNA sequences will work as input for the programmable logic devices. A storage device needs to use for making balance between these two computing systems. Multiple-valued DNA computing produces less heat than multiple-valued quantum computing. As a result, it is needed to supply extra heat from outside in the DNA circuit. It is needed to transform DNA molecules into qubits by a storage device and a transformation process by which an excited magnetic state returns to its equilibrium distribution. Quantum Computing produces more heat in circuits which can be used in further DNA computing or can transfer the heat in a cooler circuit to cool down. The DNA- Quantum FPGA implementation methods will improve the result of any logic device. Figure 12.6 shows the block diagram of DNA-Quantum FPGA .

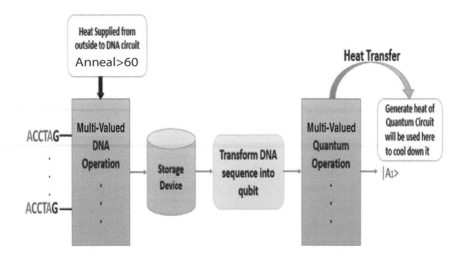

FIGURE 12.6
Block Diagram of DNA-Quantum FPGA

12.4.1   Circuit Architecture of Multiple-Valued DNA-Quantum FPGA

Ternary FPGA logic block is designed by connecting Flip-Flops, Look-up Tables (LUT), and Multiplexers. Here a simple multi-valued FPGA logic block is designed by using a ternary D Flip-Flop, a ternary Look-up Table (LUT), and a 3-to-1 multiplexer. The ternary input A and B signals are firstly decoded by the ternary decoder. Then they will be transferred to the LUT. A simple 2 input LUT is designed using four AND operations and two OR operations. The output equations is = (A0.B1 + A1.B1) + 1.(A2.B0). Note that the output of the decoder has only two logic values, i.e., "2" and "0", corresponding to logic "1" and "0" in binary logic. The output of the LUT is used as input in the D Flip-Flop and multiplexer. Ternary D Flip-Flop is a sequential circuit that consists of four ternary NAND operations and one NOT operation. Two Output (Q and Q') of the D Flip-Flop are used as input in the multiplexer. The ternary selection input S is firstly decoded by the ternary decoder. Then it will be transferred to the multiplexer as a selection input. A 3-to-1 MUX is designed using three ternary Quantum AND operations and two ternary DNA OR operation. The multiplexer generates the desired output for the FPGA logic block. Outputs of the three ternary AND will store in a Quantum cache memory. Trap Ion is used to transfer the DNA sequence into a qubit. These two qubits will perform Quantum OR operations. Heat supplied from outside to DNA circuit for operation. A heat transfer circuit is used in the quantum multiplexer circuit to transfer the extra heat produced by the Quantum circuit to the outside to cool it. Figure 12.7 shows the circuit architecture of Multi-Valued DNA-Quantum FPGA .

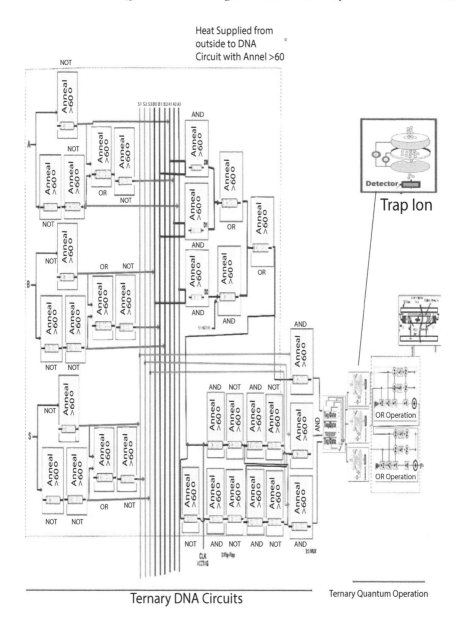

FIGURE 12.7
Multi-Valued DNA-Quantum FPGA

12.4.2 Working Procedure

A two-input ternary Look-up Tables (LUT) is used to design the FPGA control logic block. Input A and B firstly is decoded by the ternary decoder and generated the output lines A0, A1, A2 for input A, and B0, B1, B2 for input B. These outputs will go through the Look-up Tables (LUT). The LUT use four ternary AND operation and two OR operation to implement the function (A0.B1 + A1.B1) + 1. (A2.B0). During ternary DNA computing, sequence ACCTAG is considered as "0", sequence CAAGCT strands as "1" and TGGATC stands as "2". In ternary AND operation, output value depends on the minimum value of its inputs and in ternary OR operation, output value depends on the maximum value of its inputs.). The output of the decoder has only two sequences, "TGGATC (2) and "ACCTAG (0)".

1. When the input sequences A, B = ACCTAG (0), ternary decoder generate A0, A1, A2 = (2, 0, 0) and B0, B1, B2 = (2, 0, 0). None of the ternary AND operations (D0-D2) are connected to A0 and B0 lines. As a result, none of the ternary AND operation values will be true. So the output of the LUT will be ACCTAG (0).

2. When the input sequences A = ACCTAG (0) and B = CAAGCT (1), ternary decoder generate A0, A1, A2 = (2, 0, 0) and B0, B1, B2 = (0, 2, 0). D0 ternary AND operation is connected to A0 and B1 lines. As a result, D0 ternary AND operation value will be true (TGGATC). So the output of the LUT will be TGGATC.

3. When the input sequences A = ACCTAG (0) and B = TGGATC (2), ternary decoder generate A0, A1, A2 = (2, 0, 0) and B0, B1, B2 = (0, 0, 2). None of the ternary AND operations (D0-D2) are connected to A0 and B2 lines. As a result, none of the ternary AND operation values will be true. So the output of the LUT will be ACCTAG (0).

4. When the input sequences A = CAAGCT (1) and B = ACCTAG (0), ternary decoder generate A0, A1, A2 = (0, 2, 0) and B0, B1, B2 = (2, 0, 0). None of the ternary AND operations (D0-D2) are connected to A1 and B0 lines. As a result, none of the ternary AND operation values will be true. So the output of the LUT will be ACCTAG (0).

5. When the input sequences A, B = CAAGCT (1), ternary decoder generate A0, A1, A2 = (0, 2, 0) and B0, B1, B2 = (0, 2, 0). D1 ternary AND operation are connected to A1 and B1 lines. As a result D1 ternary AND operation value will be true (TGGATC). So the output of the LUT will be TGGATC.

6. When the input sequences A = CAAGCT (1), and B = TGGATC (2), ternary decoder generate A0, A1, A2 = (0, 2, 0) and B0, B1, B2 = (0, 0, 2). . None of the ternary AND operation (D0-D2) are connected to A1 and B2 line. As a result, none of the ternary AND operation values will be true. So the output of the LUT will be ACCTAG (0).

7. When the input sequences A = TGGATC (2), and B = ACCTAG (0), ternary decoder generate A0, A1, A2 = (0, 0, 2) and B0, B1, B2 = (2, 0, 0). D2 ternary AND operation is connected to A2 and B0 lines. As a result D ternary AND operation value will be true (TGGATC). Now

it will proceed ternary AND operation with CAAGCT (1) and produce minimum as CAAGCT as LUT output.

8. When the input sequences A = TGGATC (2), and B = CAAGCT (1), ternary decoder generate A0, A1, A2 = (0, 0, 2) and B0, B1, B2 = (0, 2, 0). None of the ternary AND operations (D0-D2) are connected to A2 and B1 lines. As a result, none of the ternary AND operation values will be true. So the output of the LUT will be ACCTAG (0).

9. When the input sequences A, B = TGGATC (2), ternary decoder generate A0, A1, A2 = (0, 0, 2) and B0, B1, B2 = (0, 0, 2).). None of the ternary AND operations (D0-D2) are connected to A2 and B2 lines. As a result, none of the ternary AND operation values will be true. So the output of the LUT will be ACCTAG (0).

2 inputs, one is the output of LUT and another is clock input will go through the ternary D flip-flop. The ternary D flip-flop use one NOT gate four NAND gates.

1. If the LUT output and CLK input both are ACCTAG (0), then the output of the ternary D flip-flop will remain unchanged.

2. If the LUT output is ACCTAG (0) and CLK input is CAAGCT (2), then the output of the D flip-flop will be ACCTAG (0).

3. If the LUT output is ACCTAG (0) and CLK input is TGGATC (2), then the output of the D flip-flop will be ACCTAG (0).

4. If the LUT output is CAAGCT (1) and CLK input is ACCTAG (0), then the output of the D flip-flop will remain unchanged.

5. If the LUT output and CLK input both are CAAGCT (1), then the output of the D flip-flop will be CAAGCT (1).

6. If the LUT output is CAAGCT (1) and CLK input is TGGATC (2), then the output of the D flip-flop will be CAAGCT (1).

7. If the LUT output is TGGATC (2) and CLK input is ACCTAG (0), then the output of the D flip-flop will remain unchanged.

8. If the LUT output is TGGATC (2) and CLK input is CAAGCT (1), then the output of the D flip-flop will be CAAGCT (1).

9. If the LUT output and CLK input both are TGGATC (2), then the output of the D flip-flop will be TGGATC (2).

The ternary selection input S is firstly decoded by the ternary decoder. Then it will be transferred to the multiplexer as a selection input (S0, S1, and S2). A 3-to-1 multiplexer contains three input signals (outputs of D flip-flop (Q and Q') and LUT)) and three select lines. Each selected input line and one input signal perform ternary AND operation. For ternary AND operation, Output will be the minimum value. Outputs of these AND operations will store in a DNA cache memory. Trap Ion is used to transfer the DNA sequence into a qubit. These three Quantum qubits will go through the Quantum OR operation to generate the FPGA logical block output. Heat supplied from outside to DNA circuit for operation. A heat transfer circuit is used in the quantum multiplexer circuit to transfer the extra heat produced by the Quantum circuit to the outside to cool it.

1. If all three input sequences (output of three ternary AND operations) are   · ACCTAG (0), then the output of the ternary OR operation is |0>.

2. If one of the three input sequences is TGGATC (2), then the output of the ternary Quantum OR operation is |2>.

3. If input sequences are in ACCTAG (0), and CAAGCT (1) then the output of the ternary Quantum OR operation is |1>

4. If input sequences are in ACCTAG (0), and TGGATC (2) then the output of the ternary Quantum OR operation is |2>

5. If input sequences are in CAAGCT (1), and TGGATC (2) then the output of the ternary OR operation is |2>.

12.5   Multi-Valued DNA-Quantum Complex Programmable Logic Devices

Multiple-valued Quantum computing uses Qubits (|0>, |1> and |2>) and Multiple-valued DNA Computing system uses Molecule to represent information. In multiple-valued DNA-Quantum computing CPLD design, DNA sequences will work as input for the programmable logic devices. A storage device needs to use for making balance between these two computing systems. Multiple-valued DNA computing produces less heat than multiple-valued quantum computing. As a result, it is needed to supply extra heat from outside in the DNA circuit and need to transform DNA molecules into qubits by a storage device and a transformation process by which an excited magnetic state returns to its equilibrium distribution. Quantum Computing produces more heat in circuits which can be used in further DNA computing or can transfer the heat in a cooler circuit to cool down. The DNA- Quantum FPGA implementation methods will improve the result of any logic device. Figure 12.8 shows the block diagram of DNA-Quantum CPLD .

12.5.1   Circuit Architecture of Multiple-Valued DNA-Quantum CPLD

Multi-Valued DNA-Quantum CPLD circuit is divided into two blocks. One is DNA part of the DNA-Quantum CPLD and another is Quantum part of the DNA-Quantum CPLD. DNA part consists of DNA operation of the logic function, Flip-flop and 3 and operation of the multiplexer.

Here a simple DNA part of DNA-Quantum CPLD (Figure 12.9) is designed by using a DNA D Flip-Flop, a DNA logic function and 3 AND operations of multiplexer. Three 1-to-3 decoders are used for three input A, B and S. A simple DNA

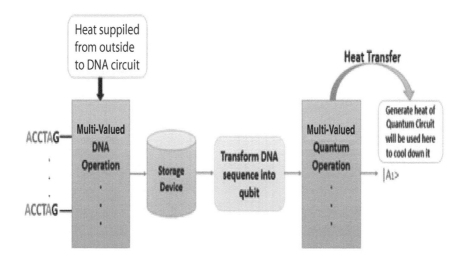

FIGURE 12.8
Block Diagram of DNA-Quantum CPLD

logic function consist of four DNA AND, and two Quantum OR operation. The output of logic function will XOR with zero. The output of the XOR will go through the DNA D Flip-Flop. The D Flip-Flop is a sequential circuit that consists of four Quantum NAND (AND+ NOT) operations and one DNA NOT operation. Heat is supplied from outside to DNA circuit for operation. Figure 12.9 depicts the DNA Part of Multi-Valued DNA-Quantum CPLD.

Another part of DNA to Quantum CPLD circuit is Quantum part CPLD. Quantum part consists of Quantum operation of Multiplexer OR operations. A simple 3-to-1 multiplexer consist of three DNA AND operations and two Quantum OR operations. Here a simple Quantum CPLD part (Figure 12.10) is designed by using two Quantum OR operation.

Multi-Valued DNA to Quantum CPLD functional block is designed by connecting DNA circuit and Quantum circuit. A simple DNA-Quantum CPLD functional block (Figure 12.11) is designed by using DNA block and Quantum block. Outputs DNA block will store in a DNA cache memory. Trap Ione used to transfer the DNA sequence into Qubit. Heat is supplied from outside to DNA circuit for operation. A heat transfer circuit is used in quantum multiplexer circuit to transfer the extra heat produced by the Quantum circuit to outside for cool it.

Ternary CPLD functional block is designed by connecting Flip-Flops, logic function and Multiplexers. Here a simple multi-valued CPLD functional block is designed by using a ternary D Flip-Flop, a logic function and a 3-to-1 multiplexer. The ternary input A, and B signals are firstly decoded by the ternary decoder. Then they will be transferred to the logic function. A simple 2 input logic function is designed using four AND operation and two OR operation. The logic function equations is =

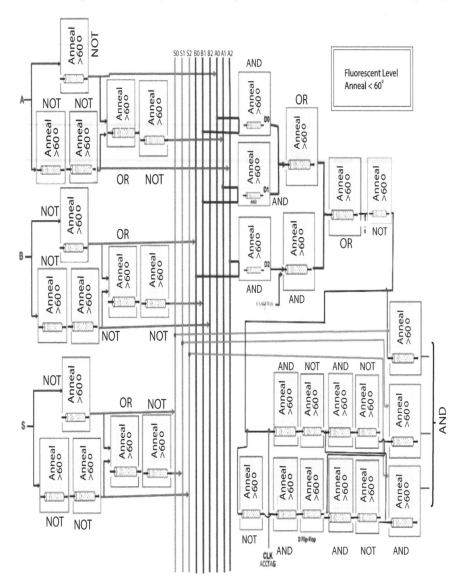

FIGURE 12.9
DNA Part of Multi-Valued DNA-Quantum CPLD

(A0.B1 + A1.B1) + 1. (A2.B0). Output of the function will XOR with 0. Note that the output of the decoder has only two logic values, i.e., "2" and "0", corresponding to logic "1" and "0" in binary logic. The output of the XOR is used as input in the D Flip-Flop and multiplexer. Ternary D Flip-Flop is a sequential circuit that consist of four ternary NAND operation and one NOT operation. Two Output of the XOR op-

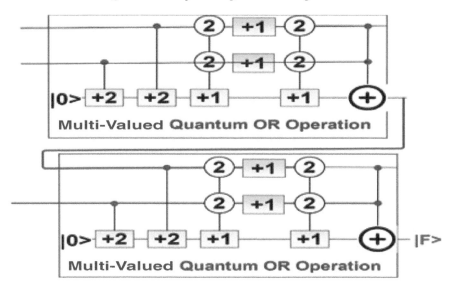

FIGURE 12.10
Quantum Part of Multi-Valued DNA-Quantum CPLD

eration and D Flip-Flop (Q and Q') are used as input in the multiplexer. The ternary selection input S firstly decoded by the ternary decoder. Then it will be transferred to the multiplexer as a selection input. A 3-to-1 MUX is designed using three ternary Quantum AND operation and two ternary DNA OR operation. The multiplexer generate the desired output for the CPLD functional block. Outputs of the three ternary

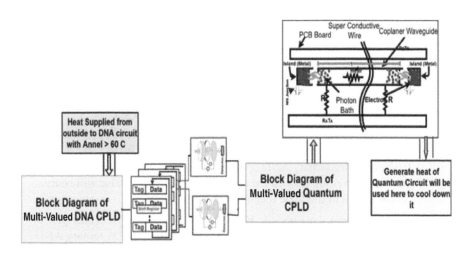

FIGURE 12.11
Block Diagram of Multi-Valued DNA-Quantum CPLD

AND will store in a Quantum cache memory. Trap Ione used to transfer the DNA sequence into qubit. These two qubit will perform Quantum OR operation. Heat is supplied from outside to DNA circuit for operation. A heat transfer circuit is used in quantum multiplexer circuit to transfer the extra heat produced by the Quantum circuit to outside for cool it.

12.5.2   Working Procedure

A two-input ternary XOR is used to design the CPLD functional block. Input A and B firstly decoded by the ternary decoder and generated the output lines A0, A1, A2 for input A, and B0, B1, B2 for input B. These outputs will go thought the logic function. The logic function use four ternary AND operation and two OR operation to implement the function (A0.B1 + A1.B1) + 1. (A2.B0). During ternary DNA computing, sequence ACCTAG is considered as "0", sequence CAAGCT strands as "1" and TGGATC stands as "2". In ternary AND operation, output value depends on the minimum value of its inputs and in ternary OR operation, output value depends on the maximum value of its inputs.). The output of the decoder has only two sequence, "TGGATC (2) and "ACCTAG (0)".

1. When the input sequences A, B = ACCTAG (0), ternary decoder generate A0, A1, A2 = (2, 0, 0) and B0, B1, B2 = (2, 0, 0). None of the ternary AND operation (D0-D2) are connected to A0 and B0 line. As a result none of the ternary AND operation value will be true. So the output of the logic function will be ACCTAG (0).

2. When the input sequences A = ACCTAG (0) and B = CAAGCT (1), ternary decoder generate A0, A1, A2 = (2, 0, 0) and B0, B1, B2 = (0, 2, 0). D0 ternary AND operation is connected to A0 and B1 line. As a result D0 ternary AND operation value will be true (TGGATC). So the output of the logic function will be TGGATC.

3. When the input sequences A = ACCTAG (0) and B = TGGATC (2), ternary decoder generate A0, A1, A2 = (2, 0, 0) and B0, B1, B2 = (0, 0, 2). None of the ternary AND operation (D0-D2) are connected to A0 and B2 line. As a result none of the ternary AND operation value will be true. So the output of the logic function will be ACCTAG (0).

4. When the input sequences A = CAAGCT (1) and B = ACCTAG (0), ternary decoder generate A0, A1, A2 = (0, 2, 0) and B0, B1, B2 = (2, 0, 0). None of the ternary AND operation (D0-D2) are connected to A1 and B0 line. As a result none of the ternary AND operation value will be true. So the output of the logic function will be ACCTAG (0).

5. When the input sequences A, B = CAAGCT (1), ternary decoder generate A0, A1, A2 = (0, 2, 0) and B0, B1, B2 = (0, 2, 0). D1 ternary AND operation is connected to A1 and B1 line. As a result D1 ternary AND operation value will be true (TGGATC). So the output of the logic function will be TGGATC.

6. When the input sequences A = CAAGCT (1), and B = TGGATC (2), ternary decoder generate A0, A1, A2 = (0, 2, 0) and B0, B1, B2 = (0, 0, 2).). None of the ternary AND operation (D0-D2) are connected to A1 and B2 line. As a result none of the ternary AND operation value will be true. So the output of the logic function will be ACCTAG (0).

7. When the input sequences A = TGGATC (1), and B = ACCTAG (0), ternary decoder generate A0, A1, A2 = (0, 0, 2) and B0, B1, B2 = (2, 0, 0). D2 ternary AND operation is connected to A2 and B0 line. As a result D ternary AND operation value will be true (TGGATC). Now it will proceed ternary AND operation with CAAGCT (1) and produce minimum as CAAGCT as LUT output.

8. When the input sequences A = TGGATC (2), and B = CAAGCT (1), ternary decoder generate A0, A1, A2 = (0, 0, 2) and B0, B1, B2 = (0, 2, 0).). None of the ternary AND operation (D0-D2) are connected to A2 and B1 line. As a result none of the ternary AND operation value will be true. So the output of the logic function will be ACCTAG (0).

9. When the input sequences A, B = TGGATC (2), ternary decoder generate A0, A1, A2 = (0, 0, 2) and B0, B1, B2 = (0, 0, 2).). None of the ternary AND operation (D0-D2) are connected to A2 and B2 line. As a result none of the ternary AND operation value will be true. So the output of the logic function will be ACCTAG (0).

Output of the logic function will XOR with be ACCTAG (0). 2 input, one is the output of logic function and another is clock input will go thought the ternary D flip-flop. The ternary D flip-flop use one NOT gate four NAND gates.

1. If the XOR output and CLK input both are ACCTAG (0), then the output of the ternary D flip-flop will remain unchanged.

2. If the XOR output is ACCTAG (0) and CLK input is CAAGCT (1), then the output of the D flip-flop will be ACCTAG (0).

3. If the XOR output is ACCTAG (0) and CLK input is TGGATC (2), then the output of the D flip-flop will be ACCTAG (0).

4. If the XOR output is CAAGCT (1) and CLK input is ACCTAG (0), then the output of the D flip-flop will remain unchanged.

5. If the XOR output and CLK input both are CAAGCT (1), then the output of the D flip-flop will be CAAGCT (1).

6. If the XOR output is CAAGCT (1) and CLK input is TGGATC (2), then the output of the D flip-flop will be CAAGCT (1).

7. If the XOR output is TGGATC (2)and CLK input is ACCTAG (0), then the output of the D flip-flop will remain unchanged.

8. If the XOR output is TGGATC (2) and CLK input is CAAGCT (1), then the output of the D flip-flop will be CAAGCT (1).

9. If the XOR output and CLK input both are TGGATC (2), then the output of the D flip-flop will be TGGATC (2).

The ternary selection input S firstly is decoded by the ternary decoder. Then it will be transferred to the multiplexer as a selection input (S0, S1, and S2). A 3 to 1 multiplexer contains three input signals (outputs of D flip-flop (Q and Q') and XOR)) and three select lines. Each selected input line and one input signal perform ternary AND operation. For ternary AND operation, Output will be the minimum value. Outputs of these AND operation will store in a DNA cache memory. Trap Ione used to transfer the DNA sequence into qubit. These three Quantum qubit will go thought the Quantum OR operation to generate the CPLD functional block output. Heat supplied from outside to DNA circuit for operation. A heat transfer circuit is used in quantum multiplexer circuit to transfer the extra heat produced by the Quantum circuit to outside for cool it.

1. If all three input sequences (output of three ternary AND operation) are ACCTAG (0), then the output of the ternary OR operation is |0>.

2. If one of the three input sequence is TGGATC (2), then the output of the ternary Quantum OR operation is |2>.

3. If input sequences are in ACCTAG (0), and CAAGCT (1) then the output of the ternary Quantum OR operation is |1>

4. If input sequences are in ACCTAG (0), and TGGATC (2) then the output of the ternary Quantum OR operation is |2>

5. If input sequences are in CAAGCT (1), and TGGATC (2) then the output of the ternary OR operation is |2>.

12.6   Summary

Quantum molecular biology is a novel concept that is discussed here. Information is represented by qubits in quantum computing, whereas information is represented by DNA molecular sequences in DNA computing. Quantum molecular biology is the result of combining these two concepts. One form of this is DNA-quantum computing. This chapter has presented four PLDs in details. The PLA, PAL, FPGA, CPLD are explained with their working principles and circuit architectures.

Bibliography

[1] Kielpinski, D., Monroe, C., & Wineland, D. J. (2002). Architecture for a large-scale ion-trap quantum computer. Nature, 417(6890), 709-711.

[2] Zadeh, R. P., & Haghparast, M. (2011). A new reversible/quantum ternary comparator. Australian Journal of Basic and Applied Sciences, 5(12), 2348-2355.

[3] Lee, J. J., & Song, G. Y. (2005, January). Design of an application-specific PLD architecture. In Proceedings of the 2005 Asia and South Pacific Design Automation Conference (pp. 1244-1247).

Part IV

Multiple-Valued Nano-Processors in Quantum Molecular Biology

Overview

Complementary metal-Oxide semiconductor (CMOS) increases integration density. As a result, CMOS technology is now facing a power scaling limit. Multi-valued logic (MVL) has been considered a promising alternative for resolving power scaling challenges. Over the last three decades, multi-valued logic has drawn the attention of researchers. They discovered that MVL logic is non-classical logic. MVL and Boolean logic are identical; the difference between them is that to- MVL can allow more than two truth values that are not possible to have in the Boolean logic gate. Truth degrees of MVL logic are treated as technical tools and intend to choose them suitably for particular applications. These distinct truth degrees are represented either "Falsum" or "Verum" like the traditional truth values. But the third value $\frac{1}{2}$ stands for "possible." The MLV circuits can store more information than conventional binary logic circuits. This part will give an apparent idea of multi-valued logic. In addition, basic definitions will also be shown here. Their design procedure, working principle, and circuit of each logic operation will also be described here.

Recent research in multi-valued logic for quantum computing has shown that this multi-valued quantum system can eradicate the power scaling limit. As a framework of quantum cryptography, a multi-valued quantum system is used. The information unit of a three-valued quantum system is known as qutrit. A qutrit can be defined as a linear superposition. Moreover, a novel design of a multi-valued quantum Nano-processor will be designed. Multi-valued logic is such a logic that consists of more than two truth variables, such as ternary logic, fuzzy logic. For example, ternary logic has three values 0, 1, and 2. A complete multi-valued Quantum Nano processor using the ternary method will be discussed. Quantum ternary logic is a promising emerging technology for future quantum computing and has potential advantages over binary ones. Multi-valued logic can reduce circuit size, is more cost-effective, simple, and more efficient.

On the other hand, quantum computing has more advantages, such as high storage capacity, ensuring data security, fast computations. Moreover, researchers have shown that quantum computing and MVL are related to each other, have a close relationship. Based on the previous research, it is apparent that it is possible to merge qubit and qutrit. Therefore, quantum computing and ternary systems have tremendous advantages, and they can incorporate; it is possible to make combinational circuits using qubit and qutrit. In addition, recent research in multi-valued logic for quantum computing has shown practical advantages for scaling up a quantum computer.

13

Multiple-Valued Quantum Nano Processor

13.1 Introduction

The Quantum computing topic is more prevalent among researchers to its potential advantages. Quantum computing can solve classical NP problems in polynomial time. It relies on quantum bits or qubits expressed by |0> and |1>. Moreover, they can lie superposition that assists a qubit to remain both |0> and |1> simultaneously. On the other hand, multi-valued logic is the non-binary valued system, in which there is more than two truth values' approach can be ternary, quaternary, or can be the higher valued system that can reduce the number of devices and system complexity. Jan Łukasiewicz is the founding father of the modern theory of multi-valued logic who argued in his numerous papers about future non-certain events belonging to the domain of many-valued logic. He identified 0 or 1 truth values with probabilities that considered statements would be factual in the classical sense in the future.

The statement perfectly fits for future non-certain events in quantum mechanics. Later many researchers have shown multi-valued logic for quantum computing. Moreover, recent studies have shown us many advantages to expanding quantum computers from qubit to multi-valued logic. Three-level quantum systems are called qutrit and are the most straightforward multi-valued system. Therefore, it is expected that qutrit-based quantum computers will be more powerful. But still, the multi-valued quantum logic field is a new and immature research area. This chapter will propose a novel quantum ternary nano-processor where each component of this nano-processor is made using qubit and qutrit to get a more powerful nano-processor . In quantum operation, qubit generates much heat when they become isolated and start computations. This heat can violate the electron position of the nano-processor, which will create chaos in this nano-processor. As a result, a heat transfer circuit is used here to pass the obtained heat from the nano-processor to the environment. And then, the generated heat of this nano-processor will be cooled down using a refrigerator. It is expected that an effective ternary quantum computer can be implemented using the procedure proposed in this chapter.

DOI: 10.1201/9781003381921-13

13.2   Basic Definitions

The computer's CPU (central process unit) is a computer that executes instructions. CPU always consists of an arithmetic logic unit, buses, a control unit, and various types of registers. All types of processors should have such a processor that contains a control unit, register, ALU, RAM, and Buses. These are the core components of a processor. Unlike other processors, the quantum ternary nano-processor has the above features like quantum ternary control unit, quantum ternary ALU (arithmetic logic unit), quantum ternary register, buses. Quantum ternary control unit consists of many selection circuits like quantum ternary multiplexers, decoders that instruct within a computer processor. The Quantum ternary register is the most miniature set of qubits that can hold any data such as instructions and registers. Quantum ternary arithmetic logic unit performs all kinds of arithmetic operations. The main motto of quantum ternary RAM is to store and access data on a short-term basis, and it is a volatile memory. Three types of buses are address bus, control bus, and data bus used to transfer data between nano-processors and other components.

13.3   Block Diagram of Complete 2-Qubit Ternary Nano Processor

The CPU (Central Processing Unit) is the heart of a computer that manipulates data and executes instructions. The following Figure is the 2-qubit ternary nano-processor. This processor is implemented using some combinational quantum ternary circuits such as Instruction register, multiplexer, program counter. A complete 2-qubit ternary nano processor diagram is shown below (Figure 13.1) with their working principles.

This is the complete 2-qubit ternary nano processor where only essential CPU components are shown. There are two inputs needed in CPUs for performing meaningful work: instruction and data. The instruction register instructs the CPU what actions need to be performed on the data. The CPU's inputs are stored in the memory. In Figure 13.1, it is seen that RAM data moves from memory to the instruction register, and the CPU functions are following a cycle of fetching an instruction. After the fetching part is completed, the ternary qubit is decoded and executed. The cycle starts when data is transferred from memory to the instruction register. Sending data from memory always uses the data bus for transferring data. First, the unique qubit patterns are extracted by selecting machine language in the IR and sent to the decoder. The main motto of the decoder is to decode coded information from one format to another format. The decoder is responsible for the second step. The decoder represents which qubit pattern will operate and activates that circuit needed to perform the actual operation.

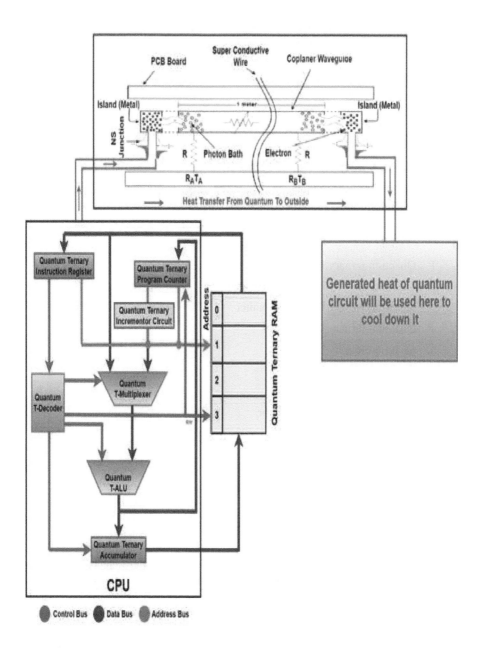

FIGURE 13.1
2-Qubit Ternary Nano-Processor

The next instruction begins to work when an operation is thoroughly performed. If an instruction is completed, the program counter is incremented by one memory location. This is the whole working procedure of the 2-qubit ternary Nano processor. Notice that a heat transfer circuit has been used as the quantum circuit produces so much heat, which can create chaos within the circuit. That is why this heat has been transferred from the circuit being cooled using the refrigerator.

13.4   Basic Components of Quantum Ternary Nano Processor

Nearly all CPUs consist of the following components that have already been discussed in quantum nano processors and DNA nano processor lessons. However, in this chapter, it will be learned with quantum ternary qubits.

The components of this CPU are

1. Quantum ternary RAM

2. Quantum ternary Instruction Register

3. Quantum ternary Program Counter

4. Quantum ternary Incrementor

5. Quantum ternary decoder

6. Quantum ternary Multiplexer

7. Quantum ternary ALU

8. Quantum ternary Accumulator.

13.4.1   Quantum Ternary RAM

Two address lines are needed for simulating 2-qubit ternary RAM . Two Quantum Ternary 1-to-3 decoder operation has been used here to achieve 2-to-9 decoder ternary quantum operation. These address lines combination will be the input of 2-to-9 decoders which consists of nine quantum ternary AND gates.

One of these outputs will be active High based on the combination of inputs present when the decoder is enabled. That means the decoder detects a particular code. The outputs of the decoder are nothing but the minterms of 'n' input variables lines when it is enabled. Each output line of the decoder will go through each quantum ternary RAM cell. Note that the word calculation of RAM will be $3 \wedge k$ where k is the address line and $3 \wedge k$ is the total words of n bit, and the decoder combination will be $k \times 3 \wedge k$. This 2-qubit ternary RAM consists of nine separate RAM cells, and each cell has 3 inputs- $|D0>$ or $|D1>$, anyone selects a line and read/ write inputs. The obtained output from nine quantum ternary RAM cells will be the input of a quantum ternary OR operation, which produces the outcome. This is the whole design procedure of 2-qubit ternary RAM.

13.4.1.1 Working Principle of Quantum Ternary RAM

This is the most crucial CPU component that stores the data quickly, a primary and volatile memory. Figure 13.2 represents the implementation of 2-qubit ternary RAM.

This quantum ternary RAM consists of nine separate "Words" of memory, and each is two cubits wide. The quantum ternary RAM Cell has three inputs and one output.

The complete circuit of a quantum ternary RAM cell is described in Figure 13.3 with proper explanation. A word consists of three ternary quantum RAM cells arranged in such a way that all qubits can be accessed simultaneously. For achieving nine terms of memory, two 1-to-3 ternary quantum decoders will be needed. Each select line input goes through a 2-to-9 decoder that selects nine worlds. The memory-enabled input enables the decoder. If the memory allows for is $|0>$, all output of the decoder will be $|0>$, and in that case, none of the memory addresses will be selected. But when the memory enables is $|1>$, one of the nine words is preferred. The read/write input determines the operation when a word is selected. The nine qubits of the word chosen pass to the quantum ternary OR gates to the output $|Y0>$ and $|Y1>$ terminals during the read operation.

To avoid complexity, the quantum ternary OR operations circuit is given in Figure 13.4. But during the write operation, the data available in the input lines are transferred into the nine quantum ternary cells of the selected word. The quantum ternary RAM cells that are chosen do not become disabled, and their previous qubit never changes. But when the memory enable input that passes into the decoder is equal to $|0>$, none of the words are selected, and then all quantum cells remain unchanged regardless of the value of the read/ write input. This is the working procedure of RAM.

The quantum ternary RAM cell is designed using a quantum ternary R-S flip flop. The total quantum ternary cells per word will be m×n, where m represents words with n bits. There are three inputs in the quantum ternary RAM-" Select," "Read/Write," and "Input," and one output line that is labeled by " Output". The "select" input is used to access either reading or writing. The cell performs the memory operation when the select line is high or $|1>$. But when the select line of the quantum ternary cell is low or $|0>$, the cell is not interested in performing a read from or written to. The following input is "Read/Write," where a system clock will conduct this input. If the clock value on the read/write line is $|0>$, this will signify "read," and when it is $|1>$, it will perform the "write" phase. Let's consider the cell that has been selected. In that case, if the clock value is $|0>$ then the cell contents are to be read, and this time the output value will depend only on the output value of the flip flop. But if the output is low, the cell output will be $|0>$, and if the output is high, the cell output will be$|1>$. It occurs because the quantum ternary AND gate added to the cell's output has three input-negated read/write, select, and Q; and both "negated read/ write" and "select are currently high.

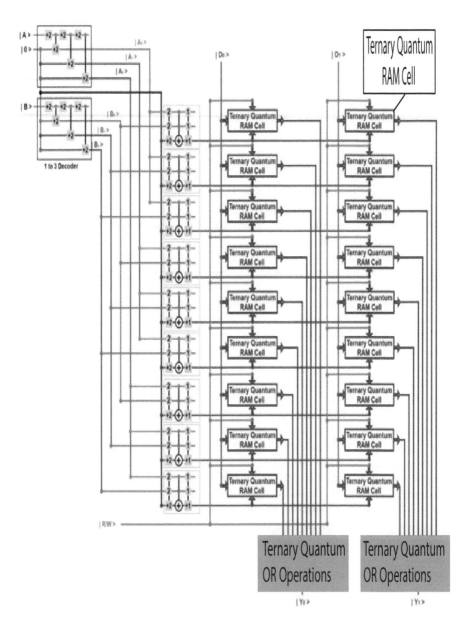

FIGURE 13.2
Ternary 2-Qubit RAM

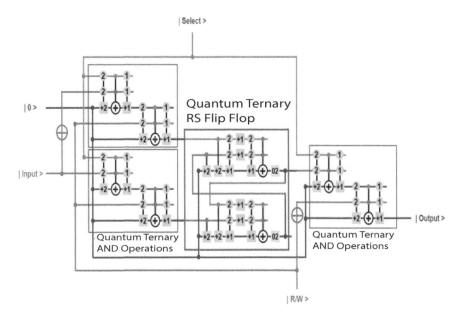

FIGURE 13.3
Ternary 2-Qubit RAM Cell

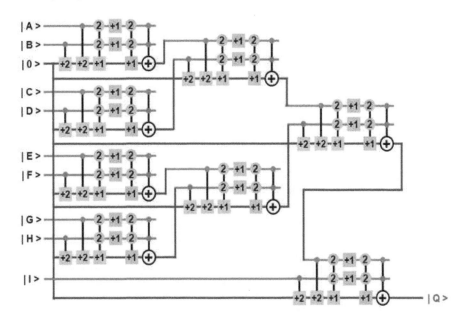

FIGURE 13.4
Ternary Quantum OR Operation

The Quantum Ternary OR operation requires the shifting operations that include (+1) and (+2) shifting. Where two (+1) operation is not controlled, and two inputs control two (+1) operation. And two (+2) operations are controlled by each input. And a two-input controlled C^2 NOT gate is also required to get the expected result.

According to Figure 13.2, it is seen that the OR gate gives the final output, which has nine inputs. First four ternary quantum OR operations are used for constructing nine inputs OR circuits and one single input. Each OR operation produces a single output. And the obtained output of four ternary quantum gates performs another two ternary quantum OR operations. Like before, those two OR operations perform another OR operation, which outputs, and the single input of this OR operation makes the final OR gate, which gives the final output. The 1st (+2) and 2nd (+2) operations of each OR circuit are controlled by the input B and A, respectively, which will open only if the input is |2>. The following (+1) is controlled by both A and B input that will only open if both inputs are |2>. The next two (+1) are not controlled. And the following (+1) is again controlled by two inputs and will open only if get |2> from both inputs. And the final XOR gate will open only if the A, B! = 0 && A! = B. The rest of the ternary quantum OR operation will work like this.

13.4.2   Quantum Ternary Instruction Register (IR)

The ternary Quantum instruction register consists of eighty-one ternary quantum AND operations. To achieve eighty-one AND operations, nine decoders are used where each decoder is 2-to-9 form. And these nine decoders are selected by another 2-to-9 ternary quantum decoder. An instruction register is a particular register that is mainly used to store the instructions currently being executed by the quantum computer. Figure 13.5 represents the ternary quantum Instruction register.

13.4.2.1   Working Procedure of Ternary Quantum Instruction Register (IR)

The purpose of the instruction register is to hold that instruction which is currently being executed. Figure 13.5 represents the ternary quantum instruction register. The number of AND gates will be eighty-one. Nine quantum decoders are used here to achieve 81 ternary quantum AND gates where each decoder will be 2-to-9 decoder forms. The first 2-to-9 decoder produces nine outputs, and each output of this decoder is fed into the other 2-to-9 decoder. The first decoder acts as an enable. Any decoder from 2-to-9 is selected according to the enable line. And the selected decoder produces nine outputs from where only one output line is activated. When the one line of any decoder becomes active, another eighty ternary quanta AND is deactivated. The IR stores the instruction word. When the CPU fetches any instruction from memory, it is temporarily stored in the instruction register. The instruction can be a qubit word or code that defines a specific operation to be performed. After that, the CPU decodes the instruction and then executes it.

The first 2-to-9 ternary quantum decoder without enabling is shown in Figure 13.6, and the rest nine decoder is with enabled input is shown in Figure 13.7.

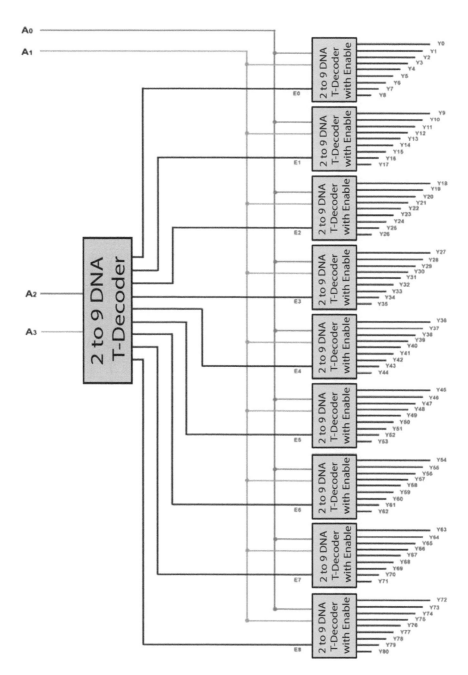

FIGURE 13.5
4-Qubit Ternary Instruction Register

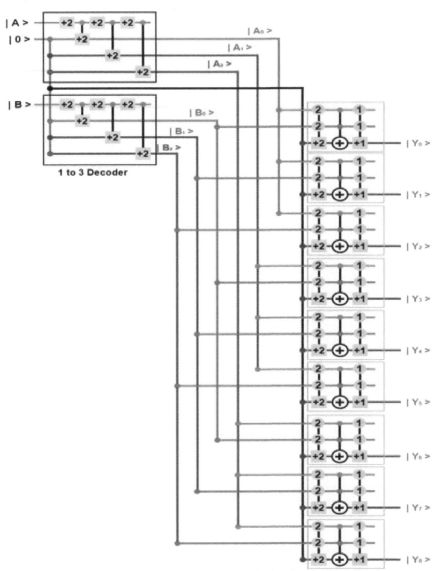

Ternary Quantum AND Operations

FIGURE 13.6
Ternary 2-to-9 Quantum Decoder

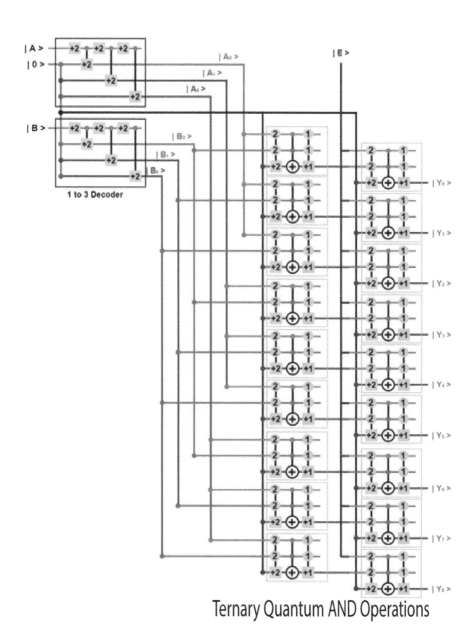

Ternary Quantum AND Operations

FIGURE 13.7
Ternary 2-to-9 Quantum Decoder with an Enable

For constructing 2-to-9 ternary quantum decoders, two 1-to-3 ternary quantum decoders are needed where each decoder has three output lines. The Quantum Ternary 1-to-3 decoder operation requires the shifting operations that include (+2) shifting only. Where three (+2) operations are not controlled, and three (+2) operations are controlled by one input. For example, for A = 0, the 1st (+2) will produce an output of 2 which will open the 1st controlled (+2) and provide 2. Thus, |D₀> will open. Others will remain closed as they will not produce the output of 2. Each decoder will work like this process, and the output lines of these two decoders are attached with nine ternary quantum AND operations—the single ternary quantum AND operation work when the output of the 1-to-3 decoder is +2.

Figure 13.7 represents the 2-to-9 decoder with an enable input shown in Figure 13.5 with the block diagrams at the right side of the 4-qubit ternary quantum instruction register. The working principle of this decoder is like a 2-to-9 ternary quantum decoder without an enable. The output combinations of 1-to-3 decoder are connected with nine ternary quantum AND operations; the output of these AND operations and an enable input make further a ternary quantum AND operation which gives the output. Only one AND gate is active, and another eight AND gates remain to deactivate.

13.4.3   Quantum Ternary Program Counter

The ternary quantum program counter consists of two ternary quantum D-flip flops where each D-flip flop has been designed using five ternary quantum NAND gates. This 2-qubit ternary program counter is only applicable for 2-qubit ternary quantum nano processors. The 2-Qubit Ternary Program Counter is shown in Figure 13.8.

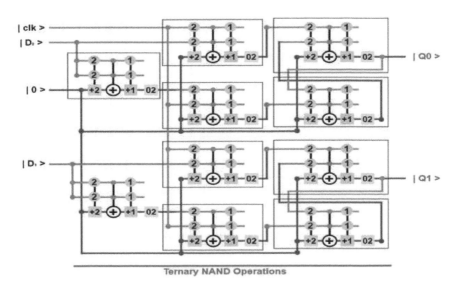

FIGURE 13.8
2-Qubit Ternary Program Counter

13.4.3.1 Working Principle of Quantum Ternary Program Counter

The program counter stores the next instruction that will be executed next. The program counter has two D-flip flops, where each D-flip flop consists of five ternary quantum NAND operations. The working procedure of NAND is described in chapter ten properly. The program counter is incremented by one if an individual instruction is completed. All instructions and data have a specific address in memory. The instructions in a program always follow the sequence memory location for storing themselves. Here $|D0>$ and $|D1>$ represent input that goes through to the D-flip flop and gives the desired output $|Q0>$ and $|Q1>$.

13.4.4 Ternary 2-Qubit Incrementor

The ternary quantum incrementor circuit is designed using two ternary quantum half adders (Quantum T-HA). Each half adder is constructed using a ternary 1-to-3 quantum decoder. The following block diagram (Figure 13.9) is 2-qubit ternary incrementor.

Quantum ternary half adder is discussed in volume 1, part 3, chapter 11 with detail description and figure. To design a ternary half adder, decoders are needed to decode the input qubit into the corresponding qubits. This carry out result will be the carry in of the second half adder circuit which produces the second output and a carry out. The second half adder has been avoided here to reduce circuit complexity.

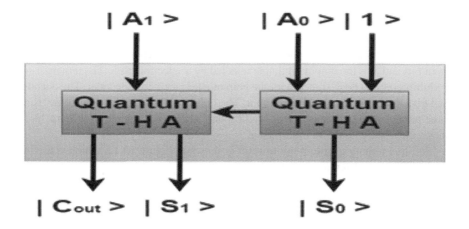

FIGURE 13.9
Block Diagram of 2-Qubit Ternary Incrementor

13.4.5   Ternary Quantum Decoder

The 2-to-9 decoder is used in 2 qubit CPU where 2-to-9 ternary quantum decoder consists of nine ternary quantum AND operations with one enable input. The decoder is a combinational ternary quantum circuit that has n input line and can produce $3 \wedge n$ output lines. Each output has one product, and for achieving this product, ternary quantum AND operations are performed here. For example, when the two input variables are $|A0\rangle$ and $|A1\rangle$, the enable input is $|1\rangle$. But if enable input is $|0\rangle$, all the decoder outputs will be equal to zero, and when enable is $|1\rangle$, one of these nine outputs will be active, i.e., $|1\rangle$.

13.4.6   Ternary Quantum ALU

This is the ternary 2 qubit ALU (Figure 13.10) which consists of fifteen ternary quantum AND operations selected by ternary quantum function selection logic. According to the function selection logic, only one logical operation is performed: addition, multiplier, or subtractor.

Every logical operation like addition and subtraction is executed with the help of a decoder. All these arithmetic operations are described in volume 1, part 3, chapter 11.

13.4.7   Ternary Quantum Multiplexer

A particular circuit device selects one of the n inputs and provides one output. It means many into one. Here 9-to-1 ternary quantum multiplexer is shown, which is implemented by a 3-to-1 ternary quantum multiplexer; Where D_0, D_1, and D_2 are input bits and Y is the output bit that is selected by a selection bit S. The ternary 3-to-1 Multiplexer can be designed with the help of a ternary 1-to-3 decoder. The decoder will give three outputs for a given input. The output will produce the outcome using three ternary AND operations followed by one ternary OR operation. Figure 13.11 shows the ternary quantum multiplexer.

A 3-to-1 ternary quantum multiplexer is shown in Figure 13.12, which is used to implement a 9-to-1 ternary quantum multiplexer.

This circuit consists of a quantum ternary 1-to-3 decoder that provides the three selection bits as output. The three quantum AND operations followed by two ternary quantum OR operations produce the expected output.

13.4.7.1   Working Procedure of Ternary Quantum Multiplexer

The multiplexer has multiple inputs and a single output. For example, a 3-to-1 multiplexer has three input signals and three select lines. Using this 3-to-1 multiplexer, a 9-to-1 ternary quantum multiplexer is implemented. To execute a 9-to-1 multiplexer, a total of 3 ternary quantum multiplexers are needed where each multiplexer has to be 3-to-1 form. Each 3-to-1 mux produces a single output connected with another

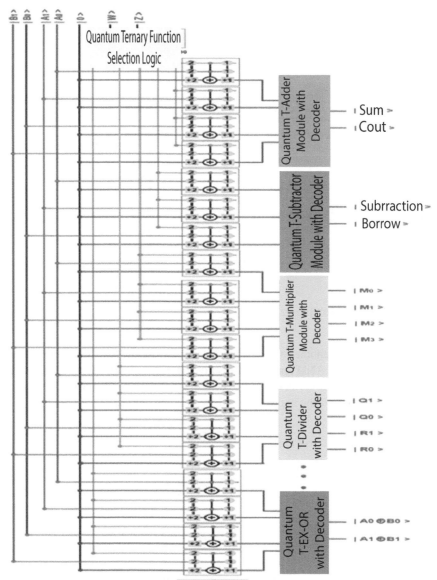

Ternary Quantum AND Operations

FIGURE 13.10
2-Qubit Ternary ALU

3-to-1 ternary quantum mux, which gives the final output. Each selected input line of 3-to-1 mux and one input signal perform ternary quantum AND operation.

The output will be the minimum value for the ternary quantum AND operation. So, the output will be that value whose input value is minimum. After this, the obtained outputs from ternary AND are attached with the ternary quantum OR opera-

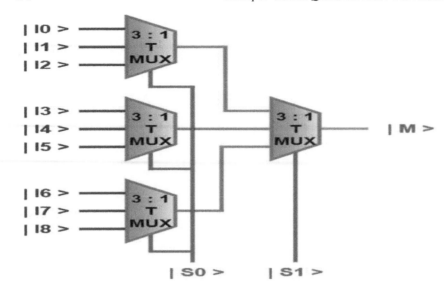

FIGURE 13.11
Ternary Quantum Multiplexer

tion as an input, and another two inputs of ternary quantum OR will come to another two ternary quantum AND operations. Finally, these three inputs give an output. For ternary OR operation, the output will always be maximum. Therefore, among the three maximum inputs the multiplexer will be the final output.

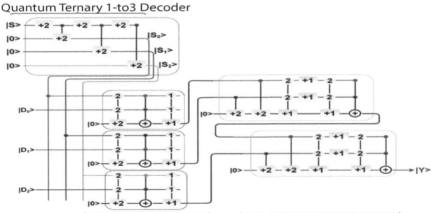

FIGURE 13.12
3-to-1 Ternary Quantum Multiplexer

Quantum Ternary AND Operations

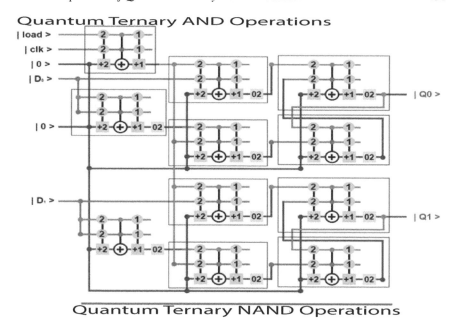

Quantum Ternary NAND Operations

FIGURE 13.13
2-Qubit Ternary Accumulator

13.4.8 Ternary Quantum Accumulator

The ternary quantum program accumulator (Figure 13.13) consists of two ternary quantum D-flip flops where each D-flip flop has been designed using five ternary quantum NAND gates. This design procedure is like a program counter; only an extra ternary quantum AND gate is used here where |LOAD> and |CLK> is the input of this operation, and the obtained output from ternary quantum AND operation will be the input of each D-flip flop. Thus, the first and second output are obtained from the first and second D-flip flop after logical execution.

13.4.8.1 Working Procedure of Ternary Quantum Accumulator

An accumulator is a special purpose register that acts as a temporary storage location and holds an intermediate value in mathematical and logical operations. Whether data will be stored or not depends on AND's output. When the ternary quantum AND's output is |0> then no data is stored in the accumulator. For example, if the AND operation's output is |1> or a high value capable of storing the data into an accumulator. But if the output of AND operation is |0> or low, the accumulator will be unable to store any data.

13.5   Applications

In recent years, more advantages are seen in integrated circuit technology, making it feasible. And it has generated more interest in electronic circuits which employ more than two discrete levels of signals. The main advantage of ternary is that they can reduce chip size as each wire can transmit more multi-valued logic. It has drawn more attention because of the less estimated interconnection cost. MVL can process large amounts of information in a small circuit.

On the other hand, it is already seen that the quantum logical circuit also has several advantages. They ensure data security and data simulations are more accessible. Moreover, quantum computing consumes less power. Another most prominent advantage is its high storage capacity, and it can solve the NP combinatorial problems in polynomial times. As MVL and quantum have a close relationship, it is possible to merge them to get more advantages. As a result, the advantages of ternary quantum logic must increase; they must be more potent than others. Recent research has shown that the main advantage of multi-valued logic for quantum computing is scaling up a quantum computer. They have also been used in the framework of quantum cryptography.

13.6   Summary

A novel multiple-valued quantum nano-processor is designed in this chapter. As quantum computers can implement algorithms in inherently multi-valued logic, it is a new approach to implement such a nano-processor that would be able to inherent advantages of both ternary logic systems and quantum computing logic together. The design of this nano-processor increases the high speed consumes low power reduces the circuit complexity. Researchers are still doing experiments to see how they are elaborated more. As they have many advantages, there are still many obstacles to implementing a real multiple-valued quantum nano-processor. It is needed to focus on solving those problems to make a helpful multiple-valued based quantum computer. The biggest challenge to further utilizing multiple-valued logic in circuit design is to create an effective computer-aided design package. Although some concepts and algorithms have already been developed, much research remains to get fruitful results.

Bibliography

[1] Yang, C. H., Leon, R. C. C., Hwang, J. C. C., Saraiva, A., Tanttu, T., Huang, W., ... & Dzurak, A. S. (2019). Silicon quantum processor unit cell operation above one Kelvin. arXiv preprint arXiv:1902.09126.

[2] M. Mukaidono, "Regular Ternary Logic Functions? Ternary Logic Functions Suitable for Treating Ambiguity" in IEEE Transactions on Computers, vol. 35, no. 02, pp. 179-183, 1986. doi: 10.1109/TC.1986.1676738

[3] Haghparast, M., Wille, R., & Monfared, A. T. (2017). Towards quantum reversible ternary coded decimal adder. Quantum Information Processing, 16(11), 1-25.

[4] Dubrova, E. (1999, November). Multiple-valued logic in VLSI: challenges and opportunities. In Proceedings of NORCHIP (Vol. 99, No. 1999, pp. 340-350).

[5] Bocharov, A., Roetteler, M., & Svore, K. M. (2017). Factoring with qutrits: Shor's algorithm on ternary and metaplectic quantum architectures. Physical Review A, 96(1), 012306.

[6] Toffano, Z., & Dubois, F. (2017). Interpolating binary and multi-valued logical quantum gates. Multidisciplinary Digital Publishing Institute Proceedings, 2(4), 152.

[7] Pykacz, J. (2021). The many-valued logic of quantum mechanics. International Journal of Theoretical Physics, 60(2), 677-686.

[8] Miller, D. M., Maslov, D., & Dueck, G. W. (2006). Synthesis of quantum multiple-valued circuits. Journal of Multiple Valued Logic and Soft Computing, 12(5/6), 431.

14

Multiple-Valued DNA Nano Processor

14.1 Introduction

In theory, people generally divide any state into two types, either "True" or "False" expressed in respectively 1 and 0 in Boolean logic. But in real life, sometimes the state of an event or things may have more than two states. For example, in a week, there are seven days. According to Boolean logic, it is not possible to represent only two days using true or false. There is no way to express the rest of the days of a week. For solving such problems, multi-valued logic is introduced. Those days can be divided into multi-valued logic, efficiently representing such situations. Xiangou ZHU et al. designed the first DNA computing model using a ternary logic system. After his design, many researchers have started to focus on this field. Like others, here also proposes multi-valued DNA logic gates. Here ternary logic systems are used to implement these operations, and ternary logic systems are examples of multi-valued logic with three states 0, 1, 2.

In ternary DNA computing, 0 is expressed as "ACCTAG," 1 is denoted by "CAAGCT" and finally, 2 is defined as "TGGATC." Based on the combinational ternary DNA computing logical operations, in this chapter, a complete ternary DNA nano processor for two-bit ternary DNA molecule sequences is proposed. In this Ternary DNA computing, the fluorescent level is used to detect the DNA sequence. Each DNA sequence will make a bond according to the Watson - Crick Model, where Adenine (A) makes a bond with Thymine (T) and Cytosine (C) sticks with Guanine (G). Heat is generated in ternary DNA computing for chemical reactions. That is why the heat for each component is measured to check how much heat is necessary for the chemical reaction that has to be provided from the outside environment. Moreover, the speed will also be shown here to measure this nano-processor motion and how much faster it is to perform any operation than any other nano-processor.

14.2 Basic Definitions

The core components of a ternary DNA nano-processor are the ternary DNA control unit, the ternary DNA register, the ternary DNA ALU (arithmetic logic unit), ternary DNA quantum RAM (random access memory), and ternary DNA buses. These com-

DOI: 10.1201/9781003381921-14

455

ponents have an individual role in the nano-processor that assists the nano-processors in performing any execution properly. All kinds of data flow and instructions are controlled by the nano-processors. This is a ternary DNA nano processor, and it consists of three, where each component works with the help of these inputs. These three inputs are ACCTAG which represents 0, CAAGCT strands as 1, and TGGATC is denoted by 2. The ternary DNA control unit provides a computer nano-processor among the core components. A ternary DNA register is the most miniature set of ternary DNA sequences that can hold the place. All mathematical and logical operations are performed in ALU. Ternary DNA RAM is another component of this nano-processor that helps to increase system performance.

The most core element of a nano-processor is buses. The buses transfer the data between nano-processors and components. A nano-processor consists of three bus data, the address, and the control bus. The main function of a data bus is to carry the data between a nano-processor and RAM. The address bus is unidirectional, carrying memory addresses from the nano-processor to other components. And the last bus is a control bus that makes sure everything flows perfectly from place to place. To accomplish meaningful work, it's all needed for a nano processor.

14.3   Block Diagram of Ternary 2-Bit DNA Nano Processor

Figure 14.1 represents the block diagram of a ternary DNA nano processor . It consists of a ternary DNA nano-arrays unit, a ternary DNA nano multiplexer, a ternary DNA nano RAM, and encodes ternary DNA sequences unit ternary DNA nano ALU. All circuits use multi-valued DNA basic logic operations as it is a ternary DNA nano processor. The red, blue, and green lines represent the control bus, data bus, and address bus. Molecular computations will work when heat is provided from outside. They cannot make any chemical reaction without heat. As a result, a source has been used for supplying additional heat. A complete DNA nano processor block diagram is shown in Figure 14.1.

The fundamental operation of this ternary DNA nano-processor is to execute a sequence of stored instructions that is called a program. These instructions are kept in some kind of memory for execution. Almost all CPUs, either a microprocessor or nano-processor, follow the three rules - fetch, decode and execute for performing their operation. These steps are collectively known as the instruction cycle. After executing an instruction, the entire process starts to work the same thing. The successive instruction cycles typically fetch the next instruction because of the incremented value in the program counter. After fetching, the instructions are decoded to various signals that control other areas of the nano-processor. Finally, the nano-processors execute the instruction. When an instruction is executed, the program counter is incremented by one memory location. This is the whole working procedure of ternary DNA nano-processors.

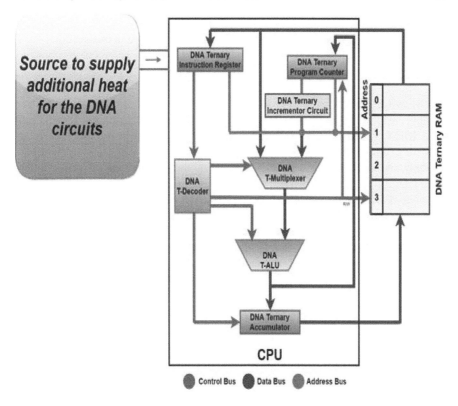

FIGURE 14.1
2-Bit Ternary DNA Nano-Processor

14.4 Basic Components of Ternary DNA Nano Processor

According to Figure 14.1, the ternary DNA nano processor consists of the following components-

1. Ternary DNA RAM
2. Ternary DNA Instruction Register
3. Ternary DNA Program Counter
4. Ternary DNA Incrementor
5. Ternary DNA decoder
6. Ternary DNA Multiplexer
7. Ternary DNA ALU
8. Ternary DNA Accumulator.

These are also essential components of the CPU that need to be accomplished for meaningful work. However, now all features will be explained in this section.

14.4.1   Ternary DNA RAM

Firstly, two address lines are needed for simulating ternary 2-bit DNA RAM, and for this, the ternary 1-to-3 DNA decoder operation is used to achieve a ternary 2-bit DNA RAM. These 1-to-3 ternary DNA decoders produce ternary 2-to-9 decoders, consisting of nine ternary AND DNA operations.

Only one result will be active High based on the combination of inputs present when the decoder is enabled. That means that at a time, the decoder detects a particular code. Each output line of the decoder is fed into each ternary DNA RAM cell. For example, if k is the address line, the word calculation of RAM will be $3 \wedge k$, and $3 \wedge k$ is the total words of n bit, and the decoder combination will be $k \times 3 \wedge k$. This ternary 2-bit DNA RAM consists of nine separate RAM cells, and each cell has 3 inputs- any ternary DNA sequence, anyone selects a line and read/ write inputs. The obtained output from nine quantum ternary RAM cells will be the input of a ternary DNA OR operation, which produces the final output. This is the whole design procedure of ternary 2-bit DNA RAM .

14.4.1.1   Working Principle of DNA RAM

RAM is a primary and volatile memory that stores the data quickly. The following circuit (Figure 14.2) shows the 2-Bit Ternary DNA RAM

This ternary DNA RAM consists of nine separate "Words" of memory. The ternary DNA RAM Cell consists of three inputs and one output. The ternary 2-bit DNA RAM cell is shown in Figure 14.3 with proper explanation.

A word consists of two ternary DNA RAM cells arranged in such a way so that all DNA sequence bits can be accessed simultaneously. Two 1-to-3 ternary DNA decoders are used to achieve nine memory words. Each selected line input goes through a 2-to-9 decoder that selects only one of the nine words at a time. The function of memory-enabled input enables the decoder. Note that, from the ternary 2-to-9 decoder, either ACCTAG or TGGATC will be obtained. There is no probability of achieving CAAGCT. If the memory enabled is ACCTAG, all output of the decoder will be ACCTAG, and in that case, none of the memory addresses will be selected. But when the memory enables TGGATC, one of the nine words is selected. The read/write input determines the operation when a word is selected. The nine DNA molecule sequence bits of the selected word pass to the ternary DNA OR operations to the output terminals during the read operation. But during the write operation, the data available in the input lines are transferred into the nine ternary DNA cells of the selected word. The ternary DNA RAM cells that are not selected have become disabled, and their previous DNA sequence bit never changes. But when the memory-enabled input that passes into the decoder is equal to ACCTAG, none of the words are selected, and then all quantum cells remain unchanged regardless of the value of the read/ write input. This is the working procedure of RAM.

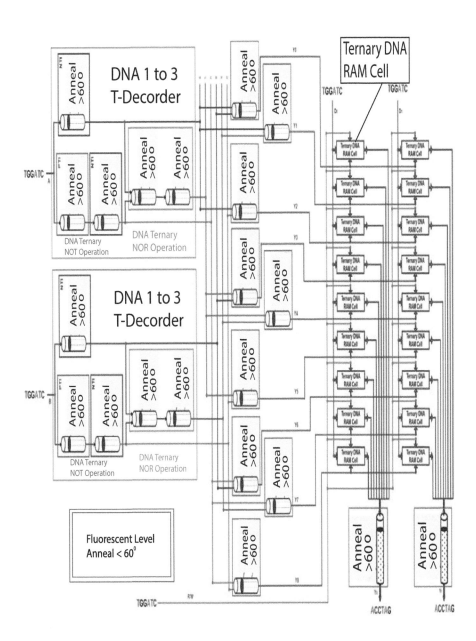

FIGURE 14.2
2-Bit Ternary DNA RAM

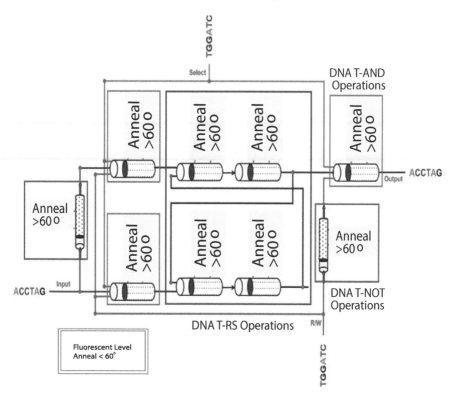

FIGURE 14.3
Ternary 2-Bit DNA RAM Cell

The ternary DNA RAM cell is designed using a ternary DNA R-S flip flop. The number of total DNA ternary cells per word will be m×n, where m represents words with n bits DNA molecule sequence. Let's consider the above example; it is seen that This ternary DNA RAM cell consists of -" Select" that is shown as TGGATC, "Read/Write" is also assumed TGGATC, and "Input" is considered as ACCTAG. Performing all operations, one output line is obtained that is ACCTAG'. The "select" input is used to access reading or writing operations. The cell performs the memory operation when the select line is high or TGGATC. But when the select line of the ternary DNA cell is low or ACCTAG, then the cell does not perform a read from or written to. The following input is "Read/Write," where a system clock will conduct this input. If the clock value on the read/write line is ACCTAG, this will signify "read," When it is TGGATC, it will perform the "write" phase. This is the fundamental working procedure.

14.4.2 Ternary 2-Bit DNA Instruction Register (IR)

The ternary DNA molecule sequence instruction register consists of eighty one ternary DNA AND operations. To achieve eighty-one DNA AND operations, the help of 2-to-9 decoders will be taken where the 2-to-9 decoder number will be a total of nine. And these nine decoders are selected by another 2-to-9 ternary DNA decoders. Figure 14.4 depicts the ternary 4-Bit Molecular Sequence Instruction Register (IR).

14.4.2.1 Working Procedure of Ternary DNA Instruction Register

An instruction register is a special register that is mainly used to store the instructions currently being executed by the quantum computer. Figure 14.4 represents the ternary DNA instruction register. The number of ternary AND gates will be eighty-one. Nine DNA decoders are used here to achieve 81 ternary DNA AND gates where each decoder will be 2-to-9 decoder form. The first 2-to-9 decoder is used here to produce nine outputs, and each output of this decoder is fed into the other 2-to-9 decoder. The first decoder acts as an enable. According to the enable line, any decoder among decoders 2-to-9 is selected. And the selected decoder yields nine outputs from which only one output line is activated at a time. When the one line of any decoder becomes active, another eighty ternary DNA AND operation is deactivated. The IR generally stores the instruction word. The instruction can be a ternary DNA bit sequence word or code that defines a specific operation to be performed. After that, the CPU decodes the instruction and then executes it.

This IR 2-to-9 decoder has been used here without enabling input and with enabling output, which is shown respectively in Figure 14.5 and Figure 14.6.

For constructing 2-to-9 ternary DNA decoders without enabling input, it is needed to use two 1-to-3 ternary DNA decoders where each decoder has three output lines. For designing 1-to-3 ternary decoders, it is needed to follow some steps. Any ternary DNA input, for example, A, first goes through NTI to produce A0, then again, the same input goes through PTI and NTI to produce A2, and finally, the result of A0 and A1 goes through ternary DNA NOR operation to produce A1. Each 1-to-3 ternary DNA decoder will work like this, and thus, the two numbers of 1-to-3 ternary decoder's output line combination produce nine ternary DNA AND operations. The single ternary DNA AND operation works when the output of the 1-to-3 decoder is TGGATC.

The working principle of the 2-to-9 decoder with enable is like a 2-to-9 ternary decoder without an enable. The output combinations of 1-to-3 ternary DNA decoder are connected with nine ternary DNA AND operations; the output of these AND operations and an enable input make further a ternary DNA AND operation, which gives the output. Only one AND gate is active, and another eight AND gates remain to deactivate.

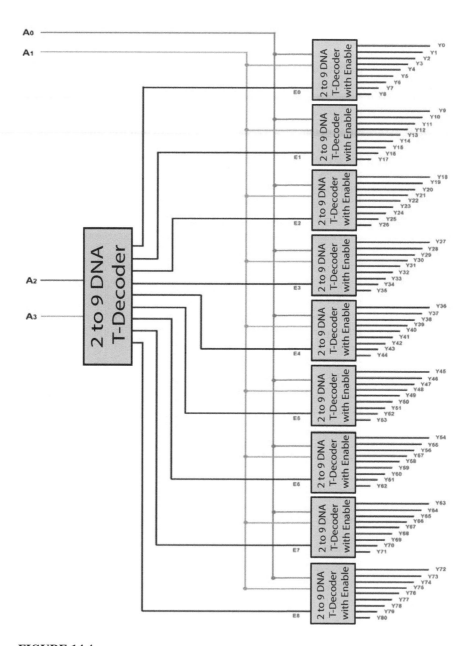

FIGURE 14.4
Ternary 4-Bit Molecular Sequence IR

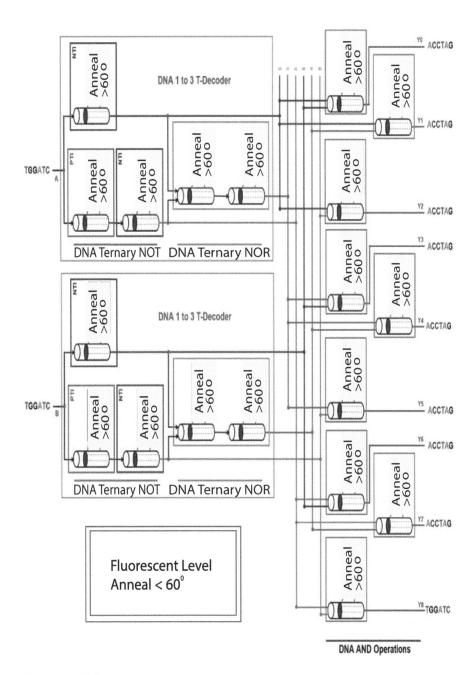

FIGURE 14.5
Ternary 2-to-9 DNA Decoder without Enable

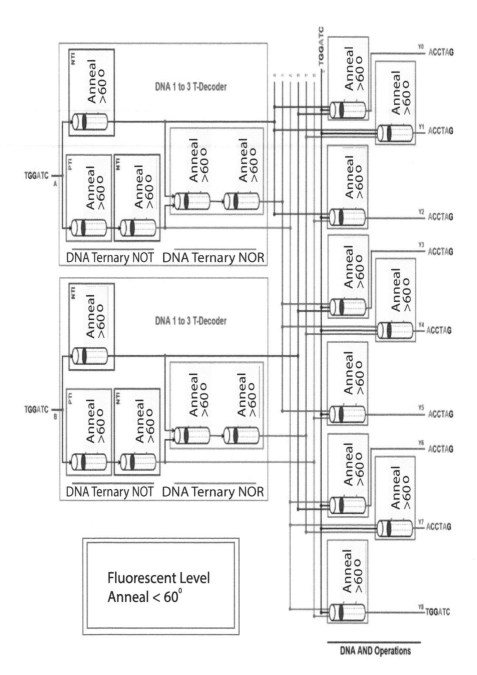

FIGURE 14.6
Ternary 2-to-9 DNA Decoder with Enable

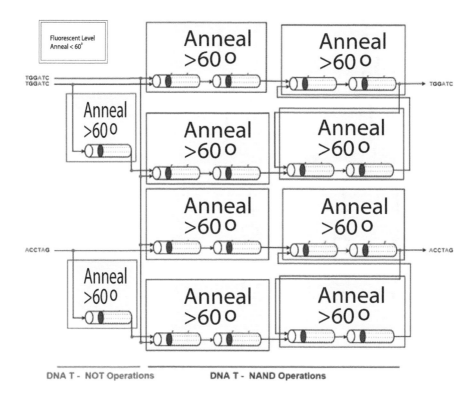

DNA T - NOT Operations   **DNA T - NAND Operations**

FIGURE 14.7
Ternary 2-Bit Molecular Sequence DNA Program Counter

14.4.3 Ternary DNA Program Counter

The ternary DNA program counter consists of two ternary DNA D-flip flops where each D-flip flop has been designed using four ternary DNA NAND gates, and each D-flip flop is connected with a ternary DNA NOT operation. Figure 14.7 represents the ternary 2-Bit Molecular Sequence DNA Program Counter .

14.4.3.1 Working Principle of Ternary DNA Program Counter

The main function of the program counter is to store the next instruction that is going to be executed next. The program counter consists of two D-flip flops where each D-flip flop consists of four ternary DNA NAND operations, and each DNA flip flip is connected with one ternary DNA NOT operation. The working procedure of ternary DNA NAND and DNA NOT are known. The program counter is incremented by one if an individual instruction is completed. All instructions and data have a specific address in memory. The instructions in a program always follow the sequence memory location for storing themselves.

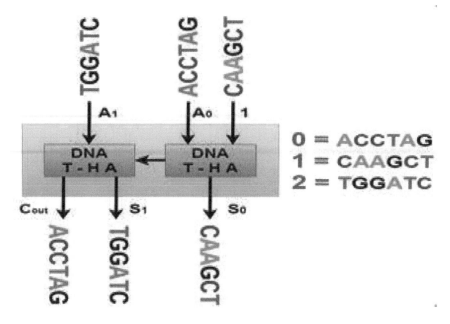

FIGURE 14.8
Ternary 2-Bit Molecular Sequence DNA Incrementor

14.4.4 Ternary 2-Bit DNA Incrementor Circuit

The ternary DNA incrementor circuit is constructed here using two ternary DNA half adders. Each half adder consists of ternary 1-to-3 decoders. Two 1-to-3 decoders are used for getting the exact output. Figure 14.8 shows the ternary 2-Bit Molecular Sequence DNA Incrementor.

This carry-out result will be the carry-in of the second half adder circuit which produces the second output and a carry-out. The second half adder has been avoided here to reduce circuit complexity. This is the working procedure of a 2-bit ternary DNA molecular program counter.

14.4.5 Ternary DNA Decoder

A decoder is another vital component of this ternary 2-bit DNA nano-processor. For designing this nano-processor, it is needed to construct a ternary 2-bit DNA decoder, and for this, a 2-to-9 decoder is used which is implemented using two 1-to-3 ternary decoders.

These 2-to-9 ternary DNA decoders consist of nine ternary DNA AND operations with one enable input described earlier. The decoder is a combinational ternary DNA circuit that has n input line and can produce 3^n output lines. Each output has one product, and for achieving this product, ternary quantum AND operations are performed here.

14.4.6 Ternary DNA ALU

An arithmetic logic unit is the core component of a nano-processor. It is a combinational ternary circuit that performs arithmetic and logic operations. The control unit indicates to ALU what process it needs to achieve on the data and stores the result in an output register. It is needed to construct a 2-bit ternary DNA ALU for a 2-bit ternary DNA nano processor. This 2-bit ternary DNA ALU consists of fifteen ternary DNA AND operations selected by ternary DNA function selection logic. This function selection logic is a 2-to-9 ternary decoder without enabling, shown in Figure 14.5. According to the function selection logic, only one logical operation is performed, such as addition, multiplier, or subtractor. Figure 14.9 depicts the Ternary 2-Bit Molecular Sequence DNA ALU.

Every logical operation like addition and subtraction is executed with the help of a decoder. All arithmetic operations are described in volume 1, part 3, chapter 13.

14.4.7 Ternary DNA Multiplexer

A multiplexer is a special component of a nano processor that selects one of the n inputs and provides it with one output. It means many to one. A ternary 9-to-1 DNA multiplexer is needed for a 2-bit ternary DNA nano-processor. 9-to-1 mux is designed using ternary three 3-to-1 mux. To perform DNA 3-to-1 multiplexer operation, it is needed to operate a 1-to-3 decoder. The purpose of using this decoder is to take three inputs. As it is a ternary multiplexer, it is needed to access three inputs: ACCTAG, CAAGCT, and TGGATC. If 1-to-3 decoders are not used, it will not be possible to access all combinations of these three inputs. Therefore, the decoder will give three outputs for a given input. The output will produce the output with the help of three ternary AND operations followed by one ternary OR operation. Figure 14.10 shows DNA Ternary Multiplexer.

14.4.8 Ternary DNA Accumulator

The ternary DNA program Accumulator is designed using two ternary DNA D-flip flops where each D-flip flop has been developed using four ternary DNA NAND operations connected ternary DNA NOT operations. This design procedure is like a program counter; only an extra ternary quantum AND gate is used here where |LOAD> and |CLK> is the input of this operation those are assumed here TGGATC and TGGATC, and the obtained output from ternary DNA AND operation will be the input of each D-flip flop. After performing logical operations, the first and second outputs will be obtained from the first and second D-flip flop. The circuit architecture of 2-Bit DNA Ternary Accumulator is shown in Figure 14.11.

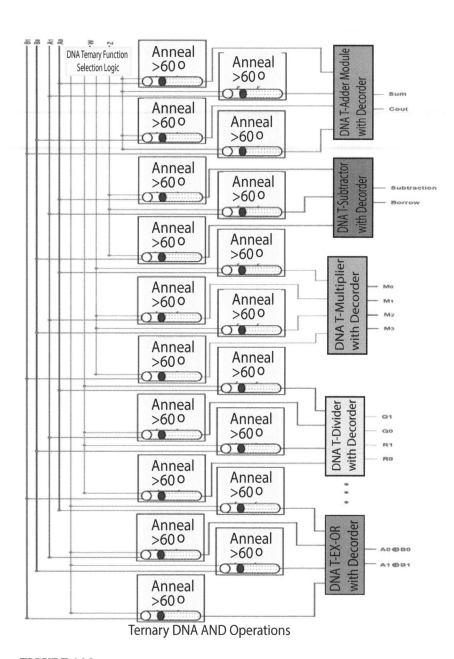

FIGURE 14.9
Ternary 2-Bit Molecular Sequence DNA ALU

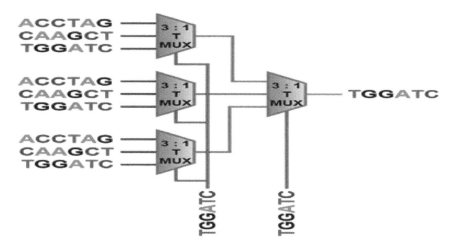

FIGURE 14.10
DNA Ternary Multiplexer

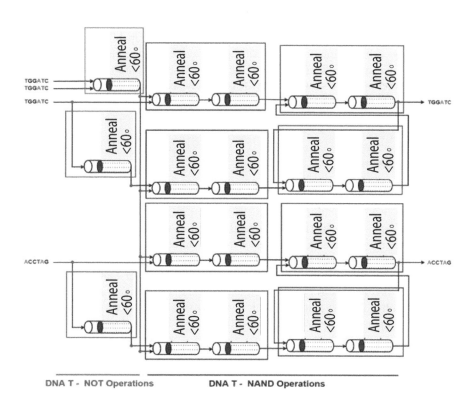

FIGURE 14.11
2-Bit DNA Ternary Accumulator

14.4.8.1 Working Procedure of Ternary DNA Accumulator

The accumulator is a special purpose register that acts as a temporary storage location in which intermediate arithmetic logic unit results are stored. Whether data will be stored or not relies on ternary DNA AND output. If the ternary DNA AND output are ACCTAG, no data is stored in the accumulator. But if the AND operation's output is TGGATC or a high value capable of storing the data into an accumulator. It is the working principle of ternary DNA accumulators.

14.5 Applications

Recently, DNA computing-based circuits have been a novel approach to data analysis. Various kinds of computation have drawn more attention from researchers than traditional calculations. DNA computing is one of them. DNA computing has several advantages, such as it can solve more powerful combinations of any search problems. The DNA-based circuit uses less energy than traditional silicon-based computing. It has reusable features. This chapter has described how to design DNA logic operations with the help of ternary logic. These circuits have the capability of high speed and low power dissipation. It will inherit superior merits such as easy synthesis, low cost, chip area, and excellent flexibility. It can play a significant role in medical science for detecting any disease quickly. It is already seen that DNA computing and multi-valued logic have several advantages

14.6 Summary

In this chapter, a novel multiple-valued DNA nano-processor is designed. DNA computing can implement algorithms inherently in multi-valued logic. Therefore, it is tried to implement such a nano-processor that would be able to inherent advantages of both ternary logic systems and DNA computing logic together. The design of this nano-processor increases the high speed consumes low power reduces the circuit complexity. It is needed to provide high temperatures for chemical reactions in this DNA computing process that is difficult to deliver outside the environment. Though this field has potential advantages, it is still murky. It is needed to pay attention in this field to acquire the best result.

Bibliography

[1] Zheng, X., Yang, J., Zhou, C., Zhang, C., Zhang, Q., & Wei, X. (2019). Allosteric DNAzyme-based DNA logic circuit: Operations and dynamic analysis. Nucleic Acids Research, 47(3), 1097-1109.

[2] Smith, K. C. (1981). The prospects for multi-valued logic: A technology and applications view. IEEE Transactions on Computers, 30(09), 619-634.

[3] Sarker, A., Hasan Babu, H. M., & Rashid, S. M. M. (2015). Design of a DNA-based reversible arithmetic and logic unit. IET Nanobiotechnology, 9(4), 226-238.

[4] Watada, J. (2008). DNA computing and its application. In Computational Intelligence: A Compendium (pp. 1065-1089). Springer, Berlin, Heidelberg.

[5] Lin, S., Kim, Y. B., & Lombardi, F. (2009). CNTFET-based design of ternary logic gates and arithmetic circuits. IEEE Transactions on Nanotechnology, 10(2), 217-225.

[6] Dhande, A. P., & Ingole, V. T. (2005, March). Design and implementation of 2 bit ternary ALU slice. In Proc. Int. Conf. IEEE-Sci. Electron., Technol. Inf. Telecommun (Vol. 17).

[7] Shan, J. Y., Yin, Z. X., Tang, X. Y., & Tang, J. J. (2014). A DNA computing model for the AND gate in three-valued logical circuit. In Applied Mechanics and Materials (Vol. 610, pp. 764-768). Trans Tech Publications Ltd.

15

Multiple-Valued Quantum-DNA Nano-Processor

15.1 Introduction

Multi-valued quantum computing and multi-valued DNA computing are now an emerging field, where quantum computing works with |0>, |1>, and |2> and multi-valued DNA computing works with ACCTAG, CAAGCT, and TGGATC. These are already described in the previous chapters of ternary DNA computing and ternary DNA nano-processor. Multi-valued logic is the non-binary valued system in which there may have more than two truth values. MVL can be any type, such as ternary, quaternary, or higher matter.

The chapter mainly concerns the design, simulations, and development of a quantum-DNA ternary nano processor where quantum computing and DNA computing consist of three values. The previous chapter only discussed quantum and DNA nano-processors for three values. But this chapter will discover the new area were to design such a nano-processor that consists of both DNA sequence and quantum qutrits together, and this sequence and qutrits valued will be three as it is a ternary quantum DNA nano-processor. All components of this nano processor will have been kept in a ternary DNA state except ternary quantum RAM and cache memory that would store ternary qutrit. As it is a loamy nano-processor, it is needed use methods where qutrit and DNA sequence can coexist without creating anarchy. In this nano-processor circuit, it will be seen that qubit passes quantum RAM to the DNA instruction register. But it is not possible to give qutrit to the instruction register because qutrit and DNA sequences are different. As a result, a media is used that converts ternary qutrit to ternary DNA sequence, and this media is NMR relaxation that needs to perform at normal room temperature.

A DNA sequence will move to RAM with a new address when an instruction is completed. But as RAM is in a quantum state, it is needed to convert the DNA sequence to the qutrit. For converting it, NMR is used at normal room temperature. A heat transfer circuit will also transfer ternary-quantum component heat to the ternary DNA component. This process will reduce costs. But this heat may not be sufficient for the ternary DNA component. That is why a source supply is used here to provide heat to ternary DNA components that help accomplish each execution properly. It is already seen that ternary quantum computing and ternary DNA computing have

DOI: 10.1201/9781003381921-15

several advantages to the computing field. Therefore, this combining nano-processor will add a new dimension and more benefits to modern technology.

15.2   Basic Definitions

Every microprocessor and nano-processor have the same components to accomplish any meaningful work of a processor. They consist of some core components like control unit, register, arithmetic logic unit (ALU), RAM (random access memory). Additionally, they mainly consist of three types of buses. Each component aids the processor in exacting tasks properly. Like other processors, a ternary quantum DNA nano-processor is a combinational circuit that executes instructions on behalf of programs. The control unit is one of the main components of a ternary quantum DNA that directs the operation of the nano processor. Registers are one kind of memory used to accept, store and transfer data and instructions used immediately by the nano process. In computing, an arithmetic logic unit is a combinational ternary quantum-DNA circuit that handles all the calculations the nano processor may need. RAM is one of the components that store information. The main goal of RAM is to store and access data on a short-term basis. Buses are high-speed internal connections that send control signals and data between the nano processor and other ternary quantum-DNA nano-processor components.

These components have already been adequately described in the previous two chapters which are called multi-valued quantum nano-processor and multi-valued DNA nano-processor.

15.3   Block Diagram of Complete Ternary 2-Qubit Quantum-DNA Nano Processor

The first and foremost component is the CPU (central processing unit), called the heart of a computer. The primary function of this is to manipulate data and execute instructions. The processor is designed using combinational circuits such as an instruction register, multiplexer, decoder, and ALU. This section will represent a ternary quantum-DNA nano-processor that also consists of some combinational circuits such as incrementor, accumulator, multiplexer, and RAM. A source has been used to supply additional heat for the DNA circuits. The following circuits (Figure 15.1) represent the ternary quantum-DNA nano-processor.

This is the combinational ternary quantum-DNA circuit where only basic components are shown. The design of this nano-processor is quite complex as it is made of both quantum qutrits and DNA sequences. There are two inputs needed in CPUs for performing meaningful work-instruction and data. The instructions register gives the

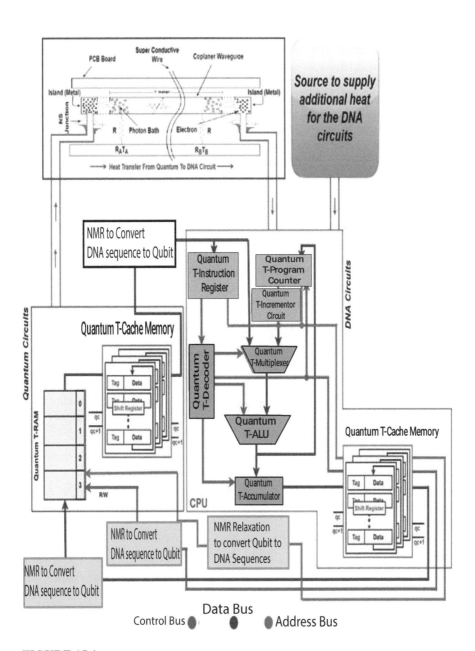

FIGURE 15.1
Ternary Quantum-DNA Nano-Processor

instructions to the CPU to operate. The CPU functions follow a cycle of fetching instructions. After fetching an instruction, it is decoded. After decoding an instruction, it determines what kind of operation the instruction should perform. Then, instruction is completed in the executed step. In this step, a computer operation is executed. And finally, it is needed to modify the instruction counter to determine the address of the next instruction. Now, let's describe the above working principle in this combinational nano-processor.

According to Figure 15.1, data as a qutrit is passed from quantum ternary RAM to DNA ternary instruction register. It is not possible to pass qubit to the DNA ternary instruction register directly. Before passing the DNA ternary instruction register, it is needed to convert the qutrit to a DNA sequence. NMR relaxation is used to convert a qutrit to a DNA sequence. Another thing is that quantum computing is faster than DNA computing. Quantum T-cash memory has been used because ternary quantum RAM sends data quickly, but DNA cannot receive the information soon. It is known that quantum qutrits generate so much heat, and on the other hand, DNA sequence needs temperatures for chemical reaction. Quantum heat is used in this nano processor for performing chemical reactions of DNA combinational circuits. That is why a heat transfer circuit has been used here that provides the obtained heat of quantum ternary RAM into a DNA ternary nano-processor.

RAM will be in the only quantum ternary form in this nano-processor, and the rest of the CPU components will be in DNA ternary form. It is executed when an instruction is fetched into the DNA ternary decoder. The fetching and execution parts are completed at the DNA ternary nano-processor. After completing the instruction, data as a DNA sequence is sent to the quantum T-RAM. But, in that case, it is not possible to pass DNA sequence to quantum T-RAM directly. Therefore, DNA sequences are needed to be stored to DNA T-cache memory, and from this cache memory, the sequence is passed into the NMR circuit to convert DNA sequence to qutrits. It is needed to consider the following steps in the nano-processor:

1. Pass data as qutrits to quantum cache memory

2. Do NMR relaxation for converting qutrit to the DNA sequence.

3. After executing an instruction, pass DNA sequence to DNA T-cache memory

4. Do only NMR for converting DNA sequence to the qubit.

Thus, the whole nano-processor will work. The working procedure of quantum ternary cache memory, NMR relaxation, NMR, and DNA ternary cache memory has been described in details in part 2 of this book.

15.4 Basic Components of Ternary Quantum-DNA Nano Processor

A complete ternary quantum-DNA nano processor has been designed using the following components shown in Figure 15.1. The members of this CPU are given below:

1. Quantum Ternary RAM
2. DNA Ternary Instruction Register
3. DNA Ternary Program Counter
4. DNA Ternary Incrementor
5. DNA Ternary Decoder
6. DNA Ternary Multiplexer
7. DNA Ternary ALU
8. DNA Ternary Accumulator.
9. Quantum cache memory
10. DNA cache memory
11. NMR relaxation
12. Heat transfer circuit and
13. NMR

Moreover, this nano-processor uses three buses -data bus, address bus, and control bus- to transfer data from one component to another. These are also essential components of the nano-processor that need to be accomplished for meaningful work.

15.4.1 Quantum Ternary RAM

As this is a 2-qubit ternary DNA nano-processor, two address lines are needed for simulating ternary 2-qubit RAM. Two Quantum Ternary 1-to-3 decoder operation is used here to achieve 2-to-9 decoder ternary quantum operation. These address lines combination will be the input of 2-to-9 decoders which consists of nine quantum ternary AND gates. Only one output will be active high based on the combination of inputs present when the decoder is enabled. This is the most crucial component of a nano-processor that stores the data quickly, a primary and volatile memory. The design and working principle of ternary quantum RAM already has been described in chapter 13.

15.4.2   Ternary DNA CPU

According to Figure 15.1, all nano-processor elements except RAM are DNA form. A ternary DNA nano-processor consists of a DNA ternary Instruction register, program counter, incrementor, multiplexer, decoder, accumulator, and ALU. To avoid complexity, the working principle and the design procedure of all these elements are skipped here and recommended to see chapter 14, where the basic components of ternary DNA Nano processor with their Design procedures and working principles have been described in detail.

Now the rest of the topics are quantum ternary cache memory, DNA ternary cache memory, NMR and NMR relaxation to convert the qubit to DNA sequences and the DNA sequences to qubit. All these topics related to quantum-DNA nano-processor are discussed in detail in volume 1, part 2.

15.5   Applications

In this chapter, a ternary quantum DNA nano-processor is designed, a new method in logical computations that will add a new dimension to computer technology. The main advantage of this proposed nano-processor is to attain high speed and storage capacity. Quantum computing is so fast that, according to IBM, the machine performed a mathematically designed calculation so complex in 200 seconds that it would take the world's most powerful supercomputer about 10,000 years to do it. This makes quantum computers about 158 million times faster than the world's fastest supercomputer.

On the other hand, the information density of DNA is remarkable because only one gram of DNA can store 215 petabytes of data. In addition, it is known that DNA resources are available in nature, so if this nano-processor is practically implemented, it will be cost-effective. Even with some disadvantages such as high error rates, heating problems can be reduced. If this nano-processor is implemented, the drug development area, artificial intelligence, and machine learning field will develop more.

15.6   Summary

The design of this nano-processor increases the high speed which consumes low power and reduces the circuit complexity. Although this nano processor can gain more and more advantages, the disadvantages are not more minor. As a result, more research are needed to be done in this field for technological progress. The proposed ternary quantum DNA nano-processor is a theoretical base but logical. Despite having more advantages, there are still many obstacles to implementing an actual ternary quantum DNA nano-processor. The environment setup will be more challenging to execute the entire operations of this nano-processor. The process that is followed here is not flexible. More human assistance is needed here. However, it is needed to solve those problems to make a functional ternary quantum DNA computer. The biggest challenge to further utilizing ternary quantum DNA logic in circuit design is developing an effective computer-aided design package. However, it is possible to be optimistic and confident that such a nano-processor-based computer plays a significant role in the entire technology system if it is implemented.

Bibliography

[1] Javadi-Abhari, C. A. K. A. (2019). A McClure DT Cross AW Temme K Nation PD Steffen M Gambetta JM Challenges and opportunities of near-term quantum computing systems. Proceedings of IEEE, 10.

[2] Shukla, S. K., & Bahar, R. I. (2004). Nano, Quantum and Molecular Computing. Kluwer Academic.

[3] Woese, C. R., Gutell, R., Gupta, R.,& Noller, H. F. (1983). Detailed analysis of the higher-order structure of 16S-like ribosomal ribonucleic acids. Microbiological Reviews, 47(4), 621-669.

[4] Dann, R., Kosloff, R., & Salamon, P. (2020). Quantum finite-time thermodynamics: insight from a single qubit engine. Entropy, 22(11), 1255.

[5] Gopal, J. K., Do, A. T., Singh, P., Chua, G. L., & Kim, T. T. H. (2014). A cantilever-based NEM nonvolatile memory utilizing electrostatic actuation and vibrational deactuation for high-temperature operation. IEEE Transactions on Electron Devices, 61(6), 2177-2185.

[6] Sharma, D., & Ramteke, M. (2021). DNA Computing: Methodologies and Challenges. DNA-and RNA-Based Computing Systems, 15-29.

16

Multiple-Valued DNA-Quantum
Nano-Processor

16.1 Introduction

In the previous chapter, a new nano-processor named ternary quantum-DNA nano-processor has been introduced where ternary quantum computing and ternary DNA computing logic have been used. The purpose of making this nano-processor is to achieve more and more advantages. This chapter will illustrate another new nano-processor named ternary DNA-quantum nano-processor. It is needed to take the help of ternary DNA computing and ternary quantum computing logic. The logical operation and working principle of ternary quantum computing and ternary DNA computing have already been expounded. With the help of this logic, a ternary DNA-quantum nano-processor will be designed. This renewed nano-processor will add a new dimension in the arena of quantum computing and DNA computing in this world. Ternary DNA computing or three-valued logic function has two inputs that can assume three states (say 0, 1, and 2) and generates one output signal that can have one of these three states. The fluorescent level is used to detect the DNA sequence in ternary computing. Fluorescence is the quick absorption of electromagnetic wavelengths from the visible light spectrum by fluorescent molecules and the subsequent light emission at a lower energy level. It is sometimes called bio-fluorescence when it befalls a living organism. This causes the emitted light to be different from the absorbed light. Stimulating light excites an electron, raising energy to an unstable level. However, now it is possible to implement a ternary DNA RAM in this nano-processor, applying the earlier logic of ternary DNA computing. On the contrary, the information unit in a three-valued quantum system is known as qutrit. The ternary quantum system represents a three-dimensional quantum system with the basis states $|0>$, $|1>$, and $|2>$. Ternary logic functions are those functions that have significance if a third value is acquainted with the binary logic. 0, 1, and 2 are false, undefined, and true, respectively. Using this logic, a ternary quantum nano-processor component will be implemented. And combining these two ternary logic systems, a ternary DNA-quantum nano-processor will be invented. The most significant advantages of this nano-processor can be high storage capacity, low power dissipation, lower error rates. But there are so many obstacles during practical implementation that need to be solved.

DOI: 10.1201/9781003381921-16   481

16.2 Basic Definitions

The core components of all nano-processors are identical. They all have the same components and functionality to accomplish any meaningful work. Each processor consists of some type of register, ALU (arithmetic logic unit), RAM (random access memory), control unit. These all are combinational circuits that have already been illustrated in each nano-processor chapter with their working principle and fundamental definitions. To avoid complexity and repetition, these explanation portions are skipped.

16.3 Block Diagram of Complete Ternary 2-Bit DNA-Quantum Nano Processor

This nano-processor has been designed using combinational circuits such as instruction register, multiplexer, decoder, ALU, incrementor, accumulator, multiplexer, RAM. Figure 16.1 represents the ternary DNA-quantum nano-processor.

This is the combinational ternary DNA quantum nano-processor where only basic components are shown. The working principle of this nano-processor is like the other nano-processors already described in each nano-processor chapter. Now, this section will describe the working principle of this nano-processor in a nutshell. First, let's explain the above working principle in this combinational nano-processor. Figure 16.1 represents a ternary 2-bit DNA-quantum circuit where RAM is the ternary DNA form, and the rest of the components are in ternary quantum form.

Notice the Figure 16.1 that data as a DNA sequence is passed from ternary DNA RAM to the ternary quantum instruction register as it is not possible to pass DNA sequence to the ternary quantum instruction register directly. It is needed to convert the DNA sequence to a qutrit before passing the quantum ternary instruction register. For converting a ternary DNA sequence to qutrit, it is needed to use the NMR method. DNA T-cash memory has been used to store DNA bit sequences for short times until an instruction is completed in the ternary quantum CPU that is already sent to the ternary quantum nano-processor. Quantum qutrits generate so much heat during computation time that they can create anarchy in the whole system. Therefore, a heat transfer circuit has been used here that brings out the obtained heat of quantum ternary logical operations and is cooled using a refrigerator. It is executed when an instruction is fetched into the quantum ternary decoder.

The fetching and executed part are completed at the quantum ternary CPU. After completing the instruction, data as a qutrit is sent to the DNA T-RAM. But, in that case, again it is not possible to pass qutrit to DNA T-RAM directly. Therefore, it is needed to use NMR relaxation to convert qutrit to the DNA sequence, and it is

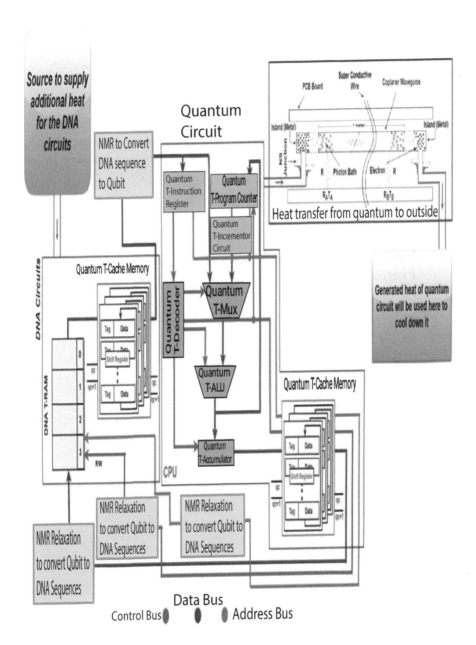

FIGURE 16.1
Ternary 2-Bit DNA-Quantum Nano-Processor

needed to store qutrit to quantum T-cache memory, and from this cache memory, the DNA sequence is passed into the NMR relaxation circuit convert qutrit to the DNA sequence. It is needed to consider the following steps in the nano-processor:

1. Pass data as DNA sequence to DNA ternary cache memory
2. Do NMR for converting DNA sequence to the qutrit.
3. After executing an instruction, pass qutrit to quantum T-cache memory
4. Do only NMR relaxation for converting qutrits to the DNA sequences.

Thus, the whole nano-processor will work. The working procedure of DNA ternary cache memory, NMR relaxation, NMR, and quantum ternary cache memory are described in detail already.

16.4   Basic Components of Ternary DNA-Quantum Nano Processor

A complete ternary DNA quantum nano processor has been designed using the following components shown in Figure 16.1. The details of this CPU are as follows:

1. DNA T- RAM
2. Quantum T-Instruction Register
3. Quantum T-Program Counter
4. Quantum T-Incrementor
5. Quantum T-decoder
6. Quantum T-Multiplexer
7. Quantum T-ALU
8. Quantum T-Accumulator.
9. Quantum ternary cache memory
10. DNA ternary cache memory
11. NMR relaxation
12. Heat transfer circuit and
13. NMR

These are the fundamental components of the nano-processor that need to be accomplished for meaningful work.

16.4.1 Ternary DNA RAM

RAM is a primary memory that stores the data for a short time. Two address lines are needed for simulating ternary 2-bit DNA RAM where it is needed to use the Ternary 1-to-3 DNA decoder operation to achieve a ternary 2-bit DNA RAM. These 1-to-3 ternary DNA decoders produce ternary 2-to-9 decoders, consisting of nine ternary AND DNA operations. The entire design procedure and the working principle have been described in chapter 14.

16.4.2 Ternary Quantum CPU

According to Figure 16.1, all nano-processor components except RAM are in ternary quantum form, and a ternary quantum CPU consists of a quantum ternary Instruction register, program counter, incrementor, multiplexer, decoder, accumulator, and ALU. The working principle and the design procedure of all these elements are already described in chapter 13.

16.5 Applications

A ternary 2-bit DNA quantum Nano-processor is demonstrated in this chapter. The field of ternary DNA and ternary quantum computing is an emerging concept in its infancy. The applications of this field are still being understood. The most significant advantages are that this Nano-processor can store high-density information and ternary quantum computing faster, i.e., they can execute any operation quickly. So, by combining these two logic systems, it is possible to achieve more advantages in a single ternary DNA quantum Nano-processor. This Nano-processor can be added to living cells to provide new detection methods in medical devices. In addition, this Nano-processor can be used to secure data, financial modeling, and cryptography for better progress in the technology world. As ternary DNA computing and ternary quantum computing work in parallel, they can achieve more advantages such as high speed, and fast computation and this Nano-processor can take the machine learning, artificial intelligence, and drug development circuits one step further.

16.6 Summary

It is expected that the design of this nano-processor will add a new dimension to the computer technology world. Although this nano processor has a higher probability of gaining more and more advantages, the disadvantages and limitations are minor. The proposed nano-processor is theoretical but logical. Like the previous nano-processor, it has the same disadvantages as setting up the environment. When combining these two systems, perhaps it is possible to face some unexpected obstacles. But if it is possible to simulate a Nano-processor according to the given instructions, undoubtedly, this will play a significant role in the following computer generation. Therefore, this field is needed to be emphasized for technological progress. The proposed ternary quantum DNA Nano-processor is a theoretical base but logical. Despite having more advantages, there are still many obstacles to implement an actual ternary quantum DNA Nano-processor. To make a functional ternary DNA-quantum computer, all those problems are needed to be solved.

Bibliography

[1] Javadi-Abhari, C. A. K. A. (2019). A McClure DT Cross AW Temme K Nation PD Steffen M Gambetta JM Challenges and opportunities of near-term quantum computing systems. Proceedings of IEEE, 10.

[2] Woese, C. R., Gutell, R., Gupta, R., & Noller, H. F. (1983). Detailed analysis of the higher-order structure of 16S-like ribosomal ribonucleic acids. Microbiological Reviews, 47(4), 621-669.

[3] Shukla, S. K., & Bahar, R. I. (2004). Nano, Quantum and Molecular Computing. Kluwer Academic.

[4] Dann, R., Kosloff, R., & Salamon, P. (2020). Quantum finite-time thermodynamics: insight from a single qubit engine. Entropy, 22(11), 1255.

[5] Freier, S. M., Kierzek, R., Jaeger, J. A., Sugimoto, N., Caruthers, M. H., Neilson, T., & Turner, D. H. (1986). Improved free-energy parameters for predictions of RNA duplex stability. Proceedings of the National Academy of Sciences, 83(24), 9373-9377.

Final Remarks

It is possible to establish a link between multi-valued quantum computing and multi-valued DNA computing which is clearly explained in this book through different operations. This book could be a milestone for the students and academicians to learn new things about multi-valued quantum and DNA computing. The combination of multi-valued quantum and DNA computing can form multi-valued quantum-DNA and DNA-quantum computing. This book describes all necessary components, design procedures and working principles with appropriate figures that are needed to establish these two new computing platforms. It also explains quantum molecular biology which could be a big topic and new dimension for researchers and postgraduate students.

The potential for multiple-valued quantum molecular biology to contribute in the development of sustainable energy solutions is widely highlighted as a driving force behind the study, especially in terms of understanding quantum-coherent excitation energy conversion in photosynthetic systems. More effective light absorption, quicker energy funneling (i.e., faster decay down the energy gradient), faster energy transfer, and more effective (irreversible) excitation trapping by the reaction center are all fundamental aspects of quantum biologically inspired design.

The characteristics of the DNA molecule aid in inducing quantum features such as superposition, tunneling, coherence, and entanglement. The innovative ideas of multiple-valued quantum-DNA computing and multiple-valued DNA-quantum computing presented in this book combine the benefits of multiple-valued quantum physics with multiple-valued molecular biology. The combination is called multi-valued quantum molecular biology. NMR is necessary for synthesizing a qubit from a DNA sequence, and NMR may be conducted at two temperatures: 0-kelvin and room temperature. In this book, multiple-valued quantum-DNA and multiple-valued DNA-quantum nano processors are built for these two temperatures. To retrieve the appropriate DNA sequence, RNMR is employed to relax the qubit. In multiple-valued quantum computing, qubits are utilized to carry out operations, whereas in multiple-valued DNA computing, DNA sequences are employed as inputs. Multiple-valued Quantum cache memory is utilized as a buffer in multiple-valued quantum-DNA computing to match the speed. The qubits for transferring data to multiple-valued DNA circuits are stored in multiple-valued quantum cache memory. Apart from that, while processing data, DNA operations need heat, and quantum operations generate heat while executing any action. Using multiple-valued quantum-DNA computing the heat generated in multiple-valued quantum computing is used in multiple-valued DNA computing. So, this feature is also another merit of multiple-valued quantum-DNA computing.

Through this book, undergraduate students can learn four common topics like Sequential circuits, Memory devices, Programmable logic devices, and Nano processors in different and new ways. They will learn the circuit construction and working principle of these four topics in multi-valued quantum, DNA, quantum-DNA, and DNA-quantum computing platforms.

Index

adenine, 67

Cache memory, 355
complex programmable logic device, 361
cytosine, 67

D flip-flop, 70
decoherence, 30
deoxyribonucleic, 3
Deoxyribose Nucleic Acid, 269
DNA computing, 116
DNA Decoder, 259
DNA NOT operation, 16
DNA Program Counter, 465
DNA Programmable Array Logic, 372
DNA ripple counter, 95
DNA shift registers, 90
DNA ternary NAND, 18
DNA ternary NOR, 20
DNA ternary OR, 19
DNA ternary XNOR, 22
DNA ternary XOR, 21
DNA-Quantum CPLD, 422
DNA-Quantum FPGA, 417, 418
DNA-Quantum PAL, 413
DNA-Quantum PLA, 410

Field Programmable Gate Array, 351

heat transfer, 134

inverter, 375

Look-Up Table, 353

molecular biology, 2

Molecular RAM Cells, 289
molecule, 71
multi-valued DNA complex programmable logic device, 384
multi-valued DNA SR Latch, 73
multi-valued FPGA, 379
multi-valued quantum computer, 30
multi-valued Quantum decoder, 219
Multi-Valued quantum Programmable Logic Array, 343
Multi-Valued Quantum PROM, 238
Multi-Valued quantum ripple counter, 55
multi-valued quantum shift registers, 50
multi-valued quantum T flip–flop, 46
multi-valued single RAM Cell, 246
Multiple-Valued Quantum PLA, 343
Multiple-valued Quantum-DNA PLA, 392
Multiple-Valued Quantum-DNA ROM, 302

NAND operation, 31
nano-processor, 435
nanoscale, 108
NMR, 303
NMR relaxation, 113

Programmable Logic Array, 367
Programmable Logic Device, 347
Programmable Quantum OR array, 343
programmable ternary quantum OR gates, 345

quanine, 67
quantum, 29
quantum cache memory, 111
Quantum computing, 397
Quantum D flip flop, 30
Quantum JK flip-flop, 42
quantum mechanics, 29
quantum MUX, 355
Quantum Nano processor, 433
quantum OR operation, 11
quantum physics, 29
Quantum SR Latch, 35
quantum technologies, 213
quantum ternary ALU, 436
quantum ternary AND operation, 9,
        10
Quantum ternary Instruction
        Register, 438
quantum ternary NAND operation,
        11
quantum ternary NOR operation, 12
Quantum ternary RAM, 248
quantum ternary shift gates, 5
quantum ternary XOR operation, 13
Quantum-DNA CPLD, 403
Quantum-DNA D flip flop, 109
Quantum-DNA JK flip-flop, 132
Quantum-DNA Ripple Counter, 145
Quantum-DNA SR Latch, 115

Qubit, 314

supercomputer, 215
superposition, 37, 215

Ternary AND, 376
ternary computer, 4
ternary decoder, 355
ternary DNA counter, 100
ternary DNA incrementor, 466
ternary DNA molecule sequence
        instruction register, 461
ternary DNA nano processor, 456
ternary DNA program Accumulator,
        467
ternary DNA RAM, 458
ternary Quantum AND Gates, 60
ternary Quantum counter, 60
ternary Quantum instruction register,
        442
ternary quantum NAND operation,
        32
ternary Quantum PLA, 351, 373
ternary quantum program
        accumulator, 451
ternary RAM, 438
ternary toffoli gate, 6
thymine, 67
Trap-Ion, 414